EDEXCEL
FURTHER MATHS
CORE PURE
YEAR 1/AS LEVEL

Series Editor
David Baker

Authors
Katie Wood, Rob Wagner, David Bowles
Brian Jefferson, Eddie Mullan, Garry Wiseman, John Rayneau, Mike Heylings

OXFORD
UNIVERSITY PRESS

UNIVERSITY PRESS

Great Clarendon Street, Oxford, OX2 6DP, United Kingdom

Oxford University Press is a department of the University of Oxford.

It furthers the University's objective of excellence in research, scholarship, and education by publishing worldwide. Oxford is a registered trade mark of Oxford University Press in the UK and in certain other countries.

© Oxford University Press 2017

The moral rights of the authors have been asserted.

First published in 2017

British Library Cataloguing in Publication Data
Data available

978 0 19 841523 7

10 9 8 7 6 5 4 3

Paper used in the production of this book is a natural, recyclable product made from wood grown in sustainable forests.
The manufacturing process conforms to the environmental regulations of the country of origin.

Printed and bound by CPI Group (UK) Ltd, Croydon, CR0 4YY

Acknowledgements

Authors
Katie Wood, Rob Wagner, David Bowles
Brian Jefferson, Eddie Mullan, Garry Wiseman, John Rayneau, Mike Heylings

Editorial team
Dom Holdsworth, Ian Knowles, Matteo Orsini Jones, Anna Gupta

With thanks also to Geoff Wake, Matt Woodford, Deb Dobson, Katherine Bird for their contribution.

Although we have made every effort to trace and contact all copyright holders before publication, this has not been possible in all cases. If notified, the publisher will rectify any errors or omissions at the earliest opportunity.

p3 p30, p54, p57, p110, p113, p141 Shutterstock; **p35** TomasSerada/iStockphoto; **p73** Scyther5/Dreamstime

In order to ensure that this resource offers high-quality support for the associated Pearson qualification, it has been through a review process for the awarding body. This process confirms that this resource fully covers the teaching and learning content of the specification or part of a specification at which it is aimed. It also confirms that it demonstrates an appropriate balance between the development of subject skills, knowledge and understanding, in addition to preparation for assessment.

Endorsement does not cover any guidance on assessment activities or processes (e.g. practice questions or advice on how to answer assessment questions) included in the resource, nor does it prescribe any particular approach to the teaching or delivery of a related course. While the publishers have made every attempt to ensure that advice on the qualification and its assessment is accurate, the official specification and associated assessment guidance materials are the only authoritative source of information and should always be referred to for definitive guidance.

Pearson examiners have not contributed to any sections in this resource relevant to examination papers for which they have responsibility. Examiners will not use endorsed resources as a source of material for any assessment set by Pearson.

Endorsement of a resource does not mean that the resource is required to achieve this Pearson qualification, nor does it mean that it is the only suitable material available to support the qualification, and any resource lists produced by the awarding body shall include this and other appropriate resources.

Contents

Year 1/AS Level

Chapter 1: Complex numbers 1
Introduction   3
1.1   Properties and arithmetic   4
1.2   Solving polynomial equations   8
1.3   Argand diagrams   14
1.4   Modulus argument form and loci   19
Summary and review   28
Exploration   30
Assessment   31

Chapter 2: Algebra and series
Introduction   35
2.1   Roots of polynomials   36
2.2   Summing powers   42
2.3   Proof by induction   48
Summary and review   52
Exploration   54
Assessment   55

Chapter 3: Integration
Introduction   57
3.1   Volumes of revolution around
      the x-axis   58
3.2   Volumes of revolution around
      the y-axis   64
Summary and review   68
Exploration   70
Assessment   71

Chapter 4: Matrices
Introduction   73
4.1   Properties and arithmetic   74
4.2   Transformations   83
4.3   Systems of linear equations   94
Summary and review   106
Exploration   110
Assessment   111

Chapter 5: Vectors
Introduction   113
5.1   The equation of a straight line   114
5.2   The scalar product   121
5.3   The equation of a plane   126
5.4   Finding distances   134
Summary and review   139
Exploration   141
Assessment   142

Mathematical formulae   144

Mathematical notation   147

Answers   150

Index   192

About this book

This book has been specifically created for those studying the Edexcel 2017 Further Mathematics AS and A Level. It's been written by a team of experienced authors and teachers, and it's packed with questions, explanation and extra features to help you get the most out of your course.

Every section starts by covering the basic **Fluency and skills** (A01).

Support for when and how to use **calculators** is available throughout this book.

Worked examples provide a model answer and commentary to realistic practice questions.

There is a Fluency and skills exercise for each section, to practise the skills before moving on to the Reasoning and problem-solving section.

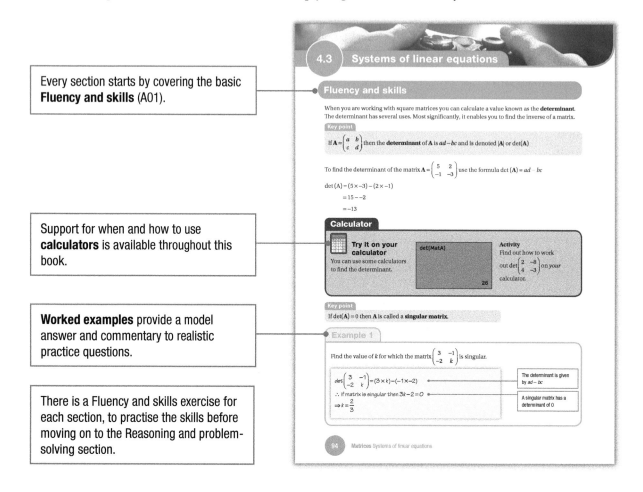

On the chapter **Introduction page**, the **Orientation box** explains what you should already know, what you will learn, and what this leads to.

At the end of every chapter, an **Exploration page** gives you an opportunity to explore the subject beyond the specification.

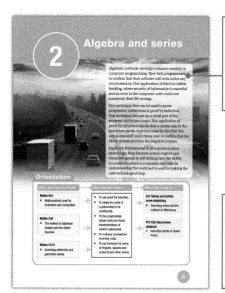

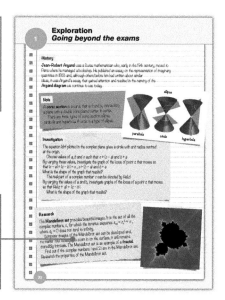

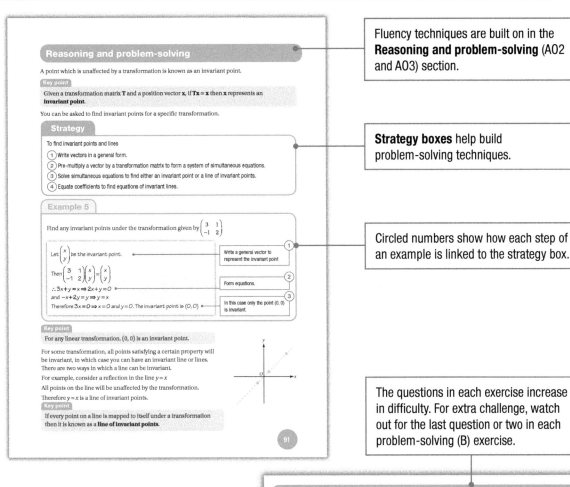

Reasoning and problem-solving

A point which is unaffected by a transformation is known as an invariant point.

Key point

Given a transformation matrix **T** and a position vector **x**, if **Tx** = **x** then **x** represents an **invariant point**.

You can be asked to find invariant points for a specific transformation.

Strategy

To find invariant points and lines

1. Write vectors in a general form.
2. Pre-multiply a vector by a transformation matrix to form a system of simultaneous equations.
3. Solve simultaneous equations to find either an invariant point or a line of invariant points.
4. Equate coefficients to find equations of invariant lines.

Example 5

Find any invariant points under the transformation given by $\begin{pmatrix} 3 & 1 \\ -1 & 2 \end{pmatrix}$

Let $\begin{pmatrix} x \\ y \end{pmatrix}$ be the invariant point. —— ① Write a general vector to represent the invariant point

Then $\begin{pmatrix} 3 & 1 \\ -1 & 2 \end{pmatrix}\begin{pmatrix} x \\ y \end{pmatrix} = \begin{pmatrix} x \\ y \end{pmatrix}$ —— ② Form equations.

$\therefore 3x + y = x \Rightarrow 2x + y = 0$ ——
and $-x + 2y = y \Rightarrow y = x$ —— ③

Therefore $3x = 0 \Rightarrow x = 0$ and $y = 0$. The invariant point is $(0, 0)$ —— In this case only the point $(0, 0)$ is invariant.

Key point

For any linear transformation, $(0, 0)$ is an invariant point.

For some transformation, all points satisfying a certain property will be invariant, in which case you can have an invariant line or lines. There are two ways in which a line can be invariant.

For example, consider a reflection in the line $y = x$

All points on the line will be unaffected by the transformation.

Therefore $y = x$ is a line of invariant points.

Key point

If every point on a line is mapped to itself under a transformation then it is known as a **line of invariant points**.

91

Fluency techniques are built on in the **Reasoning and problem-solving** (AO2 and AO3) section.

Strategy boxes help build problem-solving techniques.

Circled numbers show how each step of an example is linked to the strategy box.

The questions in each exercise increase in difficulty. For extra challenge, watch out for the last question or two in each problem-solving (B) exercise.

Exercise 4.2B Reasoning and problem-solving

1. Give the equation of the line of invariant points under the transformation given by each of these matrices.

   a $\begin{pmatrix} 1 & -2 \\ 0 & 3 \end{pmatrix}$ b $\begin{pmatrix} 5 & 2 \\ 4 & 3 \end{pmatrix}$

2. Show that the origin is the only invariant point under the transformation given by each of these matrices.

   a $\begin{pmatrix} 3 & -2 \\ 2 & 3 \end{pmatrix}$ b $\begin{pmatrix} 2 & 0 & 1 \\ 0 & 3 & -2 \\ 1 & 0 & -4 \end{pmatrix}$

3. In each of these cases, decide whether or not the order the transformations are applied affects the final image. Justify your answers.

   a A reflection in the y-axis and a stretch parallel to the y-axis.

   b A rotation about the origin and an

9. Find the equations of the invariant lines under the transformation given by each of these matrices.

   a $\begin{pmatrix} 1 & 2 \\ 2 & -1 \end{pmatrix}$ b $\begin{pmatrix} 3 & 0 \\ 0 & 3 \end{pmatrix}$ c $\begin{pmatrix} -\frac{3}{5} & \frac{4}{5} \\ \frac{4}{5} & \frac{3}{5} \end{pmatrix}$

10. a For each of these transformations, find either the invariant point or the equations of the invariant lines as appropriate.

    i Reflection in the x-axis.

    ii Rotation of 90° around the origin.

    iii Stretch of scale factor 2 parallel to the x-axis.

    iv Reflection in the line $y = -x$

    b For each of the invariant lines in part **a**.

Assessment sections at the end of each chapter test everything covered within that chapter.

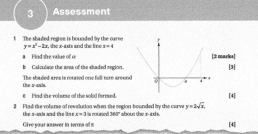

---

3 **Assessment**

1. The shaded region is bounded by the curve $y = x^2 - 2x$, the x-axis and the line $x = 4$

   a Find the value of α [2 marks]

   b Calculate the area of the shaded region.   [3]

   The shaded area is rotated one full turn around the x-axis.

   c Find the volume of the solid formed.   [4]

2. Find the volume of revolution when the region bounded by the curve $y = 2\sqrt{x}$, the x-axis and the line $x = 3$ is rotated 360° about the x-axis.

   Give your answer in terms of π [4]

10. The region R is bounded by the curve with equation $x = \frac{y-1}{\sqrt{y}}$, and the lines $y = 1$ and $y = 4$

    a Calculate the exact area of R   [4]

    b Calculate the exact volume of the solid formed when R is rotated 2π radians around the y-axis.   [5]

11. The shaded region shown is part of an ellipse with equation $\frac{x^2}{9} + \frac{y^2}{4} = 1$

    a Write the values of a and b in the diagram.   [2]

    b Calculate the volume when the shaded area is rotated π radians about the x-axis.   [5]

    c Calculate the volume when the shaded area is rotated π radians about the y-axis.   [5]

12. Part of the curve $y = \frac{1}{n}(x^3 - 20x^2 + 100x + 90)$ between $x = 0$

1 Complex numbers 1

Complex numbers are a crucial tool in the design of modern electrical components, such as motherboards. Electrical engineers use them to simplify many of their calculations. This helps them to analyse varying currents and voltages in electrical circuits. When the current flows in one direction constantly, the resistance can often be calculated using a simple formula. However, when the direction of the current is alternating, this formula doesn't work. Therefore, engineers express the quantities as complex numbers, which makes it easier to understand the processes involved and perform the necessary calculations.

Complex numbers are used in many other fields such as chemistry, economics, and statistics. Their unique properties provide powerful ways of solving and interpreting complicated equations that can be applied in a wide variety of contexts. From some of the deepest mysteries in number theory to the signal processing used in playing digital music, these so-called imaginary numbers have led to a surprising array of advances in the real world.

Orientation

What you need to know

KS4
- Use the quadratic formula.

Maths Ch1
- Solve simultaneous equations.

Maths Ch3
- Work with sine and cosine functions.

What you will learn

- To calculate with complex numbers in the form $a + bi$
- To understand and use the complex conjugate.
- To solve quadratic, cubic and quartic equations with real coefficients.
- To convert between modulus-argument form and the form $a + bi$ and calculate with numbers in modulus-argument form.
- To sketch and interpret Argand diagrams.

What this leads to

Ch6 Complex Numbers 2
- Exponential form
- The Euler formula $e^{i\theta} = \cos\theta + i\sin\theta$
- De Moivre's formula
- Roots of unity

Careers
- Electrical engineering

Fluency and skills

Some equations, including those deriving from real-world situations, have no real solutions. For example, up to this point, you have been unable to solve equations such as $x^2 = -1$. You know that the solutions are $x = \pm\sqrt{-1}$, but this is not a real number and is difficult to manipulate. In order to solve this problem, mathematicians denoted the square root of negative one by i. This means that the solutions to the equation $x^2 = -1$ can be written as $\pm i$. Although this number is known as 'imaginary', it means that all polynomial equations do indeed have solutions.

A good analogy is negative numbers: these can be hard to grasp in isolation. For example, the concept of 'minus one apple' is a difficult one, but it's useful in sums such as 3 apples − 1 apple = 2 apples. In exactly the same way, imaginary numbers are essential in many calculations and have many real-world applications.

> **Key point**
>
> The **imaginary number** i is defined as $i = \sqrt{-1}$

You can solve the equation $x^2 = -9$ by square-rooting both sides to give $x = \pm\sqrt{-9}$
Then, because $\sqrt{-9}$ can be written as $\sqrt{9}\sqrt{-1}$, you can use the fact that $i = \sqrt{-1}$ to give $x = \pm 3i$

Notice that once you have defined the square root of minus one, the square root of all other negative numbers can be written in terms of i

For example, to solve the equation $(x-3)^2 = -5$, you first square-root both sides to give $x - 3 = \pm\sqrt{-5}$, and since $\sqrt{-5} = \sqrt{5}i$, this gives the solutions $x = 3 \pm \sqrt{5}i$

Numbers with both real and imaginary parts are called **complex numbers**.

> **Key point**
>
> **Complex numbers** can be written in the form $a + bi$
> where $a, b \in \mathbb{R}$. The set of complex numbers is denoted $\mathbb{C}$

Complex numbers can be added, subtracted and multiplied by a constant in the same way as algebraic expressions.

> **Example 1**
>
> Simplify the expression $3(4-7i) - 2(3-2i)$
>
> $$3(4-7i) - 2(3-2i) = (12-21i) - (6-4i)$$
> $$= 6 - 17i$$
>
> Multiply real and imaginary parts by the constant.
>
> Simplify real parts: $12 - 6 = 6$
>
> Simplify imaginary parts: $-21i + 4i = -17i$

Example 2

Solve the equation $(x+7)^2 = -16$

$x+7 = \pm\sqrt{-16}$ ●——————————————— Square-root both sides of equation.

$x = -7 \pm \sqrt{-16}$

$ = -7 \pm \sqrt{16}\sqrt{-1}$

$ = -7 \pm 4i$ ●——————————————— Since $\sqrt{-1} = i$

Calculator

Try it on your calculator

Some calculators enable you to manipulate complex numbers.

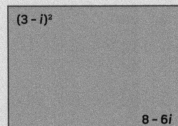

$(3 - i)^2$

$8 - 6i$

Activity

Find out how to work out $(3 - i)^2$ on *your* calculator. Use your calculator to find these expressions.

a $(2+i)(5-7i)$ **b** $4i(2-8i)$

c $(30+5i) \div (2-i)$

To rationalise the denominator in surd form such as $\dfrac{1}{a+\sqrt{b}}$, you multiply the numerator and the denominator by $a - \sqrt{b}$

You can use a similar method to simplify fractions with complex denominators. Doing this will always change the denominator into a positive real number.

Key point

To simplify $\dfrac{1}{a+bi}$, multiply the numerator and denominator by $a - bi$

Example 3

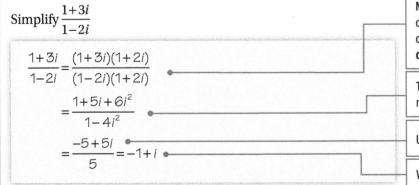

Simplify $\dfrac{1+3i}{1-2i}$

$\dfrac{1+3i}{1-2i} = \dfrac{(1+3i)(1+2i)}{(1-2i)(1+2i)}$ ●——— Multiply the numerator and denominator by the same complex number (called the **complex conjugate**).

$ = \dfrac{1+5i+6i^2}{1-4i^2}$ ●——— The imaginary parts of the denominator cancel each other out.

$ = \dfrac{-5+5i}{5} = -1+i$ ●——— Use $i^2 = -1$

  ●——— Write in the form $a + bi$

5

1 Solve these equations.

   a $x^2 = -25$ **b** $x^2 = -121$

   c $x^2 = -20$ **d** $x^2 + 8 = 0$

   e $z^2 = -9$ **f** $z^2 + 12 = 0$

2 Simplify these expressions, giving your answers in the form $a + bi$ where $a, b \in \mathbb{R}$

   a $(2 + 3i) + (5 - 9i)$

   b $(5 - 7i) - (12 + 3i)$

   c $3(6 - 9i)$ **d** $3(2 + 10i) + 5(4 - i)$

   e $4 - 9(7i + 5)$ **f** $2(6 - 2i) - 3(2i - 5)$

3 Write each of these expressions in the form $a + bi$ where $a, b \in \mathbb{R}$

   a $(2 + 3i)(i + 5)$ **b** $(7 - i)(6 - 3i)$

   c $i(8 - 3i)$ **d** $(9 - 4i)^2$

4 Fully simplify each of these expressions.

   a i^3 **b** i^4 **c** i^5

   d $(2i)^3$ **e** $(3i)^4$ **f** $2i^2(5i - 9)^2$

5 Simplify these fractions, giving your answers in the form $a + bi$ where $a, b \in \mathbb{R}$

   a $\dfrac{3}{2 + i}$ **b** $\dfrac{2i}{1 - 5i}$ **c** $\dfrac{1 + 7i}{3 - i}$

   d $\dfrac{i + 3}{2i - 1}$ **e** $\dfrac{6 + 3i}{i - \sqrt{2}}$ **f** $\dfrac{\sqrt{2}i - \sqrt{6}}{\sqrt{3} - i}$

6 You are given that $z_1 = 3i - 2$, $z_2 = 4 + i$

   Calculate these expressions, fully simplifying your answers.

   a $z_1 + z_2$ **b** $z_1 z_2$

   c $\dfrac{z_1}{z_2}$ **d** $\dfrac{z_2}{z_1}$

Reasoning and problem-solving

Strategy

To solve equations involving imaginary numbers

  (1) Write all the numbers and expressions in the form $a + bi$

  (2) Equate real parts and imaginary parts on both sides of the equation or identity.

  (3) Solve the equations simultaneously.

Example 4

Find real numbers a and b such that $(a + 5i)(2 - i) = 9 + bi$

          $(a + 5i)(2 - i) = 9 + bi$

      $2a - ai + 10i - 5i^2 = 9 + bi$

     $(2a + 5) + (10 - a)i = 9 + bi$

            $\text{Re}: 2a + 5 = 9$

                 $a = 2$

           $\text{Im}: 10 - a = b$

                 $b = 8$

Expand the brackets. **(1)**

Simplify using $i^2 = -1$ and collect together real and imaginary terms.

Equate real parts. **(2)**

Equate imaginary parts. **(2)**

Example 5

Find the complex numbers z such that $z^2 = 5 + 12i$

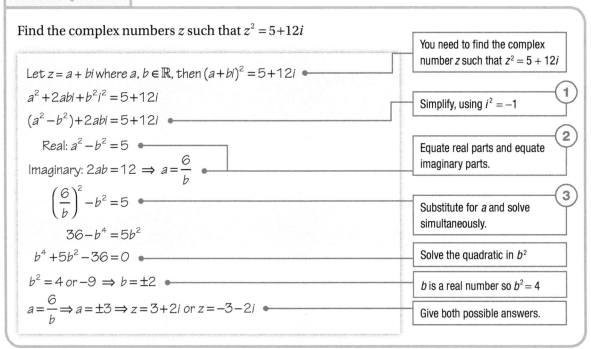

Let $z = a + bi$ where $a, b \in \mathbb{R}$, then $(a + bi)^2 = 5 + 12i$ ●————— You need to find the complex number z such that $z^2 = 5 + 12i$

$a^2 + 2abi + b^2i^2 = 5 + 12i$

$(a^2 - b^2) + 2abi = 5 + 12i$ ●————— ① Simplify, using $i^2 = -1$

Real: $a^2 - b^2 = 5$ ●

Imaginary: $2ab = 12 \Rightarrow a = \dfrac{6}{b}$ ●————— ② Equate real parts and equate imaginary parts.

$\left(\dfrac{6}{b}\right)^2 - b^2 = 5$ ●————— ③ Substitute for a and solve simultaneously.

$36 - b^4 = 5b^2$

$b^4 + 5b^2 - 36 = 0$ ●————— Solve the quadratic in b^2

$b^2 = 4$ or $-9 \Rightarrow b = \pm 2$ ●————— b is a real number so $b^2 = 4$

$a = \dfrac{6}{b} \Rightarrow a = \pm 3 \Rightarrow z = 3 + 2i$ or $z = -3 - 2i$ ●————— Give both possible answers.

Exercise 1.1B Reasoning and problem-solving

1 Find the real numbers a and b such that $(2a + i)^2 = 35 - bi$

2 Find the values $a, b \in \mathbb{R}$ that satisfy the equation $(2 - bi)(3 + 4i) = a - 13i$

3 Solve the equation

$z(2 - 5i) = 13 - 18i$ for $z \in \mathbb{C}$

4 Solve the equation

$(x + 3i)(10 - 7i) = 34 + 6i$

for a complex number x

5 Find the complex numbers z_1 and z_2 such that $z_1 + z_2 = 11 - 3i$ and $z_2 - z_1 = 5 + 7i$

6 Find the complex numbers z and w that satisfy the equations

$z + 2w = 6i$ and $3z - 4w = 20 + 23i$

7 Find $z_1, z_2 \in \mathbb{C}$ such that

$2z_1 - 3z_2 = 10 + 8i$ and $5z_1 - \dfrac{1}{2}z_2 = 4 + 6i$

8 Find the complex numbers w in each of these cases.

  a $w^2 = 30i - 16$ **b** $w^2 = -3 - 4i$

  c $w^2 - 1 = 20(1 - i)$

9 Calculate the square roots of $2 - 4\sqrt{2}i$

10 Solve these simultaneous equations, where w and z are complex numbers.

$w^2 + z^2 = 0$

$z - 3w = 10$

11 Solve each of these equations to find $z \in \mathbb{C}$

  a $z^2 - 2z = -50$ **b** $z^2(1 + i) = 7 - 17i$

12 Simplify each of these expressions, giving your answer in the form $a + bi$ in each case.

  a $(1 + 2i)^4$ **b** $(2 - 5i)^5$

  c $(3i - 1)^3(1 + i)$

13 Find $a, b \in \mathbb{R}$ such that

$(a + i)^4 = 28 + bi$

Fluency and skills

Using complex numbers, the quadratic formula $x = \dfrac{-b \pm \sqrt{b^2 - 4ac}}{2a}$ can be used to solve all quadratic equations of the form $ax^2 + bx + c = 0$, with $a, b, c \in \mathbb{R}$

When the discriminant $(b^2 - 4ac)$ is negative the solutions will contain imaginary numbers.

Because the $\pm$ in the formula gives two solutions, you can see that:

Key point

If $z = a + bi$ is a solution to an equation then $z^* = a - bi$ will also be a solution.

We call z^* the **complex conjugate** of z

Example 1

Given that $z = 5 - 3i$, find

a z^* **b** zz^* **c** $z + z^*$

a $z^* = 5 + 3i$	Write the complex conjugate of z
b $zz^* = (5 - 3i)(5 + 3i)$	
$\quad = 25 + 15i - 15i - 9i^2$	Since $i^2 = -1$
$\quad = 25 + 9$	zz^* will always be a real number.
$\quad = 34$	
c $z + z^* = 5 - 3i + 5 + 3i$	
$\quad = 10$	$z + z^*$ will always be a real number.

If you know the roots of a quadratic equation, then you can find the original equation by multiplying the factors together.

Example 2

Find the real quadratic equation that has one root of $2 + 7i$

Roots are $2 + 7i$ and $2 - 7i$	The complex conjugate will also be a root.
So the factors are $x - (2 + 7i)$ and $x - (2 - 7i)$	Ensure you know the difference between a *root* and a *factor*.
Equation is $\quad (x - (2 + 7i))(x - (2 - 7i)) = 0$	
$x^2 - x(2 - 7i) - x(2 + 7i) + (2 + 7i)(2 - 7i) = 0$	
$\qquad\qquad x^2 - 4x + 53 = 0$	Since $-49i^2 = 49$

Complex roots for any real polynomial equation will always occur in **complex conjugate pairs**. Therefore, a polynomial equation will always have an even number of complex solutions. This implies that a cubic equation will always have at least one real root (as there cannot be three complex roots). You can use these facts to help you solve cubic and quartic equations.

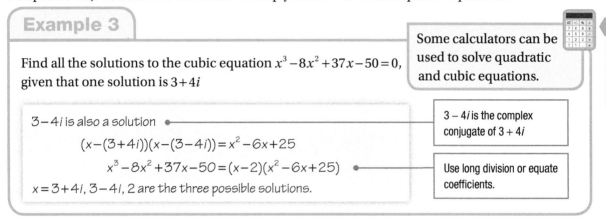

Example 3

Find all the solutions to the cubic equation $x^3 - 8x^2 + 37x - 50 = 0$, given that one solution is $3 + 4i$

Some calculators can be used to solve quadratic and cubic equations.

See Maths Ch1.4 For a reminder of using a calculator to find roots of an equation.

$3 - 4i$ is also a solution ●——— $3 - 4i$ is the complex conjugate of $3 + 4i$

$$(x - (3+4i))(x - (3-4i)) = x^2 - 6x + 25$$

$$x^3 - 8x^2 + 37x - 50 = (x-2)(x^2 - 6x + 25)$$ ●——— Use long division or equate coefficients.

$x = 3 + 4i, 3 - 4i, 2$ are the three possible solutions.

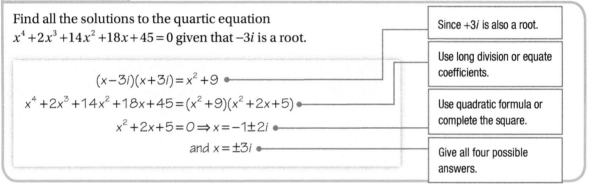

Example 4

Find all the solutions to the quartic equation $x^4 + 2x^3 + 14x^2 + 18x + 45 = 0$ given that $-3i$ is a root.

Since $+3i$ is also a root.

$$(x - 3i)(x + 3i) = x^2 + 9$$ ●——— Use long division or equate coefficients.

$$x^4 + 2x^3 + 14x^2 + 18x + 45 = (x^2 + 9)(x^2 + 2x + 5)$$ ●

$$x^2 + 2x + 5 = 0 \Rightarrow x = -1 \pm 2i$$ ●——— Use quadratic formula or complete the square.

and $x = \pm 3i$ ●——— Give all four possible answers.

Exercise 1.2A Fluency and skills

1 Write the complex conjugate of z in each case.

  a $z = 5 - 2i$ **b** $z = 8 + i$

  c $z = 5i - 6$ **d** $z = \sqrt{2} - i\sqrt{3}$

  e $z = \dfrac{1}{3} + 4i$ **f** $z = \dfrac{2}{3}i - 5$

2 Given that $z = 9 - 2i$, calculate

  a zz^* **b** $z + z^*$

  c $z - z^*$ **d** $\dfrac{z}{z^*}$

  e $(z^*)^*$ **f** $\dfrac{z^*}{z}$

3 Given that $w = -\sqrt{6} + \sqrt{2}i$, calculate

  a ww^* **b** $w + w^*$

  c $w - w^*$ **d** $\dfrac{w}{w^*}$

  e $w^2 + (w^*)^2$ **f** $(w + w^*)^2$

4 Solve each of these quadratic equations.

  a $x^2 + 5x + 7 = 0$ **b** $x^2 - 3x + 5 = 0$

  c $2x^2 + 7x + 7 = 0$ **d** $3x^2 - 10x + 9 = 0$

5 Find a quadratic equation with solutions

  a $x = 7$ and $x = -4$

  b $x = 3 + 5i$ and $x = 3 - 5i$

  c $x = -1 - 9i$ and $x = -1 + 9i$

  d $x = -5 + 4i$ and $x = -5 - 4i$

6 A quadratic equation has a solution of $z = \sqrt{3} + i$

  a Write down the other solution to the equation.

  b Find a possible equation.

7 Find a quadratic equation where one solution is given as

  a $2 + i$ **b** $4 - 3i$ **c** $7i - 1$

  d $-5 - 2i$ **e** $a + 3i$ **f** $5 - bi$

8 Find a cubic equation with solutions

  a $x = -5$, $x = 2 + 3i$ and $x = 2 - 3i$

  b $x = 2$, $x = i - 1$ and $x = -i - 1$

  c $x = 0$, $x = \sqrt{3} + 2i$ and $x = \sqrt{3} - 2i$

  d $x = 3$, $x = -\sqrt{2} - i$ and $x = -\sqrt{2} + i$

9 A cubic equation has solutions of $z = -\dfrac{1}{2}$ and $z = 6 - 2i$

  a Write down the other solution to the equation.

  b Find a possible equation that has these solutions.

10 You are given that $f(x) = x^3 + 9x^2 + 25x + 25$ and $f(-5) = 0$

  a Show that the other solutions to the equation $f(x) = 0$ satisfy $x^2 + 4x + 5 = 0$

  b Solve the equation $f(x) = 0$

11 Given that $1 + 8i$ is a root of $x^4 + 4x^3 + 66x^2 + 364x + 845 = 0$

  a Show that $x^2 + 6x + 13 = 0$

  b Find all the solutions to the equation $x^4 + 4x^3 + 66x^2 + 364x + 845 = 0$

12 Solve these cubic equations using the root given.

  a $x^3 + x^2 + 4x + 30 = 0$, $x = -3$

  b $x^3 - 5x^2 + 33x - 29 = 0$, $x = 1$

  c $x^3 - 3x^2 + 27x - 185 = 0$, $x = 6i - 1$

  d $x^3 - 8x^2 + 21x + 82 = 0$, $x = 5 + 4i$

13 Solve these quartic equations using the root given.

  a $x^4 - 2x^3 + x^2 + 8x - 20 = 0$, $x = 1 - 2i$

  b $2x^4 - 13x^3 + 75x^2 - 133x - 87 = 0$, $x = 2 + 5i$

  c $x^4 - 8x^3 + 27x^2 - 38x + 26 = 0$, $x = 1 - i$

14 Use the fact that $7i$ is a root to show that the quartic equation $x^4 - x^3 + 43x^2 - 49x - 294 = 0$ can be written in the form $(x^2 + A)(x + B)(x + C) = 0$ where A, B and C are constants to be found.

Reasoning and problem-solving

You can find a polynomial equation when given its roots.

Strategy 1

To derive polynomial equations when given its roots

  (1) Use the fact that if z is a root of $f(z) = 0$ then z^* will also be a root.

  (2) Multiply the factors together.

  (3) Simplify and write the equation in descending powers of x or z

Example 5

You can apply this result to other problems that involve quadratics.

A quadratic equation has roots α and β

Show that the equation is $x^2 - (\alpha + \beta)x + \alpha\beta = 0$

(1) If α and β are roots then $x - \alpha$ and $x - \beta$ must be factors, so multiply these together.

Roots are α and β so equation is $(x - \alpha)(x - \beta) = 0$

Which becomes $x^2 - \alpha x - \beta x + \alpha\beta = 0$

(3) Expand brackets then simplify.

$\therefore x^2 - (\alpha + \beta)x + \alpha\beta = 0$

You can also use this equation and the factor theorem to prove its factors are α and β

Example 6

Two of the roots of the equation $ax^4 + bx^3 + cx^2 + dx + e = 0$ are $3 - 2i$ and $4i - 1$

Find the values of a, b, c, d and e

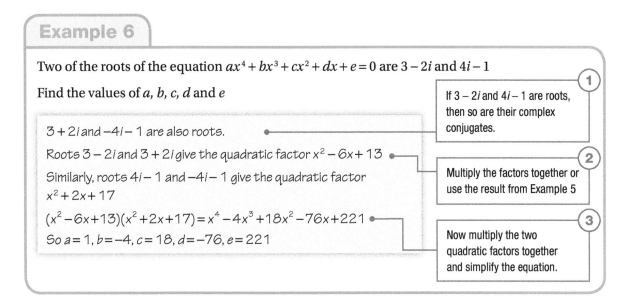

$3 + 2i$ and $-4i - 1$ are also roots.

Roots $3 - 2i$ and $3 + 2i$ give the quadratic factor $x^2 - 6x + 13$

Similarly, roots $4i - 1$ and $-4i - 1$ give the quadratic factor $x^2 + 2x + 17$

$(x^2 - 6x + 13)(x^2 + 2x + 17) = x^4 - 4x^3 + 18x^2 - 76x + 221$

So $a = 1, b = -4, c = 18, d = -76, e = 221$

1 If $3 - 2i$ and $4i - 1$ are roots, then so are their complex conjugates.

2 Multiply the factors together or use the result from Example 5

3 Now multiply the two quadratic factors together and simplify the equation.

You can solve a cubic equation by using the factor theorem to find the first solution, then dividing by the factor found and solving the remaining quadratic equation.

To find solutions to polynomial equations

1 Use the factor theorem to find one root of the equation.

2 Use the fact that if z is a root of f(z) = 0 then z^* will also be a root.

3 Divide f(z) by the factor you know.

4 Always write complex numbers in the form $a + bi$

Example 7

Solve the cubic equation $x^3 - x^2 + 5x - 14 = 0$

You could also solve this using a calculator.

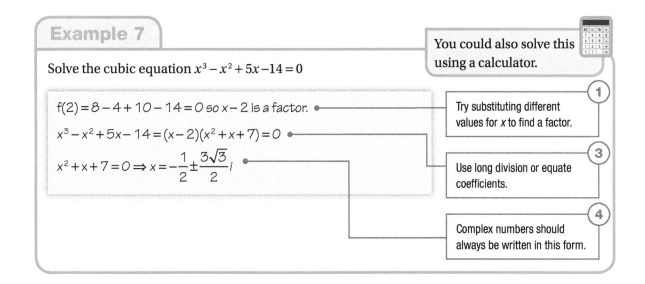

f$(2) = 8 - 4 + 10 - 14 = 0$ so $x - 2$ is a factor.

$x^3 - x^2 + 5x - 14 = (x - 2)(x^2 + x + 7) = 0$

$x^2 + x + 7 = 0 \Rightarrow x = -\dfrac{1}{2} \pm \dfrac{3\sqrt{3}}{2}i$

1 Try substituting different values for x to find a factor.

3 Use long division or equate coefficients.

4 Complex numbers should always be written in this form.

Example 8

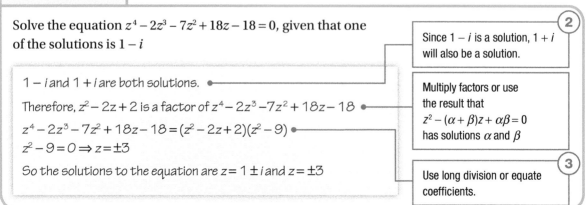

Solve the equation $z^4 - 2z^3 - 7z^2 + 18z - 18 = 0$, given that one of the solutions is $1 - i$

$1 - i$ and $1 + i$ are both solutions.

Therefore, $z^2 - 2z + 2$ is a factor of $z^4 - 2z^3 - 7z^2 + 18z - 18$

$z^4 - 2z^3 - 7z^2 + 18z - 18 = (z^2 - 2z + 2)(z^2 - 9)$

$z^2 - 9 = 0 \Rightarrow z = \pm 3$

So the solutions to the equation are $z = 1 \pm i$ and $z = \pm 3$

2 Since $1 - i$ is a solution, $1 + i$ will also be a solution.

Multiply factors or use the result that
$z^2 - (\alpha + \beta)z + \alpha\beta = 0$
has solutions α and β

3 Use long division or equate coefficients.

You can prove results involving complex numbers by manipulating them in the form $a + bi$

Strategy 3

To prove results involving complex numbers

1 Write complex numbers in the form $a + bi$

2 Use the fact that if $z = a + bi$ then $z^* = a - bi$

3 Manipulate in the usual way to prove the result.

Example 9

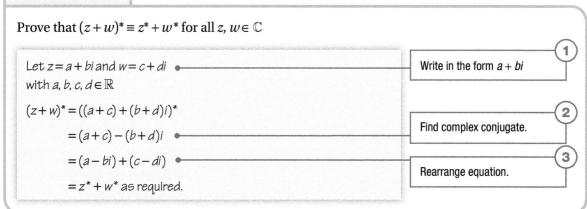

Prove that $(z + w)^* \equiv z^* + w^*$ for all $z, w \in \mathbb{C}$

Let $z = a + bi$ and $w = c + di$
with $a, b, c, d \in \mathbb{R}$

$(z + w)^* = ((a + c) + (b + d)i)^*$

$= (a + c) - (b + d)i$

$= (a - bi) + (c - di)$

$= z^* + w^*$ as required.

1 Write in the form $a + bi$

2 Find complex conjugate.

3 Rearrange equation.

Exercise 1.2B Reasoning and problem-solving

1 Prove these equations for all $z, w \in \mathbb{C}$

  a $(zw)^* = z^* w^*$

  b $(z^*)^* = z$

  c $\left(\dfrac{z}{w}\right)^* = \dfrac{z^*}{w^*}$

2 Prove that, for all $z \in \mathbb{C}$,

  a $z + z^*$ is real,

  b $z - z^*$ is imaginary,

  c zz^* is real.

3 Find the possible complex numbers z such that $z + z^* = 6$ and $zz^* = 58$

4 Find the possible complex numbers w such that $w - w^* = 18i$ and $ww^* = 85$

5 Find the complex number z such that
$z + z^* = 2\sqrt{3}$ and $\dfrac{z}{z^*} = \dfrac{1}{2} - \dfrac{1}{2}\sqrt{3}i$

6 Find the complex number w such that
$w - w^* = 4i$ and $\dfrac{w}{w^*} = \dfrac{3}{5} + \dfrac{4}{5}i$

7 Given that $x = 7$ is a solution of the equation $x^3 - 9x^2 + 31x + k = 0$

  a Find the value of k

  b Solve the equation fully.

8 Given that $x - 1$ is a factor of the equation $x^3 + 2x^2 + 5x + k = 0$, find the value of k and solve the equation.

9 Find all the solutions to the equation $x^4 - 4x^3 + 4x^2 - 4x + 3 = 0$ given that $x^2 - 4x + 3$ is a factor of $x^4 - 4x^3 + 4x^2 - 4x + 3$

10 The curve of $y = x^4 - 6x^3 - 10x^2 + 30x - 63$ is shown.

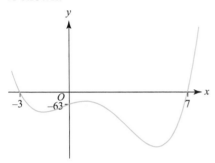

Find all the solutions to the equation $x^4 - 6x^3 - 10x^2 + 30x - 63 = 0$

11 The equation $x^4 + kx^3 + kx^2 - 110x - 111 = 0$ has a root of $i - 6$

  a Find the value of k

  b Solve the equation.

12 Given that $5 - i$ is a root of the equation $x^4 + Ax^3 + 28x^2 - 20x + 52 = 0$, find the value of A and solve the equation.

13 Find a quartic equation with repeated root

  a $1 + 3i$

  b $2i - 3$

14 $f(x) = x^3 - 19x^2 + 89x + 109$

  a Write $f(x)$ as the product of a linear and a quadratic factor.

  b Solve the equation $f(x) = 0$

  c Sketch the graph of $y = f(x)$

15 Use the fact that the quartic equation $x^4 - 4x^3 + 24x^2 - 40x + 100 = 0$ has exactly two roots to

  a Solve the equation
$$y = x^4 - 4x^3 + 24x^2 - 40x + 100$$

  b Sketch the graph of
$$y = x^4 - 4x^3 + 24x^2 - 40x + 100$$

16 The curve of a quartic equation $y = p(x)$ is shown.

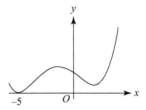

  a What does the graph tell you about the nature of the roots of $p(x) = 0$?

  Given that $p(x) = x^4 + 6x^3 - 10x^2 - 50x + 125$

  b Solve the equation $p(x) = 0$

17 If the roots of a cubic equation are a, b and c, and the coefficient of x^3 is 1, write the equation in terms of a, b and c

18 The solutions of the equation $x^3 + 5x^2 + Ax + 12 = 0$ are α, β and γ

  Write the value of

  a $\alpha + \beta + \gamma$ **b** $\alpha\beta\gamma$

19 A quartic equation has roots α, β, γ and δ

  Given that the coefficient of x^4 is 1, write an expression in terms of α, β, γ and δ for

  a The coefficient of x^3

  b The constant term.

20 Find all the solutions to

  a $x^4 = 64$

  b $x^3 = 27$

  c $x^3 = 27i$

21 Three solutions of a polynomial equation are -2, $4 - i$ and $3i$

  a State the lowest possible order of the equation.

  b Find an equation of this order with these solutions.

22 Use the result from question **1a** of this exercise to prove by induction that $(z^n)^* = (z^*)^n$ for all positive integers n

1.3 Argand diagrams

Fluency and skills

You represent real numbers visually as points on a number line, but imaginary numbers will not fit onto the real number line. Therefore, we need a new number line of imaginary numbers. Combining these two number lines together gives a plane. So complex numbers can be represented visually as points on a plane.

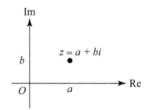

This plane is called an **Argand diagram**. Argand diagrams are used to represent complex numbers graphically.

> **Key point**
>
> The complex number $z = a + bi$ can be represented as the point (a, b) on an Argand diagram.
>
> In an Argand diagram, the horizontal axis is the **real axis** and the vertical axis is the **imaginary axis**.

Example 1

Given that $z = 1 + 3i$, show z and z^* on an Argand diagram.

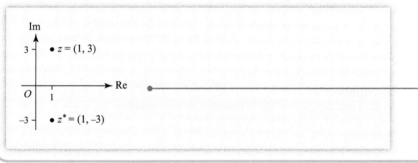

The real axis is labelled 'Re' and the imaginary axis is labelled 'Im'.

z^* is the point z reflected in the real axis.

Example 2

Complex numbers can also be shown as position vectors.

Given that $z = 4 + i$ and $w = -2 + i$, show z, w and $z + w$ on an Argand diagram.

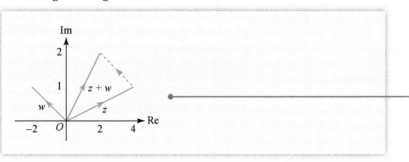

The sum can be shown as a vector addition on the Argand diagram.

Ensure you label each vector clearly.

Example 3

Given that $z = 1 + 2i$, show z, $3z$ and $-2z$ on an Argand diagram.

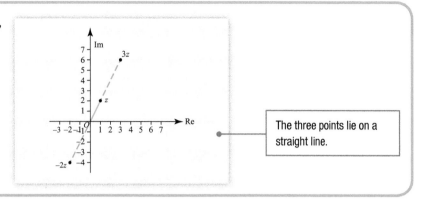

The three points lie on a straight line.

Exercise 1.3A Fluency and skills

1 Write the complex numbers represented by the vectors $\overrightarrow{OA}, \overrightarrow{OB}, \overrightarrow{OC}, \overrightarrow{OD}, \overrightarrow{OE}$ and $\overrightarrow{OF}$, as shown in the Argand diagram, in the form $a + bi$

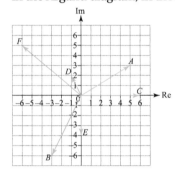

2 The Argand diagram shows the complex numbers u, v, w and z

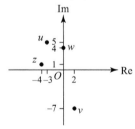

Write the complex numbers u, v, w and z in Cartesian form.

3 $z_1 = 5 - 8i$, $z_2 = 2 + 4i$

Show z_1, z_2 and $z_1 + z_2$ on an Argand diagram.

4 $z = 2 - 7i$, $w = 6i - 4$

Show z, w and $z - w$ on an Argand diagram.

5 Show the vector addition $z = (3 + 5i) + (2 - 7i)$ on an Argand diagram.

6 Show the vector subtraction $z = (6 - 2i) - (2 - 5i)$ on an Argand diagram.

7 Given that $w = \sqrt{3} + 2i$, draw w and w^* on an Argand diagram and describe the geometric relationship between them.

8 Given that $z^* = -7 + i$, draw z and z^* on the same Argand diagram and describe the geometric relationship between them.

9 Solve each of these equations and plot the solutions on an Argand diagram. Describe the geometric relationship between the two solutions in each case.

 a $x^2 = -16$ b $x^2 = -80$

10 a Solve the equation $(z + 1)^2 = -3$

 b Plot the solutions on an Argand diagram.

 c Describe the geometric relationship between the two solutions.

11 a Solve the equation $(1 - z)^2 = -25$

 b Plot the solutions on an Argand diagram.

 c Describe the geometric relationship between the two solutions.

12 For each value of z, show z and iz on an Argand diagram. What transformation maps z to iz in each case?

 a $z = -3 - 4i$ b $z = 2 - 6i$

13 For the complex numbers given, draw w and $i^2 w$ on the same Argand diagram and describe the geometrical relationship between them.

 a $w = 5 - 7i$ b $w = -2 + 3i$

Reasoning and problem-solving

To solve geometric problems involving complex numbers

(1) Draw an Argand diagram.

(2) Use rules for calculating area, lengths and angles.

(3) Use the fact that the product of gradients of perpendicular lines is −1

(4) Fully define transformations.

Example 4

$f(x) = 2x^4 + x^3 - 15x^2 + 100x - 250$

a Find all the roots of the equation $f(x) = 0$ given that $x + 5$ is a factor of $f(x)$

b Find the area of the quadrilateral formed by the points representing the four roots.

c Prove that the quadrilateral contains two right angles.

a $2x^4 + x^3 - 15x^2 + 100x - 250 = (x+5)(2x^3 - 9x^2 + 30x - 50) = 0$

Use long division or an inspection method to factorise.

$\therefore 2x^3 - 9x^2 + 30x - 50 = 0$

Solve to give roots $x = \dfrac{5}{2}, 1 \pm 3i$

Some calculators have an equation solver. This is the easiest method to use to find the three remaining roots.

b Area $= 2\left(\dfrac{1}{2} \times 7.5 \times 3\right) = 22.5$ square units

Show the roots on an Argand diagram. (1)

The quadrilateral formed is a kite which we can split into two triangles. (2)

c Gradient of $AB = \dfrac{3}{6} = \dfrac{1}{2}$

Gradient of $BC = \dfrac{-3}{1.5} = -2$

$\dfrac{1}{2} \times -2 = -1$ so $\angle ABC$ is a right angle.

The product of the gradients is −1 (3)

Similarly for $\angle ADC$

Example 5

You are given $z = 1 - 3i$

The points A, B and C are represented by the complex numbers z, z^2 and $z^2 - 4z$ respectively.

a Find the complex number $z^2 - 4z$ in the form $a + bi$

b Describe the transformation that maps line segment OA to CB

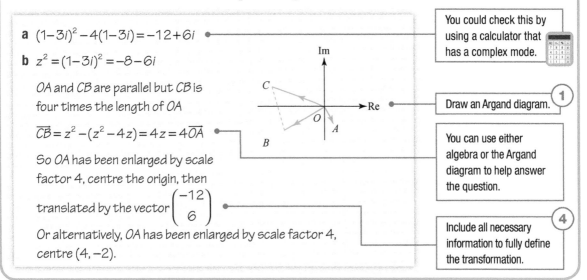

a $(1-3i)^2 - 4(1-3i) = -12 + 6i$ ●————→ You could check this by using a calculator that has a complex mode.

b $z^2 = (1-3i)^2 = -8 - 6i$

OA and CB are parallel but CB is four times the length of OA

①　Draw an Argand diagram.

$\overrightarrow{CB} = z^2 - (z^2 - 4z) = 4z = 4\overrightarrow{OA}$ ●————

So OA has been enlarged by scale factor 4, centre the origin, then

You can use either algebra or the Argand diagram to help answer the question.

translated by the vector $\begin{pmatrix} -12 \\ 6 \end{pmatrix}$ ●————

Or alternatively, OA has been enlarged by scale factor 4, centre $(4, -2)$.

④　Include all necessary information to fully define the transformation.

Exercise 1.3B Reasoning and problem-solving

1 $z_1 = 5 - 2i$, $z_2 = \dfrac{20 + 21i}{z_1}$

  a Find z_2 in the form $a + bi$

  b Show the points A and B representing z_1 and z_2 respectively on an Argand diagram.

  c Show that AOB is a right angle.

2 a Show the three roots of the equation $x^3 + x^2 + 6x - 8 = 0$ on an Argand diagram.

  b What type of triangle is formed by the points representing the three roots?

  c Find the exact area of the triangle.

3 The points A, B and C on an Argand diagram represent the solutions to the cubic equation

$x^3 - 9x^2 + 16x + 26 = 0$

Calculate the area of triangle ABC

4 a Given that the expression $x^4 + x^3 + 3x + 9$ can be written in the form $(x^2 + 3x + 3)(x^2 + Ax + B)$, find the values of A and B

  b Solve the equation $x^4 + x^3 + 3x + 9 = 0$

  c Show the roots of $x^4 + x^3 + 3x + 9 = 0$ on an Argand diagram.

  d What type of quadrilateral is formed by the points representing the roots?

  e Find the area of the quadrilateral.

5 The quartic equation

$x^4 + 8x^3 + 40x^2 + 96x + 80 = 0$

has a repeated real root.

  a Show that the repeated root is $x = -2$

  b Calculate the other solutions to the equation.

  c Show all the solutions on an Argand diagram.

6 The points A, B, C and D represent the solutions to the quartic equation

$x^4 - 3x^3 + 10x^2 - 6x - 20 = 0$

  a Use the factor theorem to find two real solutions.

  b Calculate the area of quadrilateral $ABCD$

7 The solutions of the cubic equation
$x^3 - x^2 + 9x - 9 = 0$ are represented by the
points A, B and C on an Argand diagram.

 a Prove that triangle ABC is isosceles.

 b Calculate the area of triangle ABC

8 You are given that $w = 5 - 2i$

 The points A, B and C represent the complex
numbers w, $w + iw$ and iw respectively.

 a Prove that $OABC$ is a square.

 b Find the area of the square.

9 The points P, Q and R represent the complex
numbers z, z^* and $2z + z^*$, where $z = 2 - 4i$

 a Show the points P, Q and R on an
Argand diagram.

 b Describe the transformation that maps
OP to QR

10 The points A, B and C are represented by the
complex numbers z, z^3 and $z^3 - 2z$ where
$z = 2 + i$

 a Find the complex number $z^3 - 2z$ in the
form $a + bi$

 b Describe the transformation that maps
line segment OA to CB

11 The solutions of a quartic equation
are $z = a \pm bi$ and $w = c \pm di$, where
$a, b, c, d \in \mathbb{C}, b \neq 0, d \neq 0$

 Find the area of the quadrilateral formed
by the roots on an Argand diagram.

12 Given that $z = 8 + 12i$, show the complex
numbers z, zi, zi^2 and zi^3 on an Argand
diagram.

 Describe the transformations that map z to
each of the other points.

13 The points A and B represent the complex
numbers z and iz respectively. Prove that OA
is perpendicular to OB

14 The points P and Q represent the complex
numbers $4 + 6i$ and $-3 + 4i$ respectively. Find
the area of the triangle OPQ

15 Find the area of the shape formed by the
solutions to the equation
$x^5 - x^4 + 18x^3 - 18x^2 + 81x - 81 = 0$

16 $f(z) = z^4 - 4z^3 + 56z^2 - 104z + 676$

 The equation $f(z) = 0$ has complex roots z_1
and z_2 which are represented on an Argand
diagram by the points A and B. Given that
$f(z) = 0$ has no real roots, calculate the length
of AB

17 The complex number wz is such that
$w = 3 + 2i$ and $z = 12 - 5i$

 a Show that $\dfrac{z}{w} = 2 - 3i$

 b Mark on an Argand diagram the points
A and B representing the numbers w and
$\dfrac{z}{w}$, respectively.

 c Show that triangle OAB is right-angled.

 d Hence calculate the area of triangle OAB

18 $f(z) = z^4 - az^3 + 47z^2 - bz + 290$ where $a, b \in \mathbb{R}$
are constants.

 Given that $z = 2 - 5i$ is a root of the equation
$f(z) = 0$, show all the roots of $f(z) = 0$ on an
Argand diagram.

19 $f(z) = az^4 + bz^3 + cz^2 + dz + e$ where
$a, b, c, d, e \in \mathbb{R}$ are constants.

 The solutions to $f(z) = 0$ are plotted on an
Argand diagram.

 a Describe the shape formed by the points
representing the solutions to $f(z) = 0$
when the equation has

  i Precisely two (distinct) real roots,

  ii Precisely one real root,

  iii No real roots.

 Given that the points representing the
solutions to $f(z) = 0$ form a square,

 b Calculate the values of the constants a,
b, c, d and e when

  i $z = 2i$ is a root of $f(z) = 0$,

  ii $z = 1 + 3i$ is a root of $f(z) = 0$

Modulus-argument form and loci

Fluency and skills

You have seen how the complex number $z = a + bi$ can be represented on an Argand diagram.

The length of the vector representing z is known as the **modulus** of z and written $|z|$

The angle between the positive real axis and the vector representing z is known as the **(principal) argument** of z and is written $\arg(z)$

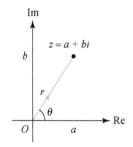

Key point

The modulus of the complex number $z = a + bi$ is given by $|z| = \sqrt{a^2 + b^2}$

Key point

Write $\arg(z) = \theta$ where $-\pi < \theta \leq \pi$

Example 1

Find the argument and modulus of the complex number $w = -3 + 3i$

$|w| = \sqrt{(-3)^2 + 3^2}$

$= 3\sqrt{2}$

Use $|w| = \sqrt{a^2 + b^2}$

$\tan\theta = \dfrac{3}{3} \Rightarrow \theta = \dfrac{\pi}{4}$

$\arg w = \pi - \dfrac{\pi}{4}$

$= \dfrac{3\pi}{4}$

An Argand diagram will help you to find the correct angle.

We always use radians on Argand diagrams: π radians is equivalent to 180°, so $\dfrac{\pi}{4}$ radians is 45°

You need the angle with the positive real axis.

Using $w = -3 + 3i$ from the previous example, if you wished to find the modulus of $-2w$ then it is clear from the diagram that $|-2w| = |-2|\,|w|$

The argument of $-2w$ can also be seen to be given by $\arg(-2w) = \arg(w) + \pi$ which is $\arg w + \arg(-2)$

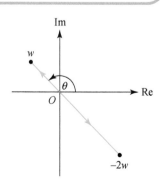

In fact, these results can be generalised for any two complex numbers.

Key point

$$|z_1 z_2| = |z_1||z_2| \text{ and } \left|\frac{z_1}{z_2}\right| = \frac{|z_1|}{|z_2|} \text{ for all } z_1, z_2 \in \mathbb{C}$$

Key point

$$\arg(z_1 z_2) = \arg(z_1) + \arg(z_2) \text{ and}$$

$$\arg\left(\frac{z_1}{z_2}\right) = \arg(z_1) - \arg(z_2) \text{ for all } z_1, z_2 \in \mathbb{C}$$

You can quote these results and will learn how to prove them in the next section.

Try it on your calculator

Some calculators can be used to find the argument of a complex number. Work out how to find the argument of $5 - 6i$ on *your* calculator.

arg(5 – 6*i*)

–0.8760580506

Example 2

Given that $z_1 = \sqrt{3} + i$ and $z_1 z_2 = -4 - 4i$, find the argument and modulus of z_2

$$|z_1| = \sqrt{(\sqrt{3})^2 + 1^2}$$
$$= 2$$

Use $|z_1| = \sqrt{a^2 + b^2}$

$$|z_1 z_2| = \sqrt{(-4)^2 + (-4)^2}$$
$$= 4\sqrt{2}$$

$$4\sqrt{2} = 2|z_2|$$

Since $|z_1 z_2| = |z_1||z_2|$

$$\Rightarrow |z_2| = 2\sqrt{2}$$

$$\arg(z_1) = \frac{\pi}{6} \text{ and } \arg(z_1 z_2) = -\frac{3}{4}\pi$$

Using complex number mode on calculator.

$$-\frac{3}{4}\pi = \frac{\pi}{6} + \arg(z_2)$$

Since $\arg(z_1 z_2) = \arg(z_1) + \arg(z_2)$

$$\Rightarrow \arg(z_2) = -\frac{11}{12}\pi$$

Instead of the form $z = a + bi$, sometimes known as Cartesian form, you can write complex numbers in **modulus** and **argument** form.

You can see in the diagram that the real component of r is $r\cos\theta$ and the imaginary component is $r\sin\theta$

Key point

The **modulus-argument form** of the complex number $z = a + bi$ is given by $z = r(\cos\theta + i\sin\theta)$ where r is the modulus of z and θ is the argument.

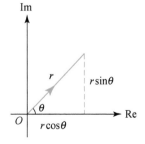

Example 3

Write the number $z = 7 - i$ in modulus-argument form.

The argument is measured in radians, where π radians is equal to 180°

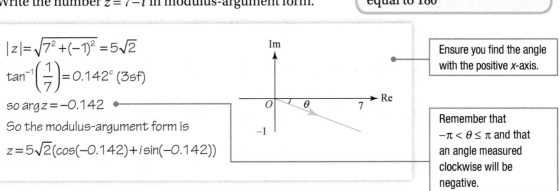

$$|z| = \sqrt{7^2 + (-1)^2} = 5\sqrt{2}$$

$$\tan^{-1}\left(\frac{1}{7}\right) = 0.142^c \text{ (3sf)}$$

so $\arg z = -0.142$

So the modulus-argument form is

$$z = 5\sqrt{2}(\cos(-0.142) + i\sin(-0.142))$$

Ensure you find the angle with the positive *x*-axis.

Remember that $-\pi < \theta \le \pi$ and that an angle measured clockwise will be negative.

Calculator

Try it on your calculator

Some calculators can be used to convert to and from modulus-argument form. Find out how to convert $\sqrt{3} + i$ to modulus-argument form on *your* calculator. Also find out how to enter a number in modulus-argument form and convert back to the form $a + bi$

$\sqrt{3} + i \blacktriangleright r\angle\theta$

$2\angle\dfrac{\pi}{6}$

You can draw **loci** in an Argand diagram. These are the set of points that obey a given rule.

Example 4

Sketch the locus of points that satisfy $|z| = 4$

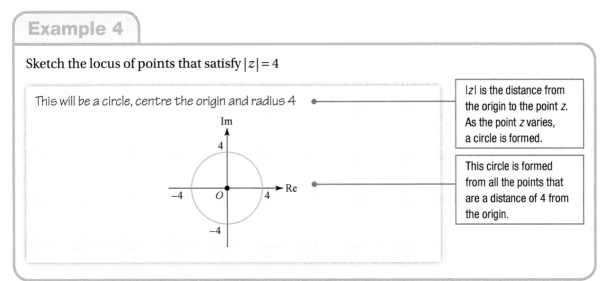

This will be a circle, centre the origin and radius 4

|z| is the distance from the origin to the point *z*. As the point *z* varies, a circle is formed.

This circle is formed from all the points that are a distance of 4 from the origin.

If we have any fixed point z_1, then

Key point

The locus of points satisfying $|z - z_1| = r$ will be a circle centre z_1 and radius r

Example 5

Sketch the locus of points that satisfy $|z-2+3i|=2$

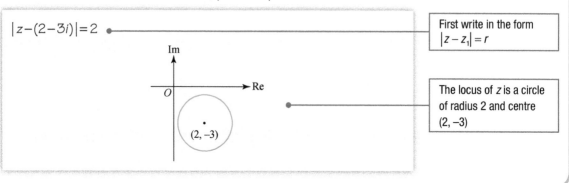

$|z-(2-3i)|=2$

First write in the form $|z-z_1|=r$

The locus of z is a circle of radius 2 and centre $(2, -3)$

Example 6

Sketch the locus of points satisfying $\arg(z)=\dfrac{2\pi}{3}$

This will be a line that makes an angle of $\dfrac{2\pi}{3}$ with the positive real axis.

$\arg(z)$ is the angle between the positive real axis and the line representing z

Notice that the line ends at the origin.

Key point

The locus of points satisfying $\arg(z-z_1)=\theta$ is a **half-line** from the point z_1 at an angle of θ to the positive real axis.

Example 7

Sketch the locus of z where $\arg(z+2+3i)=\dfrac{\pi}{6}$

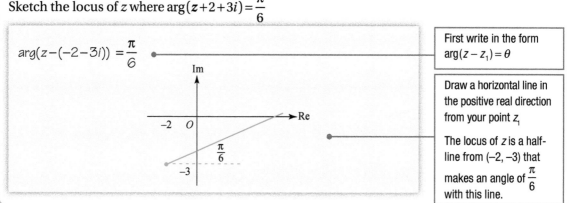

$\arg(z-(-2-3i))=\dfrac{\pi}{6}$

First write in the form $\arg(z-z_1)=\theta$

Draw a horizontal line in the positive real direction from your point z_1

The locus of z is a half-line from $(-2, -3)$ that makes an angle of $\dfrac{\pi}{6}$ with this line.

The locus of points satisfying $|z-z_1|=|z-z_2|$ is the perpendicular bisector of the line joining z_1 and z_2

This is because the locus includes all points that are equidistant from the fixed points z_1 and z_2

Example 8

Sketch the locus of points satisfying $|z-2|=|z+3i|$

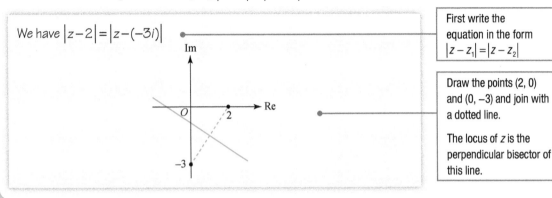

We have $|z-2|=|z-(-3i)|$

First write the equation in the form $|z-z_1|=|z-z_2|$

Draw the points (2, 0) and (0, −3) and join with a dotted line.

The locus of z is the perpendicular bisector of this line.

Exercise 1.4A Fluency and skills

1   Find the modulus and argument of each complex number.

   a $12+5i$ **b** $4-3i$

   c $-3i$ **d** $-6-8i$

   e $-1+7i$ **f** $-2-i$

   g $-\sqrt{2}-\sqrt{2}i$ **h** $\sqrt{3}-\sqrt{6}i$

2   Verify in each case that $|zw|=|z||w|$ and $\left|\dfrac{z}{w}\right|=\dfrac{|z|}{|w|}$

   a $z=1+3i,\,w=-5+2i$

   b $z=-2-i,\,w=\sqrt{5}i$

   c $z=-\sqrt{3}+6i,\,w=1-\sqrt{3}i$

3   In each case, verify that $\arg(zw)=\arg z+\arg w$ and $\arg\left(\dfrac{z}{w}\right)=\arg z-\arg w$

   a $z=1+i,\,w=3+\sqrt{3}i$

   b $z=i,\,w=2-2i$

   c $z=\sqrt{3}-3i,\,w=-2i$

4   Given that $z=1-\sqrt{3}i$ and $zw=\sqrt{3}+3i$, find the modulus and the argument of w

5   Given that $z_1=-1-i$ and $\dfrac{z_1}{z_2}=3\sqrt{2}+\sqrt{6}i$, find the modulus and the argument of z_2

6   Given that $z=-2\sqrt{3}+6i$ and $\dfrac{w}{z}=\sqrt{3}-i$, find the modulus and the argument of w

7   Write each of these complex numbers in Cartesian form.

   a $3\left(\cos\left(\dfrac{\pi}{2}\right)+i\sin\left(\dfrac{\pi}{2}\right)\right)$

   b $5(\cos(-\pi)+i\sin(-\pi))$

   c $10\left(\cos\left(\dfrac{5\pi}{6}\right)+i\sin\left(\dfrac{5\pi}{6}\right)\right)$

   d $\sqrt{3}\left(\cos\left(-\dfrac{2\pi}{3}\right)+i\sin\left(-\dfrac{2\pi}{3}\right)\right)$

8   Write each of these numbers in modulus-argument form.

   a $z=3+3i$ **b** $z=1-\sqrt{3}i$

   c $z=-2\sqrt{3}-2i$ **d** $z=-4+9i$

9   Sketch the locus of z in each case.

   a $|z|=7$ **b** $|z-2|=5$

   c $|z-i|=3$ **d** $|z-(1+2i)|=2$

   e $|z-3+5i|=5$ **f** $|z+4-2i|=4$

10 Sketch the locus of z in each case.

a $\arg z = -\dfrac{\pi}{4}$ **b** $\arg(z-3) = \dfrac{\pi}{2}$

c $\arg(z+i) = \dfrac{3\pi}{4}$ **d** $\arg(z-2i) = -\dfrac{\pi}{6}$

e $\arg(z-2+i) = \dfrac{5\pi}{6}$ **f** $\arg(z-4-i) = -\dfrac{2\pi}{3}$

g $\arg(z+5-7i) = -\dfrac{\pi}{3}$

11 Sketch the locus of z in each of these cases.

a $|z| = |z+4|$ **b** $|z-2i| = |z|$

c $|z-2i| = |z+2|$ **d** $|z+6+2i| = |z+6|$

e $|z+4-i| = |z-5+2i|$

Reasoning and problem-solving

You need to be able to prove the results: $\left|\dfrac{z}{w}\right| = \dfrac{|z|}{|w|}$, $|zw| = |z||w|$, $\arg\left(\dfrac{z}{w}\right) = \arg z - \arg w$ and $\arg(zw) = \arg z + \arg w$

Strategy 1

To prove results about modulus and argument

  1 Write complex numbers in modulus-argument form.

  2 Simplify powers of i

  3 Split into real and imaginary parts.

  4 Use the addition formulae for sine and for cosine.

Example 9

Prove that $\left|\dfrac{z}{w}\right| = \dfrac{|z|}{|w|}$ and $\arg\left(\dfrac{z}{w}\right) = \arg z - \arg w$ for all $z, w \in \mathbb{C}$

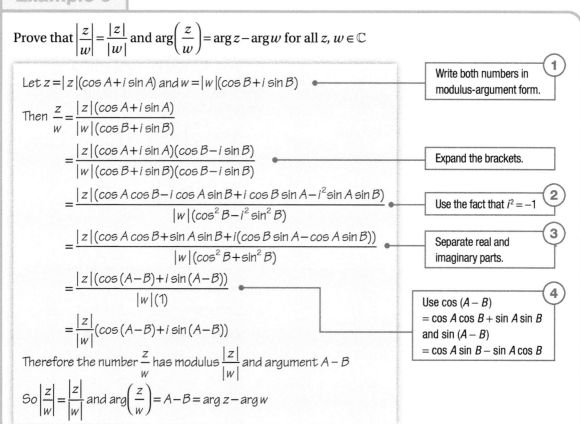

Let $z = |z|(\cos A + i \sin A)$ and $w = |w|(\cos B + i \sin B)$

> **1** Write both numbers in modulus-argument form.

Then $\dfrac{z}{w} = \dfrac{|z|(\cos A + i \sin A)}{|w|(\cos B + i \sin B)}$

$= \dfrac{|z|(\cos A + i \sin A)(\cos B - i \sin B)}{|w|(\cos B + i \sin B)(\cos B - i \sin B)}$

> Expand the brackets.

$= \dfrac{|z|(\cos A \cos B - i \cos A \sin B + i \cos B \sin A - i^2 \sin A \sin B)}{|w|(\cos^2 B - i^2 \sin^2 B)}$

> **2** Use the fact that $i^2 = -1$

$= \dfrac{|z|(\cos A \cos B + \sin A \sin B + i(\cos B \sin A - \cos A \sin B))}{|w|(\cos^2 B + \sin^2 B)}$

> **3** Separate real and imaginary parts.

$= \dfrac{|z|(\cos(A-B) + i \sin(A-B))}{|w|(1)}$

$= \dfrac{|z|}{|w|}(\cos(A-B) + i \sin(A-B))$

> **4** Use $\cos(A-B)$
> $= \cos A \cos B + \sin A \sin B$
> and $\sin(A-B)$
> $= \cos A \sin B - \sin A \cos B$

Therefore the number $\dfrac{z}{w}$ has modulus $\dfrac{|z|}{|w|}$ and argument $A - B$

So $\left|\dfrac{z}{w}\right| = \dfrac{|z|}{|w|}$ and $\arg\left(\dfrac{z}{w}\right) = A - B = \arg z - \arg w$

You know how to find the Cartesian equation of certain loci by drawing the graph and using known equations of circles or lines. However, it is possible to find the Cartesian equation of any locus by setting $z = x + iy$ and finding a relationship between x and y

See Maths Ch1.5 For a reminder of the equations of lines and circles.

Strategy 2

To find the Cartesian equation of a locus

(1) Write z as $x + iy$

(2) Calculate the modulus.

(3) Use tan to form an equation from the argument.

(4) Rearrange to the required form.

Example 10

Find the Cartesian equations of these loci.

a $\left|z - 3 + 4i\right| = \sqrt{5}$

b $\arg(z + i) = -\dfrac{\pi}{6}$

a Let $z = x + iy$ *Write in Cartesian form.* (1)

$$\left|x + iy - 3 + 4i\right| = \sqrt{5}$$

$$\sqrt{(x-3)^2 + (y+4)^2} = \sqrt{5}$$ *Find the modulus of the left-hand side.* (2)

$$(x-3)^2 + (y+4)^2 = 5$$

So the locus is a circle with centre $(3, -4)$ and radius $\sqrt{5}$

b Let $z = x + iy$ *Write in Cartesian form.* (1)

$$\arg(x + iy + i) = -\frac{\pi}{6} \text{ so } \arg(x + (y+1)i) = -\frac{\pi}{6}$$

$$\frac{y+1}{x} = \tan\left(-\frac{\pi}{6}\right) \text{ so } \frac{y+1}{x} = -\frac{\sqrt{3}}{3}$$ *Since $\tan\theta = \dfrac{\text{opposite}}{\text{adjacent}}$* (3)

$$\therefore \sqrt{3}x + 3y + 3 = 0$$ *When rearranged, you can see this is the equation of a line.* (4)

As the locus is a half-line, we only need the part where $x > 0$ and $y < -1$

Strategy 3

To find a region bounded by a locus

(1) Sketch the locus of the boundary of the region.

(2) Test a point to see if it is inside the region or not.

(3) Shade the correct area.

See Maths Ch1.7 For a reminder of set notation.

Example 11

Shade these sets of points.

a $\{z \in \mathbb{C} : |z-5| \geq 3\}$

b $\left\{z \in \mathbb{C} : 0 < \arg(z-4i) < \dfrac{3\pi}{4}\right\}$

$\{z \in \mathbb{C} : |z-5| \geq 3\}$ is the set of complex numbers z such that $|z-5| \geq 3$

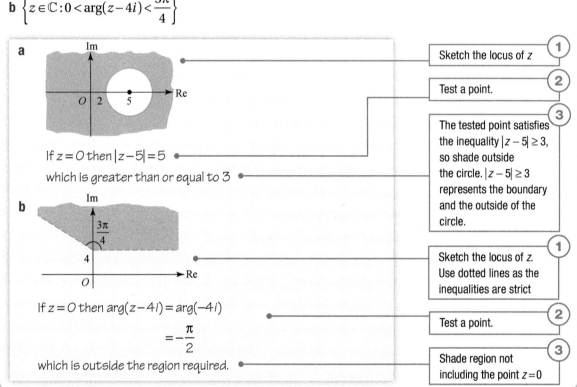

a

If $z = 0$ then $|z-5| = 5$

which is greater than or equal to 3

b

If $z = 0$ then $\arg(z-4i) = \arg(-4i)$

$= -\dfrac{\pi}{2}$

which is outside the region required.

① Sketch the locus of z

② Test a point.

③ The tested point satisfies the inequality $|z-5| \geq 3$, so shade outside the circle. $|z-5| \geq 3$ represents the boundary and the outside of the circle.

① Sketch the locus of z. Use dotted lines as the inequalities are strict

② Test a point.

③ Shade region not including the point $z=0$

Exercise 1.4B Reasoning and problem-solving

1 Prove that $|zw| = |z||w|$ and $\arg(zw) = \arg z + \arg w$ for all $z, w \in \mathbb{C}$

2 Show that the locus of points satisfying $|z+3-2i| = 4$ is a circle and sketch it.

3 a Show that the locus of points satisfying $|z-2-i| = 2$ is a circle.

The circle touches the imaginary axis at the point A and crosses the real axis at B and C

b Calculate the exact area of triangle ABC

4 Find the Cartesian equation of these loci.

a $|z-5| = |z|$ **b** $|z+2| = |z-2i|$

c $|z-4i| = |z+2|$ **d** $|z+1-i| = |z-3|$

e $|z-5+i| = |z+i-2|$

f $|z-7+4i| = |z+6-3i|$

5 In each case, find the Cartesian equation of the line on which the half-line lies.

a $\arg(z-3i) = \dfrac{\pi}{4}$ **b** $\arg(z+5) = \dfrac{\pi}{2}$

c $\arg(z+2-i) = \dfrac{\pi}{3}$ **d** $\arg(z-4+i) = \dfrac{2\pi}{3}$

6 Find the Cartesian equation of the locus of points satisfying $|z-2|=|z+3-i|$

7 Find the Cartesian equation of the locus of $\left|z-\sqrt{2}+i\right|=\left|z-1-\sqrt{2}i\right|$

8 Find the Cartesian equation of the locus of $|z-2|=|4i-z|$

9 Shade the region that satisfies

    a $|z-4i|\le 3$ **b** $|z+2-i|\ge 1$

    c $4\le |z|\le 10$ **d** $2<|z-5+2i|<5$

10 Find the area of the region that satisfies $\sqrt{7}\le |z+3-7i|\le 7$

11 Shade the region represented by each inequality.

    a $0\le \arg z\le \dfrac{\pi}{3}$

    b $-\dfrac{5\pi}{6}<\arg(z-3i)<0$

    c $\dfrac{\pi}{2}\le \arg(z-2-i)<\dfrac{3\pi}{4}$

12 Find the area of the region that satisfies $|z-5+2i|<8$ and $0\le \arg(z-5+2i)\le \dfrac{\pi}{2}$

13 Shade the region that satisfies

    a $|z-3|\ge |z+5|$

    b $|z-i|<|z-3i|$

    c $|z-2-4i|\le |z+8-4i|$

    d $|z-3-i|>|z-5+3i|$

    e $|z-1+2i|\le |z-3-2i|$ and $|z|>|z-2|$

14 Sketch and shade the region that satisfies both $|z-3+3i|\le 3$ and $-\dfrac{\pi}{4}\le \arg z\le 0$

15 Sketch and shade the region that satisfies both $|z|\le |z-8i|$ and $|z-2i|\ge 8$

16 Sketch and shade the region that satisfies both $\dfrac{\pi}{3}<\arg(z-2)<\dfrac{2\pi}{3}$ and $|z-2i|<|z-4i|$

17 Shade on an Argand diagram the set of points

$\{z\in \mathbb{C}:|z+i-2|\le 1\}$

18 Shade on an Argand diagram the set of points

$\{z\in \mathbb{C}:|z+2i|>|z-2|\}$

19 Shade on an Argand diagram the set of points

$\left\{z\in \mathbb{C}:|z|\ge 4\right\}\cap \left\{z\in \mathbb{C}:-\dfrac{\pi}{3}<\arg(z)<\dfrac{\pi}{3}\right\}$

20 Shade on an Argand diagram the set of points

$\left\{z\in \mathbb{C}:|z-3i|>|z+5i|\right\}\cap \left\{z\in \mathbb{C}:-\pi \le \arg(z)\le -\dfrac{\pi}{2}\right\}$

21 **a** Use algebra to show that the locus of points satisfying $|z+3|=2|z-6i|$ is a circle, then sketch it.

    b Shade the region that satisfies $|z+3|\le 2|z-6i|$ and $|z-1-3i|\le 20$

22 The point P represents a complex number z on an Argand diagram such that $|z-3+i|=3$

    a State the Cartesian equation of the locus of P

The point Q represents a complex number z on an Argand diagram such that $|z+2-i|=|z-1+2i|$

    b Find the Cartesian equation of the locus of Q

    c Find the complex number that satisfies both $|z-3+i|=3$ and $|z+2-i|=|z-3+2i|$, giving your answer in surd form.

23 Find the complex number that satisfies both $|z-3i|=4$ and $\arg(z-3i)=-\dfrac{\pi}{4}$

Chapter summary

- The imaginary number i is defined as $i = \sqrt{-1}$
- Complex numbers written in the form $a + bi$ can be added, subtracted and multiplied in the same way as algebraic expressions.
- Powers of i should be simplified: $i^2 = -1$, $i^3 = -i$ and so on.
- The complex conjugate of the number $z = a + bi$ is $z^* = a - bi$
- Fractions with a complex number in the denominator can be simplified by multiplying the numerator and the denominator by the complex conjugate of the denominator.
- Complex roots of polynomial equations occur in conjugate pairs.
- The complex number $z = a + bi$ can be represented by the point (a, b) on an Argand diagram.
- The modulus of a complex number $z = a + bi$ is given by $|z| = \sqrt{a^2 + b^2}$
- The (principal) argument of a complex number is the angle between the vector representing it and the positive real axis. Write $\arg z = \theta$ where $-\pi < \theta \le \pi$
- $|z_1 z_2| = |z_1||z_2|$ and $\left|\dfrac{z_1}{z_2}\right| = \dfrac{|z_1|}{|z_2|}$ for all z_1, $z_2 \in \mathbb{C}$
- $\arg(z_1 z_2) = \arg(z_1) + \arg(z_2)$ for all z_1, $z_2 \in \mathbb{C}$
- The modulus-argument form of the complex number $z = a + bi$ is given by $z = r(\cos\theta + i\sin\theta)$ where r is the modulus of z and θ is the argument.
- The locus of points satisfying $|z - z_1| = r$ will be a circle, centre z_1 and radius r
- The locus of points satisfying $\arg(z - z_1) = \theta$ is a half-line from the point z_1 at an angle of θ to the positive real axis.
- The locus of points satisfying $|z - z_1| = |z - z_2|$ is the perpendicular bisector of the line joining z_1 and z_2

Check and review

You should now be able to...	Try Questions
✔ Add, subtract, multiply and divide complex numbers in the form $a + bi$	1
✔ Understand and use the complex conjugate.	1–3
✔ Solve quadratic, cubic and quartic equations with real coefficients.	2–4
✔ Calculate the modulus and the argument of a complex number.	5
✔ Convert between modulus-argument form and the form $a + bi$	6, 7
✔ Multiply and divide numbers in modulus-argument form.	8
✔ Sketch and interpret Argand diagrams.	9
✔ Construct and interpret loci in the Argand diagram.	10–14

1 Given that $z = 5 - 4i$ and $w = -2 - 3i$, find each of these in the form $a + bi$

a $z + w$ b $3w$

c $2z - w$ d zw

e $z^2 + w^2$ f $(z + w)^2$

g z^* h w^*

i $\dfrac{z}{w}$ j $\dfrac{3w}{2z}$

k $2 \div z$ l $w^* \div 3i$

2 Solve these equations.

a $x^2 - 4x + 20 = 0$

b $x^2 + 6x + 10 = 0$

c $x^2 - 10x + 27 = 0$

3 Solve these equations using the root given.

a $x^3 - 4x^2 + 69x - 130 = 0$, $x = 2$

b $x^3 - x^2 - 20x + 50 = 0$, $x = -5$

c $x^3 - 22x^2 + 154x - 328 = 0$, $x = 9 + i$

d $x^4 + 4 = 0$, $x = 1 + i$

4 Solve the quartic equation $f(x) = 0$ in each case using the complex root given.

a $f(x) = x^4 + x^3 - 8x^2 - 3x + 35$,
$x = \dfrac{1}{2}(-5 + \sqrt{3}i)$ is a root.

b $f(x) = x^4 - 4x^3 + x - 4$, $x = \dfrac{1}{2}(1 + \sqrt{3}i)$ is a root.

5 Calculate the modulus and the argument of these complex numbers.

a $2 + 9i$ b $3 - 3i$

c $7i$ d $-2i$

e $-1 + 4i$ f $-3 - 4i$

6 Write these complex numbers in modulus-argument form.

a $8 + 6i$ b $-12 + 5i$

c $-2 - 2i$ d $\sqrt{3} - i$

e $\sqrt{5}\left(\cos\left(\dfrac{\pi}{3}\right) - i\sin\left(\dfrac{\pi}{3}\right)\right)$

7 Write these in the form $a + bi$

a $2\left(\cos\left(\dfrac{\pi}{6}\right) + i\sin\left(\dfrac{\pi}{6}\right)\right)$

b $\sqrt{3}\left(\cos\left(-\dfrac{\pi}{4}\right) + i\sin\left(-\dfrac{\pi}{4}\right)\right)$

8 Given that
$$z = \sqrt{6}\left(\cos\left(-\dfrac{\pi}{3}\right) + i\sin\left(-\dfrac{\pi}{3}\right)\right)$$
and
$$w = \sqrt{3}\left(\cos\left(\dfrac{\pi}{6}\right) + i\sin\left(\dfrac{\pi}{6}\right)\right)$$
find these in modulus-argument form.

a zw b $\dfrac{z}{w}$

c $\dfrac{w}{z}$

9 Given that $z_1 = 5 - 2i$ and $z_2 = -2 + 3i$, draw these complex numbers on the same Argand diagram.

a z_1 b z_2

c $z_1 + z_2$ d $z_1 - z_2$

10 Sketch these loci on separate Argand diagrams and give the Cartesian equation of each.

a $|z| = 7$ b $|z - 8| = 5$

c $|z + 3 - i| = 3$ d $|z - 2 - 3i| = 2$

11 Sketch these loci on separate Argand diagrams.

a $\arg z = \dfrac{\pi}{6}$ b $\arg(z - 3) = -\dfrac{\pi}{3}$

c $\arg(z + 2i) = \dfrac{2\pi}{3}$ d $\arg(z + 1 - i) = -\dfrac{\pi}{4}$

12 Sketch these loci on separate Argand diagrams and give the Cartesian equation of each.

a $|z| = |z - 6|$ b $|z + 2i| = |z - 8i|$

c $|z - 2i| = |z + 4|$ d $|z - 1 - i| = |z + 1 + i|$

13 Sketch and shade the region satisfying each inequality.

a $|z - 5 + 2i| \leq 2$ b $\dfrac{\pi}{2} \leq \arg(z - i) < \dfrac{5\pi}{6}$

c $3 \leq |z + 3 - 5i| \leq 5$ d $|z - 6i| < |z + 6|$

14 Sketch and shade the region satisfying both of these inequalities.

$|z - 7| \geq 7$ and $-\dfrac{\pi}{2} \leq \arg(z - 7) \leq 0$

Exploration
Going beyond the exams

History

Jean-Robert Argand was a Swiss mathematician who, early in the 19th century, moved to Paris where he managed a bookshop. He published an essay on the representation of imaginary quantities in 1813 and, although others before him had written about similar ideas, it was Argand's essay that gained attention and resulted in the naming of the **Argand diagram** we continue to use today.

Note

A **conic section** is a curve that is found by intersecting a plane with a double cone placed vertex to vertex.

There are three types of conic section: ellipse, parabola and hyperbola. A circle is a type of ellipse.

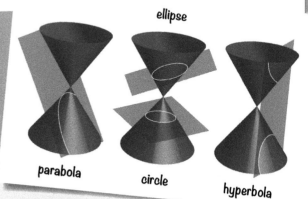

ellipse

parabola   circle   hyperbola

Investigation

The equation $|z|=1$ plotted in the complex plane gives a circle with unit radius centred at the origin.

Choose values of a, b and c such that $c > (b - a)$ and $b > a$

By varying these values, investigate the graph of the locus of point z that moves so that $|z - a| + |z - b| = c$, $c > (b - a)$ and $b > a$

What is the shape of the graph that results?

The real part of a complex number z can be denoted by $\text{Re}(z)$

By varying the values of a and b, investigate graphs of the locus of a point z that moves so that $\text{Re}(z + a) = |z - b|$

What is the shape of the graph that results?

Research

The **Mandelbrot set** provides beautiful images. It is the set of all the complex numbers, c, for which the iterative sequence $z_{n+1} = z_n^2 + c$, where $z_0 = 0$ does not tend to infinity.

Computer images of the Mandelbrot set can be developed and, no matter how closely you zoom in on the surface, it still remains incredibly intricate. The Mandelbrot set is an example of a **fractal**.

Find out if the complex numbers i and $2i$ are in the Mandelbrot set. Research the properties of the Mandelbrot set.

1 The complex numbers z and w are given as $z = 9 - 11i$ and $w = 7 + 3i$

    a Calculate these expressions, giving the answers in the form $a + bi$

        i $w - z$ **ii** wz **iii** $\dfrac{w}{z}$ **[5 marks]**

    b Calculate the modulus and the argument of z, giving your answers to 3 significant figures. **[3]**

2 The quadratic equation $x^2 + 2x + 5 = 0$ has complex roots α and β

    a Find α and β **[2]**

    b Show α and β on an Argand diagram. **[2]**

    c Describe the transformation that maps α to β **[1]**

3 $z = \dfrac{5 + 2i}{3 + i}$

    a Show that $z = a + bi$ where a and b are constants to be found. **[3]**

    b Calculate the value of

        i $z + z^*$ **ii** $z - z^*$ **iii** zz^* **[4]**

    c **i** Draw z and iz on the same Argand diagram.

        ii Describe the geometric relationship between z and iz **[3]**

4 Find the values of a and b such that $(a + bi)^2 = -15 + 8i$ **[6]**

5 Find the values of a and b such that $(a + bi)^2 = 1 + 2\sqrt{2}i$ **[6]**

6 $f(x) = x^3 - 2x^2 + 5x + 26$

    a Given that $x = 2 - 3i$ is a solution to $f(x) = 0$, find all the roots of $f(x)$ **[3]**

    b Find the area of the triangle formed by the three roots of $f(x)$ **[2]**

7 Given that $5 - i$ is a root of the equation $x^3 + ax^2 + bx - 182 = 0$

    a Find the other two roots, **[4]**

    b Calculate the values of a and b **[3]**

8 Solve the equation $z - 2z^* = -6 + 27i$ **[3]**

9 Solve the simultaneous equations

    $z - 2w = 8 + 5i$

    $2z - 3w = 13 + 3i$ **[4]**

10 Given that $z = 3 + 3\sqrt{3}i$ and $w = -\sqrt{2} + \sqrt{2}i$

    a Calculate the modulus and argument of z **[3]**

    b Calculate the modulus and argument of w **[3]**

c Hence, or otherwise, find the value of

   i $|zw|$ **ii** $\left|\dfrac{z}{w}\right|$

   iii $\arg(zw)$ **iv** $\arg\left(\dfrac{z}{w}\right)$ **[6]**

11 Given that $z = 4 + \sqrt{2}i$ and $zw = 5\sqrt{2} - 2i$

   a Find w in modulus-argument form.   **[6]**

   The points A and B represent w and zw on an Argand diagram.

   b Calculate the distance AB, giving your answer as a surd.   **[2]**

12 a Write the complex number $z = \sqrt{3} - i$ in modulus-argument form.   **[3]**

   The complex number w has argument $-\dfrac{\pi}{12}$

   b Find the argument of

   i zw **ii** $\dfrac{z}{w}$ **[4]**

   c Find $|w|$ given that $|zw| = 10$ **[2]**

13 a Sketch the locus of points that satisfy

   i $|z - 4i| = 2$ **ii** $|z + 3| = |z - 3i|$ **[4]**

   b Find the Cartesian equation of the locus of points drawn in part **a**.   **[3]**

14 a Show that the locus of points satisfying $|z + 3 - 4i| = 4$ is a circle.   **[3]**

   b Sketch the locus of points satisfying $|z + 3 - 4i| = 4$ **[2]**

   c Shade in the region that satisfies $|z + 3 - 4i| \le 4$ **[2]**

15 For the locus of points satisfying $|z + 5| = |z - i|$

   a Sketch the locus,   **[2]**

   b Find the Cartesian equation.   **[3]**

16 Find the square roots of the complex number $39 - 80i$ **[6]**

17 Solve the equation $z^2 = 4 - 2\sqrt{5}i$ **[6]**

18 The solutions of the quartic equation $x^4 - 14x^3 + 59x^2 - 14x + 58 = 0$ are represented on an Argand diagram by the points P, Q, R and S

   One solution to the equation is $w = 7 + 3i$

   Calculate the area of the quadrilateral $PQRS$ **[6]**

19 Two solutions of a cubic equation are $x = -3$ and $x = i - 4$

   a State the third solution to the equation.   **[1]**

   b Find a possible equation with these solutions.   **[3]**

20 Two solutions of a quartic equation are $z = -2 + i$ and $z = 3 - 5i$

   a State the two other solutions of the equation.   **[2]**

   b Find a possible equation with these solutions.   **[3]**

21 A quartic equation $ax^4 + bx^3 + cx^2 + dx + e = 0$ has exactly two distinct solutions, one of which is $1 + 5i$

Find the values of a, b, c, d and e **[4]**

22 Solve the simultaneous equations

$$z^2 - w^2 = -6$$
$$z + 2w = 3$$ **[6]**

23 Solve these simultaneous equations.

$$z^2 + w^2 = 30$$
$$z + 3w = 20$$ **[4]**

24 The equation $2x^5 + x^4 + 36x^3 + 18x^2 + 162x + 81 = 0$ has a repeated quadratic root.

  a Show that $x = 3i$ is a solution to the equation. **[2]**

  b Fully factorise the equation. **[3]**

25 The curve of $y = x^4 - 4x^3 - x^2 + 6x + 18$ is shown.

Find all the solutions to the equation
$x^4 - 4x^3 - x^2 + 6x + 18 = 0$ **[4]**

26 A quadratic equation has roots α and β,

where α has argument $\dfrac{5\pi}{6}$ and modulus 3

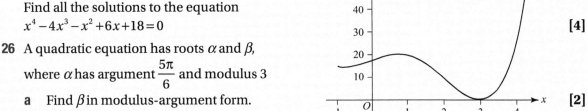

  a Find β in modulus-argument form. **[2]**

  b Calculate

    i $|\alpha\beta|$ **ii** $\left|\dfrac{\alpha}{\beta}\right|$ **iii** $\arg(\alpha\beta)$ **[6]**

  c Find the quadratic equation with roots α and β **[3]**

27 The complex numbers z_1 and z_2 are given by $z_1 = 3 + ai$ and $z_2 = 2 - i$ where a is an integer.

  a Find $\dfrac{z_1}{z_2}$ in terms of a **[3]**

Given that $\left|\dfrac{z_1}{z_2}\right| = \sqrt{18}$

  b Find the possible values of a **[5]**

28 Given that $w = \dfrac{1}{2}\left(\cos\dfrac{\pi}{3} - i\sin\dfrac{\pi}{3}\right)$

  a Write w in modulus-argument form, **[1]**

  b Find the complex number z such

    that $\arg(wz) = -\dfrac{\pi}{6}$ and $\left|\dfrac{w^2}{z}\right| = 3$

Write your answer in exact Cartesian form. **[3]**

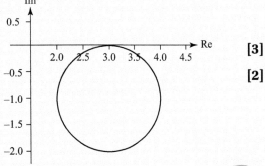

29 Describe the locus shown in terms of z **[2]**

30 a Sketch the locus of points that satisfy $\arg(z-2i)=\dfrac{\pi}{4}$ **[3]**

b Find the Cartesian equation of the locus drawn in part **a**. **[3]**

31 Sketch and shade the region satisfying

$0 < \arg(z+1+3i) \le \dfrac{2\pi}{3}$ **[5]**

32 Sketch and shade the region satisfying
$|z-i| \ge 1$ and $|z+i| \le 2$ **[4]**

33 The complex numbers $4-3i$ and $-2-i$ represent the points A and B respectively.

Find the area of the triangle OAB **[5]**

34 The locus of the complex number z is a half-line as shown.

Describe the locus in terms of z **[3]**

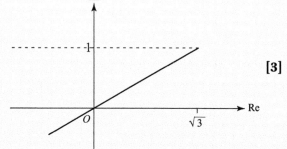

35 The point P represents a complex number z on an Argand diagram such that

$|z+2| = 3|z-2i|$

Show that, as z varies, the locus of P is a circle and state the radius of the circle and the coordinates of the centre. **[6]**

36 The point P represents a complex number z on an Argand diagram such that
$|z-1| = 1$

The point Q represents a complex number z on an Argand diagram such that
$\arg(z-i) = -\dfrac{\pi}{4}$

a Sketch the loci of P and Q on the same axes. **[4]**

b Find the complex number that satisfies both

$|z-1| = 1$ and $\arg(z-i) = -\dfrac{\pi}{4}$ **[6]**

37 a Find the possible complex numbers, z, that satisfy both $|z+3i| = |z-i|$ and $|z-2+i| = 4$ **[4]**

b Sketch the region that satisfies both $|z+3i| > |z-i|$ and $|z-2+i| \ge 4$ **[5]**

38 a Shade the region that satisfies $1 \le |z-3i| \le 3$ **[4]**

b Find the exact area of the shaded region. **[2]**

39 a Shade the region that satisfies both

$|z+i-3| \le 2$ and $-\dfrac{\pi}{2} \le \arg(z+i) \le 0$ **[4]**

b Find the exact area of the shaded region. **[2]**

2
Algebra and series

Algebraic methods can help to ensure certainty in computer programming. They help programmers to confirm that their software will work under any circumstances. One application of this is in online banking, where security of information is essential and an error in the computer code could cost somebody their life savings.

One technique that can be used to prove programme correctness is proof by induction. This technique focuses on a small part of the program which uses loops. This application of proof by induction checks that a certain step in the procedure works, and then uses the fact that this step is repeated many times over to confirm that the whole system provides the required outputs.

Algebra is fundamental to all aspects modern technology: from internet search engines and computer games to self-driving cars, the ability to model situations is a valuable tool both for understanding the world as it is and for making the next technological leap.

Orientation

What you need to know	What you will learn	What this leads to
Maths Ch1 • Mathematical proof by deduction and exhaustion.	• To use proof by induction. • To relate the roots of a polynomial to its coefficients. • To find polynomials whose roots are linear transformations of another polynomial. • To evaluate expressions involving roots. • To use formulae for sums of integers, squares and cubes to sum other series.	**Ch7 Series and further curve sketching** • Summing series and the method of differences.
Maths Ch2 • The method of algebraic division and the factor theorem.		**FP2 Ch6 Recurrence relations** • Inductive proofs of closed forms.
Maths Ch13 • Summing arithmetic and geometric series.		

Fluency and skills

See Maths Ch 2.3

For a reminder of the factor theorem.

You already know how to find the roots of polynomials using the factor theorem and long division. You are now going to learn how to transform one polynomial into another polynomial with roots that are related in some way.

Quadratic equations

Let the quadratic equation $ax^2 + bx + c = 0$ have roots $x = \alpha$ and $x = \beta$

Dividing by a gives

$$x^2 + \frac{b}{a}x + \frac{c}{a} = 0$$

Since $x = \alpha$ and $x = \beta$ are the roots of this quadratic, you can write the equation in the form

$$(x - \alpha)(x - \beta) = 0$$

Expanding the brackets gives

$$x^2 - (\alpha + \beta)x + \alpha\beta = 0$$

Comparing the two versions of the quadratic equation gives

$$x^2 + \frac{b}{a}x + \frac{c}{a} \equiv x^2 - (\alpha + \beta)x + \alpha\beta = 0$$

So, comparing the coefficients for x and the constant gives $(\alpha + \beta) = -\frac{b}{a}$ and $\alpha\beta = \frac{c}{a}$

Key point

For a quadratic equation:

The sum of the roots $= \alpha + \beta = -\frac{b}{a}$ and the product of the roots $= \alpha\beta = \frac{c}{a}$

This also shows that all quadratics can be written in the form
$x^2 - (\text{sum of the roots})x + (\text{product of the roots}) = 0$

Example 1

The roots of $x^2 - 7x + 12 = 0$ are $x = \alpha$ and $x = \beta$. Without finding the values of α and β separately,

a Write down the values of $\alpha + \beta$ and $\alpha\beta$

b Hence find the quadratic equations whose roots are   **i** α^2 and β^2 **ii** $\frac{1}{\alpha}$ and $\frac{1}{\beta}$

a $a = 1, b = -7$ and $c = 12$, so

$\alpha + \beta = -(-7) = 7$ and $\alpha\beta = 12$ • $(\alpha + \beta) = -\frac{b}{a}$ and $\alpha\beta = \frac{c}{a}$

(Continued on the next page)

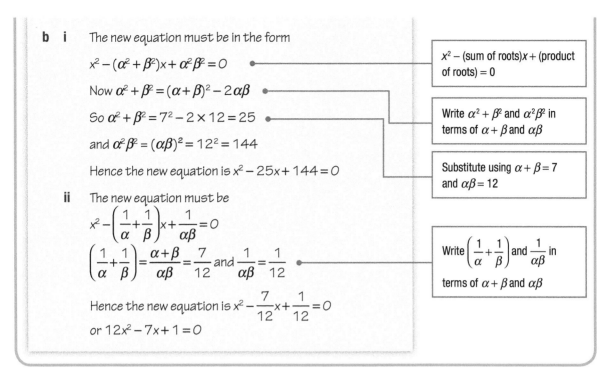

b i The new equation must be in the form

$$x^2 - (\alpha^2 + \beta^2)x + \alpha^2\beta^2 = 0$$

x^2 − (sum of roots)x + (product of roots) = 0

Now $\alpha^2 + \beta^2 = (\alpha + \beta)^2 - 2\alpha\beta$

So $\alpha^2 + \beta^2 = 7^2 - 2 \times 12 = 25$

and $\alpha^2\beta^2 = (\alpha\beta)^2 = 12^2 = 144$

Write $\alpha^2 + \beta^2$ and $\alpha^2\beta^2$ in terms of $\alpha + \beta$ and $\alpha\beta$

Hence the new equation is $x^2 - 25x + 144 = 0$

Substitute using $\alpha + \beta = 7$ and $\alpha\beta = 12$

ii The new equation must be

$$x^2 - \left(\frac{1}{\alpha} + \frac{1}{\beta}\right)x + \frac{1}{\alpha\beta} = 0$$

$$\left(\frac{1}{\alpha} + \frac{1}{\beta}\right) = \frac{\alpha + \beta}{\alpha\beta} = \frac{7}{12} \text{ and } \frac{1}{\alpha\beta} = \frac{1}{12}$$

Write $\left(\frac{1}{\alpha} + \frac{1}{\beta}\right)$ and $\frac{1}{\alpha\beta}$ in terms of $\alpha + \beta$ and $\alpha\beta$

Hence the new equation is $x^2 - \frac{7}{12}x + \frac{1}{12} = 0$

or $12x^2 - 7x + 1 = 0$

Cubic equations

There are similar relationships between the coefficients of x and the roots of the equation for higher-order equations (cubics, quartics, and so on).

Let the cubic equation $ax^3 + bx^2 + cx + d = 0$ have roots $x = \alpha$, $x = \beta$ and $x = \gamma$

Let $ax^3 + bx^2 + cx + d = 0 \equiv (x - \alpha)(x - \beta)(x - \gamma) = 0$

$$\equiv (x^2 - \alpha x - \beta x + \alpha\beta)(x - \gamma) = 0$$

$$\equiv x^3 - \alpha x^2 - \beta x^2 + \alpha\beta x - x^2\gamma + \alpha\gamma x + \beta\gamma x - \alpha\beta\gamma = 0$$

$$\equiv x^3 - (\alpha + \beta + \gamma)x^2 + (\alpha\beta + \beta\gamma + \gamma\alpha)x - \alpha\beta\gamma = 0$$

Dividing $ax^3 + bx^2 + cx + d = 0$ by a gives $x^3 + \frac{b}{a}x^2 + \frac{c}{a}x + \frac{d}{a} = 0$

Hence

$$(\alpha + \beta + \gamma) = -\frac{b}{a} \quad (\alpha\beta + \beta\gamma + \gamma\alpha) = \frac{c}{a} \quad \text{and } \alpha\beta\gamma = -\frac{d}{a}$$

Example 2

The roots of the equation $x^3 - 7x^2 + 3x + 2 = 0$ are α, β and γ. Find the values of

a $\alpha^2 + \beta^2 + \gamma^2$ **b** $\alpha^3 + \beta^3 + \gamma^3$

a $(\alpha + \beta + \gamma)^2 = \alpha^2 + 2\alpha\beta + 2\alpha\gamma + \beta^2 + 2\beta\gamma + \gamma^2$

Consider the expression $(\alpha + \beta + \gamma)^2$ to find a way to write $\alpha^2 + \beta^2 + \gamma^2$

$$= (\alpha^2 + \beta^2 + \gamma^2) + 2(\alpha\beta + \beta\gamma + \alpha\gamma)$$

$$7^2 = (\alpha^2 + \beta^2 + \gamma^2) + 2(3)$$

$$\alpha^2 + \beta^2 + \gamma^2 = 49 - 6 = 43$$

Substitute $\alpha + \beta + \gamma = 7$ and $\alpha\beta + \beta\gamma + \alpha\gamma = 3$ and rearrange.

(Continued on the next page)

b α is a root of the equation, which means

$\alpha^3 - 7\alpha^2 + 3\alpha + 2 = 0$, so by rearranging, $\alpha^3 = 7\alpha^2 - 3\alpha - 2$

The same can be done with the other roots.

$\beta^3 = 7\beta^2 - 3\beta - 2$ $\gamma^3 = 7\gamma^2 - 3\gamma - 2$

Now find the sum of these expressions.

$\alpha^3 + \beta^3 + \gamma^3 = 7\alpha^2 - 3\alpha - 2 + 7\beta^2 - 3\beta - 2 + 7\gamma^2 - 3\gamma - 2$

$= 7(\alpha^2 + \beta^2 + \gamma^2) - 3(\alpha + \beta + \gamma) - 6$ ●————

$= 7(43) - 3(7) - 6 = 274$

> Substitute $\alpha^2 + \beta^2 + \gamma^2 = 43$ and $\alpha + \beta + \gamma = 7$

Quartic equations

The same method can also be used to find the relationships between the roots, α, β, γ and δ, and the coefficients of a quartic equation $ax^4 + bx^3 + cx^2 + dx + e = 0$

Key point

$(\alpha + \beta + \gamma + \delta) = -\dfrac{b}{a}$ i.e. $\Sigma\alpha = -\dfrac{b}{a}$

$(\alpha\beta + \alpha\gamma + \alpha\delta + \beta\gamma + \beta\delta + \gamma\delta) = \dfrac{c}{a}$ i.e. $\Sigma\alpha\beta = \dfrac{c}{a}$

$(\alpha\beta\gamma + \beta\gamma\delta + \gamma\delta\alpha + \delta\alpha\beta) = -\dfrac{d}{a}$ i.e. $\Sigma\alpha\beta\gamma = -\dfrac{d}{a}$

$\alpha\beta\gamma\delta = \dfrac{e}{a}$

> $\alpha + \beta + \gamma + \delta$ is often abbreviated to $\Sigma\alpha$
>
> $\alpha\beta + \beta\gamma + \gamma\delta + \delta\alpha$ is often abbreviated to $\Sigma\alpha\beta$
>
> $\alpha\beta\gamma + \beta\gamma\delta + \gamma\delta\alpha + \delta\alpha\beta$ is often abbreviated to $\Sigma\alpha\beta\gamma$

Example 3

a   Find the cubic equation whose roots are one less than the roots of the equation

$x^3 - 3x^2 - 22x + 24 = 0$

b   Solve your new equation.

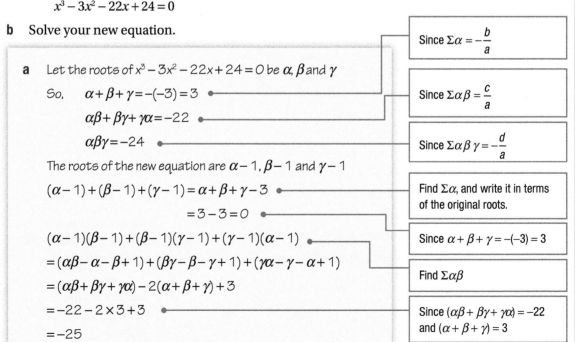

a   Let the roots of $x^3 - 3x^2 - 22x + 24 = 0$ be α, β and γ

So, $\alpha + \beta + \gamma = -(-3) = 3$ ●———— Since $\Sigma\alpha = -\dfrac{b}{a}$

$\alpha\beta + \beta\gamma + \gamma\alpha = -22$ ●———— Since $\Sigma\alpha\beta = \dfrac{c}{a}$

$\alpha\beta\gamma = -24$ ●———— Since $\Sigma\alpha\beta\gamma = -\dfrac{d}{a}$

The roots of the new equation are $\alpha - 1$, $\beta - 1$ and $\gamma - 1$

$(\alpha - 1) + (\beta - 1) + (\gamma - 1) = \alpha + \beta + \gamma - 3$ ●———— Find $\Sigma\alpha$, and write it in terms of the original roots.

$= 3 - 3 = 0$ ●———— Since $\alpha + \beta + \gamma = -(-3) = 3$

$(\alpha - 1)(\beta - 1) + (\beta - 1)(\gamma - 1) + (\gamma - 1)(\alpha - 1)$ ●————

$= (\alpha\beta - \alpha - \beta + 1) + (\beta\gamma - \beta - \gamma + 1) + (\gamma\alpha - \gamma - \alpha + 1)$

$= (\alpha\beta + \beta\gamma + \gamma\alpha) - 2(\alpha + \beta + \gamma) + 3$ ●———— Find $\Sigma\alpha\beta$

$= -22 - 2 \times 3 + 3$ ●———— Since $(\alpha\beta + \beta\gamma + \gamma\alpha) = -22$ and $(\alpha + \beta + \gamma) = 3$

$= -25$

(Continued on the next page)

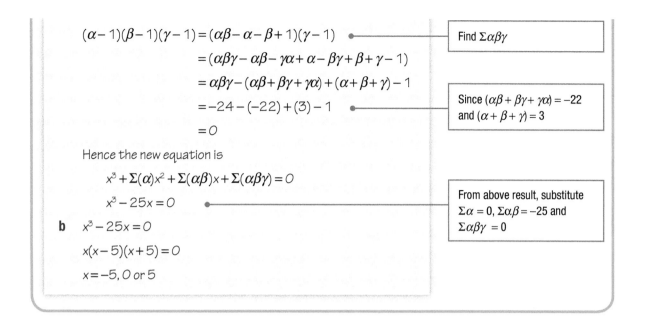

$(\alpha - 1)(\beta - 1)(\gamma - 1) = (\alpha\beta - \alpha - \beta + 1)(\gamma - 1)$

Find $\Sigma\alpha\beta\gamma$

$= (\alpha\beta\gamma - \alpha\beta - \gamma\alpha + \alpha - \beta\gamma + \beta + \gamma - 1)$

$= \alpha\beta\gamma - (\alpha\beta + \beta\gamma + \gamma\alpha) + (\alpha + \beta + \gamma) - 1$

$= -24 - (-22) + (3) - 1$

Since $(\alpha\beta + \beta\gamma + \gamma\alpha) = -22$ and $(\alpha + \beta + \gamma) = 3$

$= 0$

Hence the new equation is

$x^3 + \Sigma(\alpha)x^2 + \Sigma(\alpha\beta)x + \Sigma(\alpha\beta\gamma) = 0$

$x^3 - 25x = 0$

From above result, substitute $\Sigma\alpha = 0$, $\Sigma\alpha\beta = -25$ and $\Sigma\alpha\beta\gamma = 0$

b $x^3 - 25x = 0$

$x(x - 5)(x + 5) = 0$

$x = -5, 0 \text{ or } 5$

Exercise 2.1A Fluency and skills

1 Find the sum and product of the roots of these equations.

a $x^2 - 5x + 9$ **b** $x^2 + 6x + 7$

c $x^2 - 8x - 12$ **d** $x^2 + 10x - 5$

e $3x^2 - 12x + 8$ **f** $4x^2 + x + 6$

2 These quadratic equations have roots α and β

a $x^2 - 4x + 3 = 0$ **b** $x^2 + 4x - 5 = 0$

c $x^2 + x + 1 = 0$ **d** $2x^2 - 7x - 8 = 0$

e $5x^2 + 3x + 12 = 0$ **f** $4x^2 - 2x + \dfrac{1}{4} = 0$

Without finding α and β, find the equations whose roots are

i α^2 and β^2 **ii** $\dfrac{1}{\alpha}$ and $\dfrac{1}{\beta}$

iii $\alpha + 2$ and $\beta + 2$ **iv** α^3 and β^3

v $\dfrac{\alpha}{\beta}$ and $\dfrac{\beta}{\alpha}$

3 Write down the sum, the sum of the products in pairs and the product of the roots of these equations.

a $x^3 + 4x^2 - 9x - 14 = 0$

b $x^3 - 7x^2 - 11x + 12 = 0$

c $x^3 - 13x^2 + 22x - 26 = 0$

d $2x^3 + 5x^2 + 17x - 21 = 0$

e $4x^3 - x^2 + 3x + 8 = 0$

f $\dfrac{1}{2}x^3 + \dfrac{3}{8}x^2 - \dfrac{3}{4}x + \dfrac{5}{16} = 0$

4 These equations have roots α, β and γ

a $x^3 + 2x^2 - 4x + 3 = 0$

b $x^3 - 7x^2 + 9x + 11 = 0$

c $3x^3 - 7x^2 - 12x - 8 = 0$

d $5x^3 + 12x^2 - 14x - 5 = 0$

Without finding α, β and γ, find the equations whose roots are

i α^2, β^2 and γ^2 **ii** $\dfrac{1}{\alpha}, \dfrac{1}{\beta}$ and $\dfrac{1}{\gamma}$

iii $\alpha + 2$, $\beta + 2$ and $\gamma + 2$

5 These equations have roots α, β, γ and δ

a $x^4 - 13x^2 + 36 = 0$

b $x^4 - 4x^3 - 7x^2 + 22x + 24 = 0$

Find the values of

i $\alpha^2 + \beta^2 + \gamma^2 + \delta^2$ **ii** $\dfrac{1}{\alpha} + \dfrac{1}{\beta} + \dfrac{1}{\gamma} + \dfrac{1}{\delta}$

6 By substituting y for x^2, solve the equation $x^4 - 89x^2 + 1600 = 0$

7 Show that the equation $x^4 - 8x^2 - 9 = 0$ has only two real roots, and find them.

8 Prove that the equation $x^4 + 5x^2 + 4 = 0$ has no real roots.

9 Solve the equation $3x^3 - 14x^2 + 32 = 0$, given that one root is half one of the other roots.

Reasoning and problem-solving

Sometimes it is useful to transform an equation into another whose roots are related in a simple way to the roots of the original equation.

Key point

If the roots are transformed in a linear way, so that $y = mx + c$,

then you transform the equation by substituting $x = \dfrac{y-c}{m}$

If the new roots are reciprocals, so that $y = \dfrac{1}{x}$, then you transform the equation by substituting $x = \dfrac{1}{y}$

Strategy

To solve a question involving the transformation of one polynomial into another

(1) Rewrite the transformation $y = mx + c$ as $x = \dfrac{y-c}{m}$

(2) Substitute $\dfrac{y-c}{m}$ for x in the original polynomial and simplify to produce the transformed equation.

Example 4

Find the equation whose roots are

a One less than,   **b** Double   the roots of $x^3 - 4x^2 - 11x + 30 = 0$

a The new equation has roots one less than the first,

so $y = x - 1 \Rightarrow x = y + 1$

$(y+1)^3 - 4(y+1)^2 - 11(y+1) + 30 = 0$

$\Rightarrow \quad y^3 + 3y^2 + 3y + 1 - 4y^2 - 8y - 4 - 11y - 11 + 30 = 0$

$\Rightarrow \quad y^3 - y^2 - 16y + 16 = 0$

b The new root $y = 2x \Rightarrow x = \dfrac{y}{2}$

$\left(\dfrac{y}{2}\right)^3 - 4\left(\dfrac{y}{2}\right)^2 - 11\left(\dfrac{y}{2}\right) + 30 = 0 \Rightarrow y^3 - 8y^2 - 44y + 240 = 0$

> (1) Let y be the new root. Write y in terms of x and then rearrange to give x in terms of y

> (2) Substitute for x in the original equation.

> Multiply out and simplify.

Example 5

The roots of the equation $x^3 - 2x^2 - x + 2 = 0$ are α, β and γ

a Write the value of $\alpha\beta\gamma$

b **i** Use the substitution $y = x - 2$ to find a new cubic equation in y with integer coefficients.

ii By considering the product of the roots of this new equation, find the value of
$\alpha\beta\gamma(1 + \beta\gamma)(1 + \gamma\alpha)(1 + \alpha\beta)$

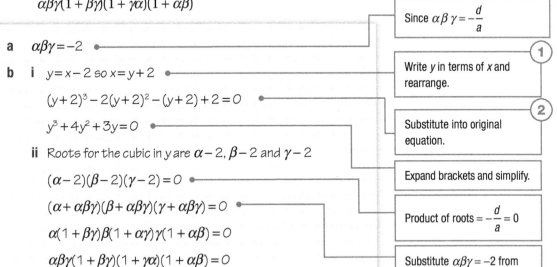

a $\alpha\beta\gamma = -2$ ────────────── Since $\alpha\beta\gamma = -\dfrac{d}{a}$ ①

b **i** $y = x - 2$ so $x = y + 2$ ────────────── Write y in terms of x and rearrange.

$(y + 2)^3 - 2(y + 2)^2 - (y + 2) + 2 = 0$ ────────────── Substitute into original equation. ②

$y^3 + 4y^2 + 3y = 0$ ────────────── Expand brackets and simplify.

ii Roots for the cubic in y are $\alpha - 2$, $\beta - 2$ and $\gamma - 2$

$(\alpha - 2)(\beta - 2)(\gamma - 2) = 0$ ────────────── Product of roots $= -\dfrac{d}{a} = 0$

$(\alpha + \alpha\beta\gamma)(\beta + \alpha\beta\gamma)(\gamma + \alpha\beta\gamma) = 0$

$\alpha(1 + \beta\gamma)\beta(1 + \alpha\gamma)\gamma(1 + \alpha\beta) = 0$ ────────────── Substitute $\alpha\beta\gamma = -2$ from part **a**

$\alpha\beta\gamma(1 + \beta\gamma)(1 + \gamma\alpha)(1 + \alpha\beta) = 0$

Exercise 2.1B Reasoning and problem-solving

1 Find the polynomial whose roots are
  a Reciprocals of, **b** Triple
  the roots of $x^3 + 3x^2 + 5x + 1 = 0$

2 A cubic equation is $x^3 - 2x^2 - x + 2 = 0$
  a Use the substitution $y = x - 2$ to find a cubic equation in y with integer coefficients.
  b Solve this cubic equation in y
  c Use the substitution $y = x - 2$ to find the roots of the original equation in x

3 Transform $x^3 - 7x^2 + 9x + 11 = 0$ using the substitution $y = \dfrac{1}{2}x - 1$

4 Transform $x^4 + 6x^3 + 7x + 8 = 0$ using the substitution $y = 2x - 1$

5 Solve the equation $x^3 - 9x^2 + 6x + 56 = 0$ using the substitution $y = x - 4$

6 Solve the equation $x^3 + 6x^2 - x - 30 = 0$ by making the substitution $y = x + 2$

7 Solve $2x^3 - 3x^2 - 23x + 12 = 0$ by making the substitution $y = x + 3$

8 Solve $x^4 + 4x^3 - 11x^2 - 30x = 0$ by increasing the roots by one.

9 Find the value of b if the equations $x^2 + 6x + b = 0$ and $x^2 + 4x - b = 0$ have a common root that is not equal to 0

10 α, β and γ are the roots of $x^3 = 4x + 3$
  By substituting $x = \alpha$, $x = \beta$ and $x = \gamma$ into the equation, prove that $\alpha^3 + \beta^3 + \gamma^3 = 9$

11 α, β and γ are the roots of $2x^3 - 2x^2 - 6x - 3 = 0$. By considering the sum of the roots, find a suitable substitution to transform this equation into a polynomial with roots $\alpha + \beta$, $\beta + \gamma$ and $\gamma + \alpha$

12 The equation $x^3 - 4x^2 - x + 7 = 0$ has roots α, β and γ
  a Write down the value of $\alpha + \beta + \gamma$
  b Use the substitution $x = 4 - y$ to find a cubic equation in y
  c By considering the product of the roots for this equation, find the value of $(\beta + \gamma)(\alpha + \gamma)(\alpha + \beta)$

Fluency and skills

See Maths Ch13.3

For a reminder of sigma notation.

If you want to write the sum of a series of n terms $u_1, u_2, \ldots u_r, \ldots u_{n-1}, u_n$, you can use sigma notation, Σ

$$\sum_{r=1}^{r=n} u_r \text{ or } \sum_{1}^{n} u_r \text{ means 'find the sum of all the terms from } u_1 \text{ to } u_n\text{'}$$

> Σ means 'find the sum of all these terms'.

u_r is the general term and, if all terms are defined algebraically, u_r is a function of the variable.

Key point

If $S_n = u_1 + u_2 + \ldots + u_r + \ldots + u_{n-1} + u_n$, then $S_n = \sum_{1}^{n} u_r$

> S_n means 'the sum of these n terms'.

If $u_r = r^2$, then $S_n = 1^2 + 2^2 + 3^2 + \ldots + r^2 \ldots + n^2 = \sum_{1}^{n} r^2$

You need to know the formulae for the sums of integers, squares and cubes.

Finding the sum of integers from 1 to n

The sum of integers from 1 to n can be written out as shown.

$$\sum_{1}^{n} r = \quad 1 \quad + \quad 2 \quad + \quad 3 \quad + \ldots + (n-2) + (n-1) \ + \ n$$

Writing the same sequence in reverse:

$$\sum_{1}^{n} r = \quad n \ + \ (n-1) + (n-2) + \ldots + \quad 3 \quad + \quad 2 \quad + \ 1$$

Adding these two sequences together gives a sequence where every term is $(n+1)$

$$2\sum_{1}^{n} r = (n+1) + (n+1) + (n+1) + \ldots + (n+1) + (n+1) + (n+1)$$

There are n lots of $(n+1)$, so $2\sum_{1}^{n} r = n(n+1)$ and so $\sum_{1}^{n} r = \dfrac{n(n+1)}{2}$

Key point

$$\sum_{1}^{n} r = 1 + 2 + \ldots + n = \frac{n(n+1)}{2}$$

> You will be expected to remember and be able to quote, without proof, the formulae for Σr

Finding the sum of a constant from 1 to n

$$\sum_{1}^{n} x = x + x \ldots + x$$

There are n lots of x and so $\sum_{1}^{n} x = nx$

Key point

$$\sum_{1}^{n} x = x + x + \ldots + x = nx$$

Example 1

Find the value of $2 + 4 + 6 + 8 + \ldots + 40$

$$2 + 4 + 6 + 8 + \ldots + 40 = 2(1 + 2 + 3 + 4 + \ldots + 20)$$

Rewrite as a sum of integers.

$$= 2 \sum_{1}^{20} r$$

$$= 2 \times \frac{20(20+1)}{2} = 2 \times \frac{20 \times 21}{2}$$

Substitute $n = 20$ into the formula $S_n = \frac{n(n+1)}{2}$

$$= 420$$

Sum of squares

Consider the expression $(2r + 1)^3 - (2r - 1)^3 \equiv (8r^3 + 12r^2 + 6r + 1) - (8r^3 - 12r^2 + 6r - 1)$

$$\equiv 24r^2 + 2$$

so

$$r^2 \equiv \frac{1}{24}[(2r+1)^3 - (2r-1)^3 - 2]$$

Substituting for values of r gives

$$(1)^2 \equiv \frac{1}{24}[(2(1)+1)^3 - (2(1)-1)^3 - 2] \qquad \equiv \frac{1}{24}[3^3 - 1^3 - 2]$$

$$(2)^2 \equiv \frac{1}{24}[(2(2)+1)^3 - (2(2)-1)^3 - 2] \qquad \equiv \frac{1}{24}[5^3 - 3^3 - 2]$$

$$(3)^2 \equiv \frac{1}{24}[(2(3)+1)^3 - (2(3)-1)^3 - 2] \qquad \equiv \frac{1}{24}[7^3 - 5^3 - 2]$$

$$\downarrow \qquad\qquad \downarrow \qquad\qquad\qquad \downarrow \quad \downarrow$$

$$(n)^2 \equiv \frac{1}{24}[(2(n)+1)^3 - (2(n)-1)^3 - 2] \qquad \equiv \frac{1}{24}[(2n+1)^3 - (2n-1)^3 - 2]$$

When you add all the terms together, many cancel out, leaving

$$\sum_{1}^{n} r^2 \equiv \frac{1}{24}[(2n+1)^3 - 1^3 - 2n]$$

$$\equiv \frac{1}{24}[8n^3 + 12n^2 + 6n + 1 - 1 - 2n]$$

$$\equiv \frac{1}{24}[8n^3 + 12n^2 + 4n]$$

$$\equiv \frac{n}{6}[2n^2 + 3n + 1] \equiv \frac{n}{6}(n+1)(2n+1)$$

This derivation uses the method of differences, but you do not need to know this method for your qualification.

Key point

$$\sum_{1}^{n} r^2 \equiv \frac{n(n+1)(2n+1)}{6}$$

Sum of cubes

Compare sums of r and r^3 for different values of n

n	Σr	Σr^3
1	1	1
2	$1+2=3$	$1+8=9$
3	$1+2+3=6$	$1+8+27=36$
4	$1+2+3+4=10$	$1+8+27+64=100$
5	$1+2+3+4+5=15$	$1+8+27+64+125=225$

For each value of n, $\Sigma r^3 = (\Sigma r)^2$

You can use the method of differences to prove that

$$\Sigma n^3 = (\Sigma n)^2 = \left(\frac{n(n+1)}{2}\right)^2 = \frac{n^2(n+1)^2}{4}$$

Key point

$$\sum_1^n r^3 = \frac{n^2(n+1)^2}{4}$$

Example 2

a Find a formula for the sum to n terms of the series

$$1 \times 2 \times 4 + 2 \times 3 \times 5 + 3 \times 4 \times 6 + \dots$$

b Find the sum of the first 10 terms and check your formula.

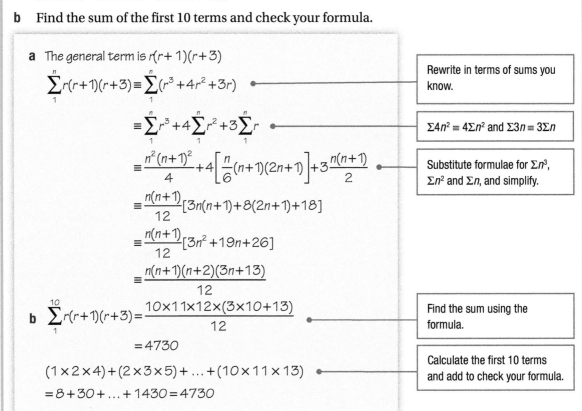

a The general term is $r(r+1)(r+3)$

$$\sum_1^n r(r+1)(r+3) \equiv \sum_1^n (r^3 + 4r^2 + 3r)$$

Rewrite in terms of sums you know.

$$\equiv \sum_1^n r^3 + 4\sum_1^n r^2 + 3\sum_1^n r$$

$\Sigma 4n^2 \equiv 4\Sigma n^2$ and $\Sigma 3n \equiv 3\Sigma n$

$$\equiv \frac{n^2(n+1)^2}{4} + 4\left[\frac{n}{6}(n+1)(2n+1)\right] + 3\frac{n(n+1)}{2}$$

Substitute formulae for Σn^3, Σn^2 and Σn, and simplify.

$$\equiv \frac{n(n+1)}{12}[3n(n+1) + 8(2n+1) + 18]$$

$$\equiv \frac{n(n+1)}{12}[3n^2 + 19n + 26]$$

$$\equiv \frac{n(n+1)(n+2)(3n+13)}{12}$$

b $$\sum_1^{10} r(r+1)(r+3) = \frac{10 \times 11 \times 12 \times (3 \times 10 + 13)}{12}$$

Find the sum using the formula.

$$= 4730$$

$$(1 \times 2 \times 4) + (2 \times 3 \times 5) + \dots + (10 \times 11 \times 13)$$

Calculate the first 10 terms and add to check your formula.

$$= 8 + 30 + \dots + 1430 = 4730$$

If a series has alternate positive and negative terms, $\Sigma f(x)$ can be written with a $(-1)^n$ multiple in it.

For example, $-1+2-3+4-...$ for n terms is expressed as $\sum_{1}^{n}(-1)^n r$

1 Write out these series.

a $\displaystyle\sum_{1}^{5}r^3$

b $\displaystyle\sum_{4}^{n}r^2$

c $\displaystyle\sum_{1}^{n}(r^2-2r)$

d $\displaystyle\sum_{1}^{n}\frac{1}{m+2}$

e $\displaystyle\sum_{1}^{6}(-1)^r r^3$

f $\displaystyle\sum_{n-3}^{n}r(r+1)$

2 Use Σ notation to write these sums.

a $1+3+5+...+31$

b $1^5+2^5+3^5+...+n^5$

c $1+\dfrac{1}{2}+\dfrac{1}{3}+\dfrac{1}{4}+...+\dfrac{1}{n+1}$

d $3-6+9-12+...-42$

e $1\times3+2\times4+3\times5+...$ for n terms.

f $\dfrac{1\times4}{3}-\dfrac{3\times5}{4}+\dfrac{5\times6}{5}-\dfrac{7\times7}{6}+...$ for n terms.

3 a How many terms are there in the series $\displaystyle\sum_{1}^{3n}(r^2-r+3)$?

b Write and simplify the $(2n+1)$th term.

4 Find the sums of these series.

a $1+2+3+...+3n$

b $1^2+2^2+3^2+...+(2n-1)^2$

c $1^3+2^3+3^3+...+(2n-1)^3$

d $1+3+5+...+(2n-1)$

e $1\times2+2\times3+3\times4+...+(n)(n+1)$

f $0+2+6+12+...+(n^2-n)$

5 a Find a formula for the sum to n terms of the series

$0\times1+1\times2+2\times3+...+(n-1)n$

b Find the sum of the first 10 terms and check your formula.

6 Find the rth term and the sum to n terms of these series.

a $1\times2\times3+2\times3\times4+3\times4\times5+...$

b $3\times-2+5\times-1+7\times0+...$

c $1\times3\times5+2\times4\times6+3\times5\times7+...$

d $2+6+10+16+...$

e $2+5+10+17+...$

Reasoning and problem-solving

Strategy

To solve problems involving sums of series

① Rearrange the expression into multiples of integers, squares and cubes.

② Substitute standard formulae or terms as required.

③ Simplify the resulting expression for the sum.

Example 3

Evaluate $\displaystyle\sum_{5}^{n} r(r+4)$

Write out some of the sequence to help you understand it. This is the same as $\displaystyle\sum_{1}^{n} r(r+4)$ without the first four terms.

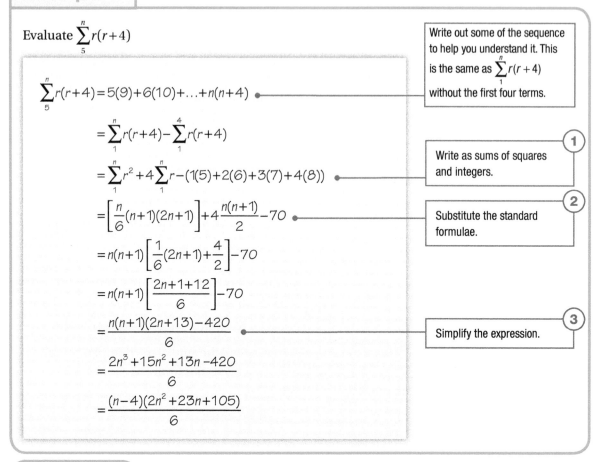

$$\sum_{5}^{n} r(r+4) = 5(9) + 6(10) + \ldots + n(n+4)$$

$$= \sum_{1}^{n} r(r+4) - \sum_{1}^{4} r(r+4)$$

1 Write as sums of squares and integers.

$$= \sum_{1}^{n} r^2 + 4\sum_{1}^{n} r - (1(5) + 2(6) + 3(7) + 4(8))$$

2 Substitute the standard formulae.

$$= \left[\frac{n}{6}(n+1)(2n+1)\right] + 4\frac{n(n+1)}{2} - 70$$

$$= n(n+1)\left[\frac{1}{6}(2n+1) + \frac{4}{2}\right] - 70$$

$$= n(n+1)\left[\frac{2n+1+12}{6}\right] - 70$$

3 Simplify the expression.

$$= \frac{n(n+1)(2n+13) - 420}{6}$$

$$= \frac{2n^3 + 15n^2 + 13n - 420}{6}$$

$$= \frac{(n-4)(2n^2 + 23n + 105)}{6}$$

Example 4

a Use the standard result for $\displaystyle\sum_{r=1}^{n} r$ to show that the sum of the first n odd numbers is a perfect square.

b Find the smallest value of n where the sum of these odd numbers is a cube number.

a $1 + 3 + 5 + \ldots + (2r - 1) + \ldots$

$$= \sum_{r=1}^{n} 2r - 1 = 2\sum_{r=1}^{n} r - \sum_{r=1}^{n} 1$$

2 Substitute for the sum of integers.

$$\equiv \frac{2n(n+1)}{2} - n$$

$$\equiv n(n+1) - n$$

3 Simplify the expression.

$$\equiv n^2 + n - n \equiv n^2$$

b The first non-trivial square number which is also a cube number is $64 \equiv 8^2$

$1 + 3 + 5 + 7 + 9 + 11 + 13 + 15 = 64 = 4^3$

So the smallest n where the sum is a cube is $n = 8$

1 Evaluate

a $\displaystyle\sum_{11}^{90}(r^2+r)$

b $\displaystyle\sum_{5}^{n}(2r+3)$

c $\displaystyle\sum_{3}^{n-2}(r^2-r)$

d $\displaystyle\sum_{n}^{2n+1}(r^3+3r)$

e $\displaystyle\sum_{n-4}^{2n-5}r^3$

2 The sum to n terms of a series is $2n^2+4n$

a By considering S_1, S_2 and S_3 find the first three terms of the series.

b By considering S_{n-1} and S_n, find the nth term.

3 a How many terms are there in the series $\displaystyle\sum_{1}^{2n}(r^2-r-2)$?

b Find the $(2n-1)$th term.

4 Use standard results to show that

$\displaystyle\sum_{1}^{n}(2r+1)^2\equiv\frac{n(4n^2+an+b)}{3}$ and find the

values of a and b

5 a Use standard results to evaluate $\displaystyle\sum_{1}^{n}2r(r^2-1)$

b Evaluate $\displaystyle\sum_{1}^{10}2r(r^2-1)$

c Use your answer to prove that all terms in the sequence are multiples of 12

6

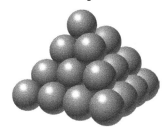

Admiral Nelson's ship has a triangular pile of cannonballs stacked on deck. The top cannonball is supported by three cannonballs in the layer underneath it. These three are supported by six cannonballs in the

layer underneath them, and so on. There are n layers of cannonballs.

a How many cannonballs are there in the bottom layer?

b How many cannonballs are there altogether?

c Calculate the weight of the pile of 'six pounder' cannon balls contained in a stack of 10 layers. A 'six pounder' cannonball weighs approximately 2.722 kg.

7

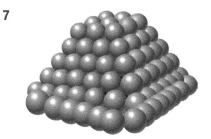

The Spanish Armada's cannon balls are in square pyramids. The base is a square with $2n$ cannonballs on each side, the layer above has $(2n-1)$ cannonballs on each side, and so on. There are n layers in total. How many cannonballs are there in this pyramid?

8 Write the general term in the series $S=1(m)+2(m-1)+3(m-2)+...+(m)(1)$ and find S

9 a Use standard results to find a formula for the sum of the series $1^2+3^2+5^2+...+(2n-1)^2$

b Confirm your result by finding the formula in a different way.

10 a Find a formula for the sum of the series $9+33+81+...+(6n^2+3^n)$

b Find the smallest value of r where $3^r>6r^2$

Proof by induction

Fluency and skills

See Maths Ch1.1 For a reminder of methods of proof.

There are a number of different methods of directly proving a mathematical statement. **Proof by induction** is a method that is generally used to prove a mathematical statement for all natural numbers (positive integers).

Other methods of direct proof are **proof by deduction** and **proof by exhaustion**.

It is a powerful method that can be used in many different contexts. The principle behind proof by induction is that if you prove a statement is true when $n = 1$, and if you can prove it is true for $n = k + 1$ by assuming it is true for any $n = k$, then you can deduce that the statement must be true for all $n \in \mathbb{N}$

$\mathbb{N}$ means the set of natural numbers: these are the positive integers 1, 2, 3, ...

Key point

The three key steps to a proof by induction are:

1 Prove the statement is true for $n = 1$
2 Assume the statement is true for $n = k$ and use this to prove the statement is true for $n = k + 1$
3 Write a conclusion.

You must always explain the steps fully and end your proof with a conclusion.

Example 1

Prove by induction that $\sum_{r=1}^{n} 4^{r-1} = \frac{1}{3}(4^n - 1)$ for all $n \in \mathbb{N}$

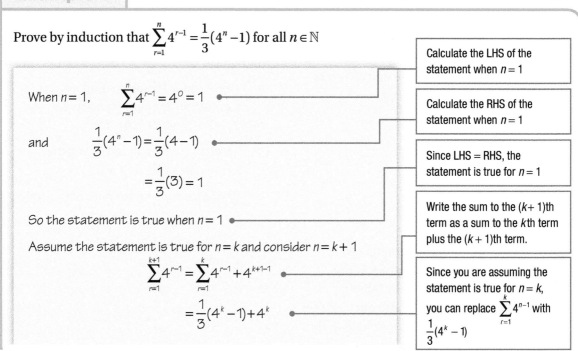

When $n = 1$, $\sum_{r=1}^{n} 4^{r-1} = 4^0 = 1$

Calculate the LHS of the statement when $n = 1$

and $\frac{1}{3}(4^n - 1) = \frac{1}{3}(4 - 1)$

Calculate the RHS of the statement when $n = 1$

$= \frac{1}{3}(3) = 1$

Since LHS = RHS, the statement is true for $n = 1$

So the statement is true when $n = 1$

Assume the statement is true for $n = k$ and consider $n = k + 1$

$\sum_{r=1}^{k+1} 4^{r-1} = \sum_{r=1}^{k} 4^{r-1} + 4^{k+1-1}$

Write the sum to the $(k + 1)$th term as a sum to the kth term plus the $(k + 1)$th term.

$= \frac{1}{3}(4^k - 1) + 4^k$

Since you are assuming the statement is true for $n = k$, you can replace $\sum_{r=1}^{k} 4^{n-1}$ with $\frac{1}{3}(4^k - 1)$

(Continued on the next page)

$$= \frac{1}{3}(4^k - 1 + 3(4^k)) = \frac{1}{3}(4(4^k) - 1)$$

Collect like terms and then use index laws.

$$= \frac{1}{3}(4^{k+1} - 1)$$

This is $\frac{1}{3}(4^n - 1)$ with n replaced by $k + 1$

So the statement is true when $n = k + 1$

The statement is true for $n = 1$ and by assuming it is true for $n = k$ it is shown to be true for $n = k + 1$

Therefore, by mathematical induction, it is true for all $n \in \mathbb{N}$

You must always write a conclusion.

Example 2

Prove by induction that $\displaystyle\sum_{r=1}^{n} r^2 = \frac{1}{6}n(n+1)(2n+1)$ for all $n \in \mathbb{N}$

Calculate the LHS of the statement when $n = 1$

When $n = 1$, $\displaystyle\sum_{r=1}^{n} r^2 = 1^2 = 1$

Calculate the RHS of the statement when $n = 1$

and $\frac{1}{6}n(n+1)(2n+1) = \frac{1}{6} \times 1 \times (1+1)(2 \times 1 + 1) = \frac{1}{6}(2)(3)$

$$= 1$$

Since LHS = RHS, the statement is true for $n = 1$

So the statement is true when $n = 1$

Assume statement is true for $n = k$ and substitute $n = k + 1$ into the formula:

$$\sum_{r=1}^{k+1} r^2 = \sum_{r=1}^{k} r^2 + (k+1)^2$$

Write the sum to the $(k+1)$th term as sum to the kth term plus the $(k+1)$th term.

$$= \frac{1}{6}k(k+1)(2k+1) + (k+1)^2$$

Since you are assuming the statement is true for $n = k$, you can replace $\displaystyle\sum_{r=1}^{k} r^2$ with $\frac{1}{6}k(k+1)(2k+1)$

$$= \frac{1}{6}(k+1)[k(2k+1) + 6(k+1)]$$

$$= \frac{1}{6}(k+1)(2k^2 + k + 6k + 6)$$

$$= \frac{1}{6}(k+1)(2k^2 + 7k + 6)$$

Look for common factors – avoid multiplying out brackets unless necessary.

$$= \frac{1}{6}(k+1)(k+2)(2k+3)$$

$$= \frac{1}{6}(k+1)((k+1)+1)(2(k+1)+1)$$

This is $\frac{1}{6}n(n+1)(2n+1)$ with n replaced by $k+1$

So statement is true when $n = k + 1$

The statement is true for $n = 1$ and by assuming it is true for $n = k$ it is shown to be true for $n = k + 1$

Therefore, by mathematical induction, it is true for all $n \in \mathbb{N}$

You must always write a conclusion.

1 Use proof by induction to prove these statements for all $n \in \mathbb{N}$

a $\displaystyle\sum_{r=1}^{n} 1 = n$ **b** $\displaystyle\sum_{r=1}^{n} r = \frac{1}{2}n(n+1)$

c $\displaystyle\sum_{r=1}^{n}(2r+3) = n(n+4)$

d $\displaystyle\sum_{r=1}^{n} r(r+1) = \frac{1}{3}n(n+1)(n+2)$

e $\displaystyle\sum_{r=1}^{n}(r-1)^2 = \frac{1}{6}n(n-1)(2n-1)$

f $\displaystyle\sum_{r=1}^{n}(r+1)(r-1) = \frac{1}{6}n(2n+5)(n-1)$

2 Use induction to prove these sums for all natural numbers, n

a $4+9+16+25+\dots+(n+1)^2$
$= \frac{1}{6}n(2n^2+9n+13)$

b $1+5+25+125+\dots+5^{(n-1)} = \frac{1}{4}(5^n-1)$

3 Prove by induction that $\displaystyle\sum_{r=1}^{n} r^3 = \frac{1}{4}n^2(n+1)^2$ for all $n \in \mathbb{N}$

4 Prove by induction that $\displaystyle\sum_{r=1}^{2n} r = n(2n+1)$ for all $n \in \mathbb{N}$

5 Prove by induction that
$\displaystyle\sum_{r=1}^{2n} r^2 = \frac{1}{3}n(2n+1)(4n+1)$ for all $n \in \mathbb{N}$

6 Use induction to prove these statements for all $n \in \mathbb{N}$

a $\displaystyle\sum_{r=1}^{n} 2^r = 2(2^n-1)$ **b** $\displaystyle\sum_{r=1}^{n} 3^r = \frac{3}{2}(3^n-1)$

c $\displaystyle\sum_{r=1}^{n} 4^r = \frac{4}{3}(4^n-1)$ **d** $\displaystyle\sum_{r=1}^{n} 2^{r-1} = 2^n-1$

e $\displaystyle\sum_{r=1}^{n} 3^{r-1} = \frac{1}{2}(3^n-1)$

f $\displaystyle\sum_{r=1}^{n}\left(\frac{1}{2}\right)^r = 1-\left(\frac{1}{2}\right)^n$

7 Prove by induction that $\displaystyle\sum_{r=1}^{n} \frac{1}{r(r+1)} = \frac{n}{n+1}$ for all $n \in \mathbb{N}$

8 Prove by induction that $\displaystyle\sum_{r=2}^{n} \frac{1}{r(r-1)} = \frac{n-1}{n}$ for all $n \in \mathbb{N}$

9 Prove by induction that
$\displaystyle\sum_{r=1}^{n} \frac{1}{r^2+2r} = \frac{n(3n+5)}{4(n+1)(n+2)}$ for all $n \in \mathbb{N}$

Reasoning and problem-solving

See Ch4.1

To see how to apply induction to matrix proofs.

As well as proving sums of series, you can use proof by induction in other contexts such as to prove an algebraic expression is divisible by a given integer.

Strategy

To prove an expression is divisible by a particular integer

(1) Substitute $n = 1$ into the expression and show the result is divisible by the integer you are using.

(2) Assume the expression is divisible by the integer for $n = k$ and substitute in $n = k+1$

(3) Separate the part you know to be divisible by the integer given.

(4) Write the conclusion.

Example 3

Prove that $n^3 + 2n$ is divisible by 3 for all $n \in \mathbb{N}$

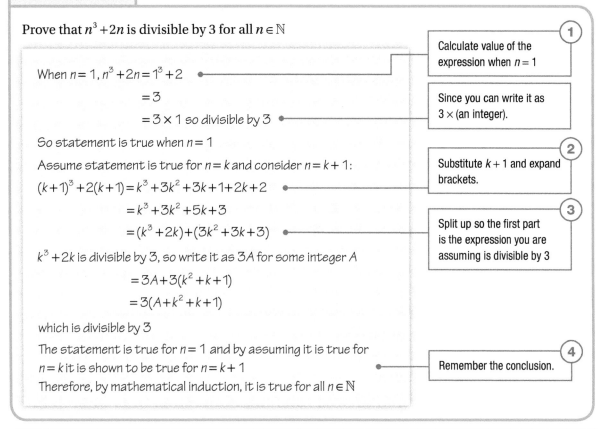

When $n = 1$, $n^3 + 2n = 1^3 + 2$

$\qquad = 3$

$\qquad = 3 \times 1$ so divisible by 3

So statement is true when $n = 1$

Assume statement is true for $n = k$ and consider $n = k + 1$:

$(k+1)^3 + 2(k+1) = k^3 + 3k^2 + 3k + 1 + 2k + 2$

$\qquad = k^3 + 3k^2 + 5k + 3$

$\qquad = (k^3 + 2k) + (3k^2 + 3k + 3)$

$k^3 + 2k$ is divisible by 3, so write it as $3A$ for some integer A

$\qquad = 3A + 3(k^2 + k + 1)$

$\qquad = 3(A + k^2 + k + 1)$

which is divisible by 3

The statement is true for $n = 1$ and by assuming it is true for $n = k$ it is shown to be true for $n = k + 1$

Therefore, by mathematical induction, it is true for all $n \in \mathbb{N}$

Callouts:

1. Calculate value of the expression when $n = 1$

Since you can write it as $3 \times$ (an integer).

2. Substitute $k + 1$ and expand brackets.

3. Split up so the first part is the expression you are assuming is divisible by 3

4. Remember the conclusion.

Exercise 2.3B Reasoning and problem-solving

1 Use proof by induction to prove these statements for all $n \in \mathbb{N}$

 a $n^2 + 3n$ is divisible by 2

 b $5n^2 - n$ is divisible by 2

 c $8n^3 + 4n$ is divisible by 12

 d $11n^3 + 4n$ is divisible by 3

2 Prove by induction that $7n^2 + 25n - 4$ is divisible by 2 for all $n \in \mathbb{N}$

3 Prove by induction that $n^3 - n$ is divisible by 3 for all $n \geq 2$, $n \in \mathbb{N}$

4 Prove by induction that $10n^3 + 3n^2 + 5n - 6$ is divisible by 6 for all $n \in \mathbb{N}$

5 Use proof by induction to prove these statements for all $n \in \mathbb{N}$

 a $6^n + 9$ is divisible by 5

 b $3^{2n} - 1$ is divisible by 8

 c $2^{3n+1} - 2$ is divisible by 7

6 Prove by induction that $5^n - 4n + 3$ is a multiple of 4 for all $n \in \mathbb{N}$

7 Prove by induction that $3^n + 2n + 7$ is a multiple of 4 for all $n \in \mathbb{N}$

8 Prove by induction that $7^n - 3n + 5$ is a multiple of 3 for all $n \in \mathbb{N}$

9 Use proof by induction to prove these statements for all $n \in \mathbb{N}$

 a $\displaystyle\sum_{r=n+1}^{2n} r^2 = \frac{1}{6}n(2n+1)(7n+1)$

 b $\displaystyle\sum_{r=n}^{2n} r^3 = \frac{3}{4}n^2(5n+1)(n+1)$

 c $8^n - 5^n$ is divisible by 3 for all $n \in \mathbb{N}$

Chapter summary

- For the quadratic equation $ax^2 + bx + c = 0$, the sum of the roots $= -\dfrac{b}{a}$ and the product of the roots $= \dfrac{c}{a}$

- If the roots of the cubic equation $ax^3 + bx^2 + cx + d = 0$ are α, β and γ, then
$$(\alpha + \beta + \gamma) = -\frac{b}{a}, \qquad (\alpha\beta + \beta\gamma + \gamma\alpha) = \frac{c}{a} \qquad \text{and} \qquad \alpha\beta\gamma = -\frac{d}{a}$$

- If the roots of the quartic equation $ax^4 + bx^3 + cx^2 + dx + e = 0$ are α, β, γ and δ, then
$$(\alpha + \beta + \gamma + \delta) = -\frac{b}{a}, \qquad (\alpha\beta + \beta\gamma + \gamma\delta + \delta\alpha) = \frac{c}{a}, \qquad (\alpha\beta\gamma + \beta\gamma\delta + \gamma\delta\alpha + \delta\alpha\beta) = -\frac{d}{a}$$
and $\alpha\beta\gamma\delta = \dfrac{e}{a}$

- If roots are transformed by a linear amount so $y = mx + c$, then substitute $x = \dfrac{y - c}{m}$

- If the new roots are the reciprocals of the original roots so $y = \dfrac{1}{x}$, then substitute $x = \dfrac{1}{y}$

- $\displaystyle\sum_{1}^{n} u_r$ means 'find the sum of all the terms u_1, u_2, ..., u_r, ... u_{n-1}, u_n'

- $\displaystyle\sum_{1}^{n} r \equiv 1 + 2 + 3 + ... + n = \dfrac{n(n+1)}{2}$

- $\displaystyle\sum_{1}^{n} r^2 \equiv \dfrac{n}{6}(n+1)(2n+1)$

- $\displaystyle\sum_{1}^{n} r^3 \equiv \dfrac{n^2(n+1)^2}{4}$

- If a series has alternate positive and negative terms, include $(-1)^r$ in $\Sigma\mathrm{f}(r)$
  For example, $1 - 2 + 3 - 4 + ... + n$ can be written as $\displaystyle\sum_{1}^{n}(-1)^r r$

- To prove a statement by induction you must first prove the result for $n = 1$ then assume the statement is true for $n = k$ and use this fact to prove the statement is true for $n = k + 1$. Always remember to write a conclusion.

Check and review

You should now be able to...	Try Questions
✓ Relate the roots of a polynomial to its coefficients.	1, 5
✓ Find a new polynomial whose roots are a linear transformation of the roots of another polynomial.	2–5
✓ Evaluate expressions involving the roots.	6
✓ Use the formulae for the sums of integers, squares and cubes to sum other series.	7–11
✓ Prove an expression for the sum of a series using mathematical induction.	12–14
✓ Use proof by induction to prove divisibility.	15–17

1 Write down the sum, the sum of the products in pairs and the product of the roots of these equations.

a $x^3 + 5x^2 + 2x + 1 = 0$

b $x^3 + 9x^2 - 11x - 8 = 0$

2 Find the polynomial whose roots are double the roots of $x^3 - 4x^2 + 6x - 2 = 0$

3 Solve the equation $x^3 - 4x^2 - x + 4 = 0$ by making the substitution $y = x - 1$

4 Find the roots of the equation which are 5 less than the roots of the equation $x^3 - 12x^2 + 20x = 0$

5 The roots of the equation $x^3 - x^2 - 3x + 1 = 0$ are α, β and γ

  a Write down the value of $\alpha\beta\gamma$

  b i Use the substitution $y = x - 1$ to find a cubic equation in y with integer coefficients.

    ii By considering the product of the roots of this new equation, find the value of $\alpha\beta\gamma(1 + \beta\gamma)(1 + \gamma\alpha)(1 + \alpha\beta)$

6 The equation $2x^3 - 6x^2 + 3x - 1 = 0$ has roots α, β and γ

Find the values of

  a $\dfrac{1}{\alpha} + \dfrac{1}{\beta} + \dfrac{1}{\gamma}$

  b $(2 - \alpha)(2 - \beta)(2 - \gamma)$

  c $\alpha^2 + \beta^2 + \gamma^2$

7 Write out these sequences.

  a $\displaystyle\sum_{3}^{10} r$ b $\displaystyle\sum_{n}^{n+4} r^2$

  c $\displaystyle\sum_{1}^{n}(r^2 + 5r)$ d $\displaystyle\sum_{1}^{n}\dfrac{1}{m-3}$

  e $\displaystyle\sum_{2}^{10}(-1)^r r^2$ f $\displaystyle\sum_{n-1}^{n+3} 2r(3r - 2)$

8 Find the sums of the series with these general terms, to n terms. In each case, the series starts with $r = 1$

  a $3 - 2r$ b $2r^2 - 4r$ c $5r + r^3$

9 Sum the series $1^2 + 3^2 + 5^2 + \ldots + (99)^2$

10 a Find $\displaystyle\sum_{n}^{2n+1}(r^3 - 3r + 2)$

  b Verify your solution using $n = 6$

11 The rth term of a series is $r^2(r - 1)$. Find the sum of the series

  a To n terms,

  b To $2n + 1$ terms.

12 Prove these results by induction.

  a $\displaystyle\sum_{r=1}^{n} 3r^2 = \frac{1}{2}n(n+1)(2n+1)$ for all $n \in \mathbb{N}$

  b $\displaystyle\sum_{r=1}^{n} r^2(r-1) = \frac{1}{12}n(n+1)(3n+2)(n-1)$ for all $n \in \mathbb{N}$

  c $\displaystyle\sum_{r=1}^{n}(r+3)(2r-1) = \frac{1}{6}n(4n^2 + 21n - 1)$ for all $n \in \mathbb{N}$

13 Use proof by induction to prove that $\displaystyle\sum_{r=1}^{2n} r^2 = \frac{1}{3}n(2n+1)(4n+1)$ for all $n \in \mathbb{N}$

14 Prove these results by induction.

  a $\displaystyle\sum_{r=1}^{n} 5^r = \frac{5}{4}(5^n - 1)$ for all $n \in \mathbb{N}$

  b $\displaystyle\sum_{r=1}^{n} 2^{r+1} = 4(2^n - 1)$ for all $n \in \mathbb{N}$

  c $\displaystyle\sum_{r=1}^{2n} 5^{r-1} = \frac{1}{4}(25^n - 1)$ for all $n \in \mathbb{N}$

15 Prove by induction that $3^{2n+1} + 1$ is divisible by 4 for all $n \in \mathbb{N}$

16 Prove by induction that $2^{2n} - 3n + 2$ is divisible by 3 for all $n \in \mathbb{N}$

17 Prove by induction that $6^n - 1$ is divisible by 5 for all $n \in \mathbb{N}$

Did you know?

The **Riemann zeta function** is an infinite summation formula. It is given by the formula

$\zeta(s) = \sum_{n=1}^{\infty} \frac{1}{n^s}$ where s is any complex number with $\text{Re}(s) > 1$

The Riemann zeta function is a famous function in mathematics because it has a link to prime numbers. The **Riemann hypothesis**, proposed by Bernhard Riemann, is a conjecture that makes statements about the zeros of the zeta function. The hypothesis implies results about how the prime numbers may be distributed.

The Riemann hypothesis is yet to be proved, but there is $1 million for the first person who manages it.

ζ is the Greek letter "zeta"

Research

The **Basel problem** was first posed by **Pietro Mengoli** in 1644. The Basel problem simply asks for the precise summation of the reciprocals of the squares of the natural numbers. In other words, $\zeta(2)$

Many famous mathematicians failed to solve the problem until **Leonhard Euler**, at the age of 28, solved it in 1734.

Euler found the solution to be $\sum_{n=1}^{\infty} \frac{1}{n^2} = \frac{\pi^2}{6}$

Find out how Euler solved the problem.

After Euler, **Augustin-Louis Cauchy** came up with an alternative proof for the result.
Find out Cauchy's approach to the problem.

Leonhard Euler

Information

The sum of the natural numbers is a divergent sequence. This can be shown by considering increasingly longer finite sums of the natural numbers:

$\sum_{n=1}^{100} \frac{1}{n} = 5050$ $\sum_{n=1}^{1000} \frac{1}{n} = 500\,500$ $\sum_{n=1}^{10000} \frac{1}{n} = 50\,005\,000$

As the length of the sequence increases, so does the sum. Therefore as the length tends to infinity, so does the sum.

Research

In 1913, the Indian mathematician **Ramanujan** claimed to be able to show that

$\sum_{n=1}^{\infty} \frac{1}{n} = -\frac{1}{12}$

Find out about the proof he offered, and how it related to the Riemann zeta function.

1. The quadratic equation $3x^2 + 4x + 2 = 0$ has roots α and β. Work out the value of $\alpha + \beta$ **[1 mark]**

2. The quartic equation $x^4 - 4x^2 + x - 2 = 0$ has roots α, β, γ and δ. Work out the value of $\sum \alpha\beta\gamma$ **[1]**

3. Express $\displaystyle\sum_{r=1}^{n}(6r^2 + 2r)$ as a polynomial in n **[3]**

4. A cubic equation is given by $4x^3 + 3x = 1 + 6x^2$

   a. Use the substitution $u = 2x - 1$ to form a cubic in u **[3]**

   b. Solve this cubic equation in u **[1]**

   c. Hence, or otherwise, solve the original equation in x **[1]**

5. a. Express $\displaystyle\sum_{r=1}^{n} 3r^2 - 2r - 1$ as a polynomial in n **[4]**

   b. Hence evaluate the sum to $n = 16$ **[1]**

6. The rth term of a sequence is defined by $u_r = (2r-1)(r+3)$

   a. Express the sum to n terms of the sequence as a function of n **[3]**

   b. Evaluate the sum of the first 15 terms. **[2]**

   c. Find the sum of the 16th to the 20th terms. **[3]**

7. a. Prove that $3^n + 2n + 3$ is divisible by 4 for all integers $n \geq 0$ **[5]**

   b. Prove that $3^n + 2n + 3$ is not divisible by 8 for all integers $n \geq 0$ **[2]**

8. Prove by induction that $\displaystyle\sum_{r=1}^{n} 1 + 3r = \frac{1}{2}n(3n+5)$ for all $n \in \mathbb{N}$ **[6]**

9. Prove by induction that $\displaystyle\sum_{r=1}^{n}(r-1)^3 = \frac{1}{4}(n-1)^2 n^2$ for all $n \in \mathbb{N}$ **[6]**

10. a. Form a polynomial function, f(x), whose roots are 2, 3 and −4 and whose curve passes through the point $(1, 5)$ **[3]**

    b. The function f(x) undergoes the transformation described by $y = \frac{1}{2}x + 1$ Find the equation of the transformed polynomial, g(x) **[3]**

11. The roots of the cubic $2x^3 - x^2 - 3x + 2 = 0$ are α, β and γ

    a. Write the values of $\alpha\beta\gamma$, $\sum \alpha\beta$ and $\sum \alpha$ **[3]**

    b. Find the cubic equation with roots $2\alpha - 1$, $2\beta - 1$ and $2\gamma - 1$ **[4]**

12. The roots of the equation $x^3 - 5x^2 - kx + 4 = 0$ are α, β and γ

    a. Write down the value of $\alpha + \beta + \gamma$ **[1]**

    b. Work out the value of $\alpha^2 + \beta^2 + \gamma^2$ in terms of k **[4]**

13. Given that $\displaystyle\sum_{r=1}^{n} 1 = n$, that $\displaystyle\sum_{r=1}^{n} r = \frac{n}{2}(n+1)$ and that $(r+1)^3 - r^3 = 3r^2 + 3r + 1$,

    find an expression for $\displaystyle\sum_{r=1}^{n} r^2$ in terms of n **[6]**

14 Use the formulae $\sum_{r=1}^{n} r = \frac{1}{2}n(n+1)$ and $\sum_{r=1}^{n} r^2 = \frac{1}{6}n(n+1)(2n+1)$ to show that

    a $\sum_{r=1}^{n} 3r^2 - 2r = \frac{1}{2}n(n+1)(2n-1)$ **[3]**

    b $\sum_{r=1}^{n} (r+1)^2 = \frac{1}{6}n(2n^2 + 9n + 13)$ **[4]**

15 a   Use standard results to show that $\sum_{r=1}^{2n} r^2(r+2) = \frac{1}{3}n(2n+1)(6n^2 + 11n + 2)$ **[4]**

    b   Hence evaluate the sum $3 + 16 + 45 + \ldots + 640$ you must show your working.   **[3]**

16 Prove by induction that $\sum_{r=1}^{n} 4^{r-1} = \frac{1}{3}(4^n - 1)$ for all $n \in \mathbb{N}$ **[6]**

17 Prove by induction that $4n^3 + 8n$ is divisible by 12 for all $n \in \mathbb{N}$ **[6]**

18 Prove by induction that $4^{2n+1} - 1$ is a multiple of 3 for all $n \in \mathbb{N}$ **[6]**

19 The roots of the equation $4x^2 + kx + 3(k+1) = 0$ are α and β

    a   Show that $\alpha^2 + \beta^2 = \frac{1}{16}(k^2 - 24k - 24)$ **[4]**

    b   **i**   Use the substitution $y = x + 3$ to find a new quadratic equation in y

            ii   Write down the roots of this equation in terms of α and β **[4]**

20 The quartic equation $x^4 + x + 1 = 0$ has four roots α, β, γ and δ

    a   Find the quartic equation with integer coefficients which has roots 3α, 3β, 3γ and 3δ **[3]**

    b   By considering the substitution $u = x^2$, or otherwise, find the quartic equation
            with roots α^2, β^2, γ^2 and δ^2 **[3]**

21 a   Show that $\sum_{r=n}^{2n} r(r-1) = An(n+1)(7n-4)$ where A is a constant to be found.   **[6]**

    b   Hence evaluate the sum $15 \times 16 + 16 \times 17 + 17 \times 18 + \ldots + 31 \times 32$ **[3]**

22 A sequence is given as $2\ln 3$, $3\ln 9$, $4\ln 27$, $5\ln 81$, $\ldots$

    a   State an expression for the nth term of the sequence.   **[1]**

    b   Show that the sum of the first n terms of the sequence is

     $S_n = \frac{n}{3}(n+1)(n+2)\ln 3$ **[4]**

23 A sequence is given by the iterative formula $u_{n+1} = \frac{1+u_n}{2}$ where $u_1 = 2$

     Prove by induction that $u_n = \frac{1+2^{n-1}}{2^{n-1}}$ for all $n \in \mathbb{N}$ **[5]**

24 A sequence is given by the iterative formula $u_{n+1} = 3u_n - 7$ where $u_1 = 5$

     Prove by induction that $u_n = \frac{1}{2}(7 + 3^n)$ for all $n \geq 1$ **[4]**

25 a   Prove by induction that $\sum_{r=2}^{n} \frac{1}{r^2 - 1} = \frac{3n^2 - n - 2}{4n(n+1)}$ for all $n \geq 2 \in \mathbb{N}$ **[8]**

    b   Explain why $\sum_{r=1}^{n} \frac{1}{r^2 - 1}$ does not exist.   **[1]**

3 Integration

Integral calculus allows us to calculate the **volume of revolution** of a two-dimensional shape about a fixed axis, which can be used to support the design of objects in all manner of industries. For example, some of our most iconic buildings have been designed using such techniques. In this case the cross-section, or parts of the cross-section, of the building can be modelled by a mathematical function, which can then be considered to be rotated about a vertical axis to generate the three-dimensional outline of the building. Developing this further, calculus can be used to find the volume inside the building.

Architecture is just one area where integral calculus has been employed. The underlying principles have a wide range of applications from the design of cars to planning public sculptures.

Orientation

What you need to know	What you will learn	What this leads to
Maths Ch1 • Finding solutions of simultaneous equations. **Maths Ch4** • Finding definite integrals. **Maths Ch12** • Parametric equations of curves **Maths Ch16** • Methods of integration • Integrating rational functions	• To find the area enclosed by a curve and lines. • To calculate the volume of revolution when rotated around the x-axis or the y-axis. • To calculate more complicated volumes of revolution by adding or subtracting volumes.	**Ch8 Integration and differentiation** • Volumes of revolution for parametric functions. **FP2 Ch7 Further methods in calculus 2** • Surface area of a volume of revolution

Fluency and skills

See Maths Ch4.7

For a reminder of finding the area under a curve.

Calculus can be used to find the area under a curve, but you can also use calculus to find the volumes of 3D shapes and to generate the formulae for shapes like cones and spheres.

Look at this graph of $y = f(x)$. If the section of this curve between $x = a$ and $x = b$ is rotated 360° around the x-axis, it will form a solid.

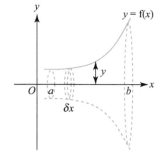

Since the curve has rotated through 360°, the cross-section of this solid will always be a circle with variable radius y.

Now consider thin strips of thickness δx. Each of these strips is approximately a cylinder, so they will have a volume, δV of $\pi y^2 \delta x$.

The volume of the whole solid is the sum of these thin strips.

Using integration to sum these strips gives the formula

$$V = \int_a^b \pi y^2 \, dx$$

Key point

The volume of the solid formed by rotating the curve $y = f(x)$ between $x = a$ and $x = b$ a full turn around the x-axis is given by $V = \int_a^b \pi y^2 \, dx$

Because this is an integral with respect to x, you need the expression in terms of x only, so replace the y^2 with the function in terms of x.

Example 1

See Maths Ch14.1

For a reminder of radians.

Find the volume formed when the area enclosed by the curve with equation $y = \dfrac{x^3}{3}$, the x-axis and the lines $x = -3$ and $x = 3$, is rotated 2π radians around the x-axis.

$$V = \pi \int_{-3}^{3} \left(\frac{x^3}{3} \right)^2 dx$$

$$= \pi \int_{-3}^{3} \frac{x^6}{9} \, dx$$

$$= \pi \left[\frac{x^7}{63} \right]_{-3}^{3}$$

$$= \pi \left(\frac{243}{7} - -\frac{243}{7} \right) = \frac{486}{7} \pi \text{ cubic units}$$

Using $V = \int_a^b \pi y^2 \, dx$ where $y = \dfrac{x^3}{3}$

Remember to square the $\dfrac{1}{3}$ as well.

Example 2

Find the volume formed when the area enclosed by the curve with equation $y = x^3 + 1$, the coordinate axes and the line $x = 2$ is rotated 180° around the x-axis.

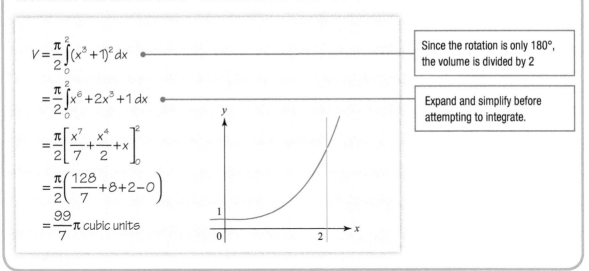

$V = \dfrac{\pi}{2} \displaystyle\int_0^2 (x^3 + 1)^2 \, dx$ •————— Since the rotation is only 180°, the volume is divided by 2

$= \dfrac{\pi}{2} \displaystyle\int_0^2 x^6 + 2x^3 + 1 \, dx$ •————— Expand and simplify before attempting to integrate.

$= \dfrac{\pi}{2} \left[\dfrac{x^7}{7} + \dfrac{x^4}{2} + x \right]_0^2$

$= \dfrac{\pi}{2} \left(\dfrac{128}{7} + 8 + 2 - 0 \right)$

$= \dfrac{99}{7} \pi$ cubic units

Exercise 3.1A Fluency and skills

1 Find the volume formed when the region bounded by the line $y = 2x$, the x-axis and the lines $x = \dfrac{1}{2}$ and $x = 2$ is rotated 360° around the x-axis.

2 Find the volume formed when the region bounded by the curve $y = 3x^2$, the x-axis and the lines $x = 1$ and $x = 4$ is rotated 360° around the x-axis.

3 Find the volume formed when the region bounded by the curve $y = x^3 + 5x^2$, the x-axis and the lines $x = -2$ and $x = -1$ is rotated 360° around the x-axis.

4 Find the volume formed when the region bounded by the curve $y = \dfrac{1}{9}(x^4 + 21x)$, the x-axis and the lines $x = 0$ and $x = 3$ is rotated 360° around the x-axis.

5 The region R is enclosed by the curve $y = 8 - x^3$ and the coordinate axes. This region is shown after part **b**.

    a   Show that the area of R is 12 square units.

The region R is rotated 360° around the x-axis.

b Calculate the volume of the solid formed.

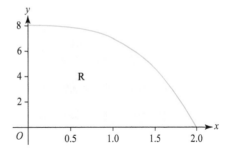

6 The shaded region is enclosed by the curve $y = x^4 - x^3$ and the x-axis.

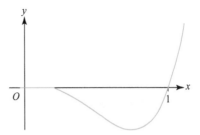

    a   Find the area of the shaded region.

    b   Calculate the volume of the solid formed when the shaded region is rotated 2π radians around the x-axis.

7 Find the volume of the solid formed when the region bounded by the curve $y = 4x^2 - 5$, the x-axis and the lines $x = 0.5$ and $x = 1$ is rotated 180° around the x-axis.

8 Show that the volume of the solid formed when the region bounded by the curve $y = \dfrac{2}{\sqrt{x^3}}$, the x-axis and the lines $x = a$ and $x = 2$ for $a < 2$ is rotated 2π radians around the x-axis is $\dfrac{(4 - a^2)\pi}{2a^2}$

9 The region R is enclosed by the curve $y = \dfrac{1}{x^2}$, the x-axis and the lines $x = \dfrac{1}{2}$ and $x = 1$.
Show that the volume of the solid formed when R is rotated 2π radians around the x-axis is $\dfrac{7}{3}\pi$

10 The region A is enclosed by the curve $y = \dfrac{2 + \sqrt{x}}{5x^2}$, the x-axis and the lines $x = 1$ and $x = 3$

   a Show that the area of A is $\dfrac{10 - 2\sqrt{3}}{15}$

   b Calculate the volume of the solid formed when the region A is rotated

     i 360° around the x-axis,

     ii 90° around the x-axis.

Give your answers to 3 significant figures.

11 The shaded region is enclosed by the curve with equation $y = 8\sqrt{x} - \dfrac{x^2}{8}$ and the x-axis.

   a Work out the x-intercepts of the curve.

   b Calculate the area of the shaded region.

The region is rotated 360° around the x-axis.

   c Calculate the volume of the solid formed. Give your answer to 3 significant figures.

12 Calculate the volume of the solid formed when the area bounded by the curve with equation $y = 4 - x^2$ and the lines $x = 0$ and $y = 0$ is rotated 360° around the x-axis.

13 Find the exact volume of the solid formed when the region in the first quadrant bounded by the curve $y = 3x - x^3$ and the x-axis is rotated 180° around the x-axis.

Reasoning and problem-solving

Strategy

To find a volume of revolution:

1. Find any necessary points of intersection.
2. Sketch a graph.
3. Use the formula for a volume of revolution, $V = \int_a^b \pi y^2 \, dx$ and the correct limits.
4. Add or subtract another volume where necessary.

See Maths Ch1.6

For a reminder of how to find points of intersection.

Example 3

The region R is enclosed by the curve $y = +\sqrt{x}$ and the lines $y = x - 6$ and $y = 0$

Find the volume of the solid formed when R is rotated 360° around the x-axis.

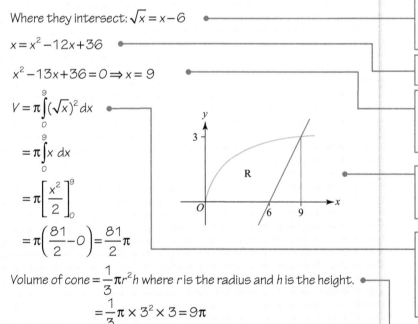

Where they intersect: $\sqrt{x} = x - 6$

> The point of intersection is where the two curves are equal.

$x = x^2 - 12x + 36$

> Square both sides.

$x^2 - 13x + 36 = 0 \Rightarrow x = 9$

> When $x = 4$, the line intersects $y = -\sqrt{x}$, so this solution isn't relevant.

$V = \pi \int_0^9 (\sqrt{x})^2 \, dx$

> ② Since it's a more complicated situation, it is helpful to sketch a graph.

$= \pi \int_0^9 x \, dx$

$= \pi \left[\dfrac{x^2}{2} \right]_0^9$

> ③ Find the volume formed when the curve is rotated. Use the x-coordinate of point of intersection for the upper limit.

$= \pi \left(\dfrac{81}{2} - 0 \right) = \dfrac{81}{2} \pi$

Volume of cone $= \dfrac{1}{3} \pi r^2 h$ where r is the radius and h is the height.

> You can derive this from the volume of revolution formula.

$= \dfrac{1}{3} \pi \times 3^2 \times 3 = 9\pi$

Volume required $= \dfrac{81}{2} \pi - 9\pi$

> ④ Subtract the volume of the cone to find the volume required.

$= \dfrac{63}{2} \pi$

You can use volumes of revolution to prove some standard volume formulae.

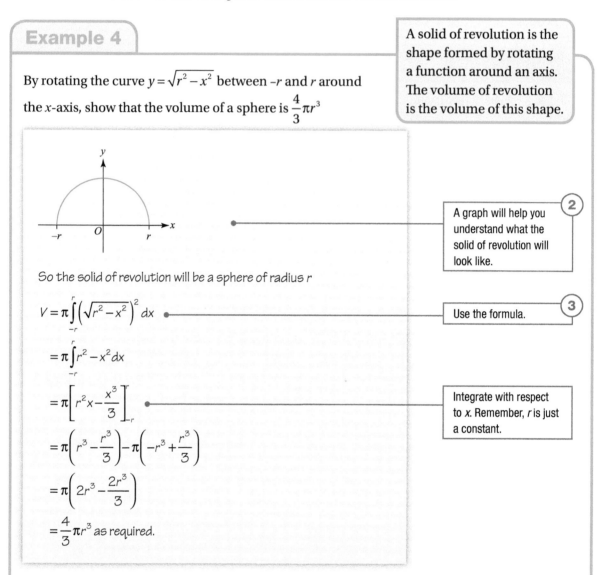

Example 4

By rotating the curve $y = \sqrt{r^2 - x^2}$ between $-r$ and r around the x-axis, show that the volume of a sphere is $\dfrac{4}{3}\pi r^3$

A solid of revolution is the shape formed by rotating a function around an axis. The volume of revolution is the volume of this shape.

So the solid of revolution will be a sphere of radius r

$$V = \pi \int_{-r}^{r} \left(\sqrt{r^2 - x^2}\right)^2 dx$$

$$= \pi \int_{-r}^{r} r^2 - x^2\, dx$$

$$= \pi \left[r^2 x - \frac{x^3}{3} \right]_{-r}^{r}$$

$$= \pi \left(r^3 - \frac{r^3}{3} \right) - \pi \left(-r^3 + \frac{r^3}{3} \right)$$

$$= \pi \left(2r^3 - \frac{2r^3}{3} \right)$$

$$= \frac{4}{3}\pi r^3 \text{ as required.}$$

2 A graph will help you understand what the solid of revolution will look like.

3 Use the formula.

Integrate with respect to x. Remember, r is just a constant.

Exercise 3.1B Reasoning and problem-solving

1 The shaded region is bounded by the curve $y = 9x - 8x^2$, the x-axis and the line $y = 2x - 1$

  a Calculate the area of the shaded region.

  b Calculate the volume of the solid formed when the shaded region is rotated 360° around the x-axis. Give your answer in terms of π

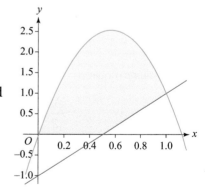

2 The shaded region shown is bounded by the curve $y = 10 - x^3$, the y-axis and the line $y = 2$

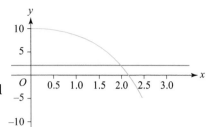

  a Calculate the area of the shaded region.

  b Calculate the volume of the solid formed when the shaded region is rotated 360° around the x-axis. Give your answer in terms of π

3 The region R is bounded by the curve $y = \sqrt{x}$, the line $y = x - 2$ and the x-axis.

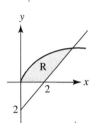

  a Calculate the area of R.

  R is rotated 180° around the x-axis

  b Calculate the volume of the solid formed.

4 The region R is bounded by the curve with equation $y = 16 - x^4$, the line $y = 7$ and the y-axis.

  a Show that the area of R is $\dfrac{36}{5}\sqrt{3}$ square units.

  b Calculate the volume of the solid formed when R is rotated 360° around the x-axis.

5 The region A is bounded by the curve with equation $y = 8 + 2x - x^2$, the line $y = x - 4$ and the y-axis.

  a Show that the area of A is $\dfrac{104}{3}$ square units.

  b Calculate the volume of the solid formed when A is rotated 180° around the x-axis.

6 By rotating the line $y = ax$ between $x = 0$ and $x = b$, 360° around the x-axis, show that the volume of a cone is $\dfrac{1}{3}\pi r^2 h$, where r is the radius of the base and h is the height.

7 Use a volume of revolution to show that the volume of a cylinder is $\pi r^2 h$, where r is the radius and h is the height.

8 Use a volume of revolution to derive a formula for the volume of a hemisphere of radius r

9 The parabola $y^2 = 28x$ intersects the line $x = k$ at the points A and B.

  The area bounded by the parabola and the line segment AB is $130\dfrac{2}{3}$ square units.

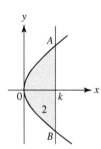

  a Find the coordinates of A and B.

  The shaded region is rotated 180° around the x-axis.

  b Calculate the volume of the solid formed.

10 A region, R, is bounded by the curve $y = (x+1)(x-1)(5-x)$, the line $y = 4(x+1)$ and the positive x and y-axes.

  a Calculate the area of R.

  b Calculate the volume of the solid formed when R is rotated 2π radians around the x-axis. Give your answer in terms of π.

Fluency and skills

You could also be asked to calculate the volume when a curve is rotated around the *y*-axis. In this case, the roles of *x* and *y* are swapped. You can think about it by using the *y*-axis as the base and rotating the area "under" the curve and "above" the *y*-axis.

Therefore you need the function to be $x = f(y)$ and the limits to be

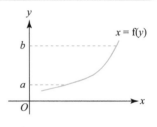

$y = a$ and $y = b$. The volume is given by the formula $V = \int_a^b \pi x^2 \, dy$

Key point

The volume of the solid formed by rotating the curve $x = f(y)$
between $y = a$ and $y = b$ a full turn around the *y*-axis is given by $V = \int_a^b \pi x^2 \, dy$

This is an integral with respect to *y*, so replace the x^2 with the function in terms of *y*

Example 1

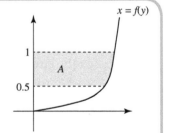

The region A is enclosed by the curve with equation $y = x^5$, the *y*-axis and the lines $y = 0.5$ and $y = 1$

a Calculate the area of region A.

b Calculate the volume of the solid formed when region A is rotated 2π radians around the *y*-axis.

a Rearranging gives you $x = y^{\frac{1}{5}}$

> Always rearrange into the form $x = f(y)$.

$$\text{Area} = \int_{0.5}^{1} y^{\frac{1}{5}} \, dy$$

> Since the area between a curve $x = f(y)$, the *y*-axis and the lines $y = a$ and $y = b$ is given by $\int_a^b f(y) \, dy$

$$= \left[\frac{5}{6} y^{\frac{6}{5}} \right]_{0.5}^{1}$$

$$= \frac{5}{6}(1 - 0.4353) = 0.471 \text{ square units}$$

b $$V = \int_{0.5}^{1} \pi \left(y^{\frac{1}{5}} \right)^2 \, dy$$

> Using $V = \int_a^b \pi x^2 \, dy$, recall that 2π radians is the same as $360°$.

$$= \pi \int_{0.5}^{1} y^{\frac{2}{5}} \, dy$$

$$= \pi \left[\frac{5}{7} y^{\frac{7}{5}} \right]_{0.5}^{1}$$

$$= \pi \left(\frac{5}{7} - 0.2707 \right) = 1.39 \text{ cubic units}$$

> Answer to 3 significant figures.

1   Calculate the volume when the curve
    $x = \dfrac{y^2}{2}$ between $y = 0$ and $y = 5$ is rotated $360°$
    around the y-axis.

2   Calculate the volume when the segment
    of the line $y = 5x$ between $x = 0$ and $x = \dfrac{1}{5}$ is
    rotated $360°$ around

    a   The x-axis,   **b**   The y-axis.

3   Calculate the volume of the solid formed
    when the region bounded by the curve
    $x = 2y^3 - y$ and the lines $y = 1$ and $y = 2$ is
    rotated 2π radians around the y-axis.

4   Calculate the volume of the solid formed
    when the region bounded by the curve
    $x = y - \dfrac{y^2}{2}$ the coordinate axes and the line
    $y = 3$ is rotated 2π radians around the y-axis.

5   Calculate the volume of the solid formed
    when the region bounded by the curve
    $x = \dfrac{4}{y^2}$ between $y = 1$ and $y = 2$ is rotated $180°$
    around the y-axis.

6   The region R is bounded by the curve with
    equation $y = (x-2)^2$, the y-axis and the line
    $y = 1$

    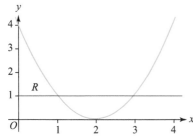

    a   Write x as a function of y

    b   Calculate the area of region R

    c   Find the volume of the solid formed when
          R is rotated 2π radians around the y-axis.

7   The shaded region is bounded by the curve
    $y = 5 - x^2$, the line $y = 2$

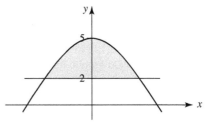

    a   Calculate the area of the shaded region.

    The shaded region is rotated π radians
    around the y-axis.

    b   Calculate the volume of revolution.

8   The shaded region is bounded by the curve
    with equation $y = x^2 - 3$ and the coordinate
    axes. Find the volume of revolution when
    the shaded region is rotated $360°$ around

    a   The x-axis,   **b**   The y-axis.

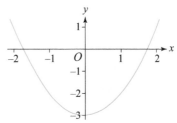

9   Calculate the volume of the solid formed
    when the area bounded by the curve with
    equation $y = 4 - x^2$ and the lines $x = 0$ and
    $y = 0$ is rotated $360°$ around the y-axis.

10 The area bounded by the curve with
    equation $y = 4x^4$ and the lines $y = 2$ and
    $y = 8$ is rotated $360°$ around the y-axis. Show
    that the volume of the solid formed is $A\sqrt{2}\pi$
    where A is a constant to be found.

11 The section between $x = 1$ and $x = 3$ of the
    curve with equation $y = \dfrac{Ax}{\sqrt{x}}$ is rotated 2π
    radians around the x-axis and the volume of
    the solid formed is $\dfrac{8}{3}\pi$

    a   Find the value of A,

    b   Calculate the volume when the same
          section of the curve is rotated 2π radians
          around the y-axis.

Reasoning and problem-solving

Strategy

To find a volume of revolution around the *y*-axis:

1. Find any necessary points of intersection.
2. Sketch a graph.
3. Use the formula for a volume of revolution, $V = \int_a^b \pi x^2 \, dy$ with the correct limits.
4. Add or subtract another volume where necessary.

Example 2

The region R is enclosed by the curve with equation $y^2 = \dfrac{18}{x}$, $y \geq 0$ and the lines $y = \dfrac{3}{2}x$
and $y = 5$

Calculate the volume when the shaded region is rotated by 360° around the *y*-axis.
Give your answer in terms of π.

$y^2 = \dfrac{18}{x}$ and $y = \dfrac{3}{2}x$ intersect when $\left(\dfrac{3}{2}x\right)^2 = \dfrac{18}{x} \Rightarrow x^3 = 8$

> **1** Find the point of intersection between the curve $y^2 = \dfrac{18}{x}$ and the line $y = \dfrac{3}{2}x$

$$\Rightarrow x = 2$$
$$\Rightarrow y = \dfrac{3}{2} \times 2 = 3$$

$y^2 = \dfrac{18}{x}$ rearranges to $x = \dfrac{18}{y^2}$, so the volume
of revolution for the blue region is given by

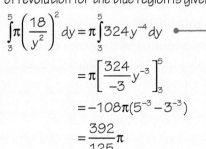

> **2** Sketch the graph, from this you can see that two separate volumes are required

$$\int_3^5 \pi \left(\dfrac{18}{y^2}\right)^2 dy = \pi \int_3^5 324 y^{-4} dy$$

> **3** Use the formula
> $$V = \int_a^b \pi x^2 \, dy$$

$$= \pi \left[\dfrac{324}{-3} y^{-3}\right]_3^5$$

$$= -108\pi(5^{-3} - 3^{-3})$$

$$= \dfrac{392}{125}\pi$$

The green region will form a cone when rotated 360° around the *y*-axis.
So its volume is given by $\dfrac{1}{3}\pi r^2 h = \dfrac{1}{3}\pi \times 2^2 \times 3$

$$= 4\pi$$

Therefore, the total volume is $V = \dfrac{392}{125}\pi + 4\pi$

$$= \dfrac{892}{125}\pi \text{ cubic units}$$

> **4** Add the two volumes together.

1 The shaded region shown is bounded by the curve $y = 10 - x^3$, the y-axis and the line $y = 2$.

Calculate the volume of the solid formed when the shaded region is rotated 180° around the y-axis. Give your answer in terms of π

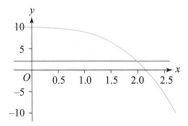

2 The region R is bounded by the curve $y = 8 - x^2$, the line $x = 2$ and the coordinate axes.

Calculate the volume of revolution when R is rotated 2π radians around

a The x-axis,

b The y-axis.

Give your answers in terms of π.

3 The region A is bounded by the curve with equation $y = 2\sqrt{x}$ and the line $y = x$.

Calculate the volume of revolution when the shaded region is rotated 2π radians around

a The x-axis, **b** The y-axis.

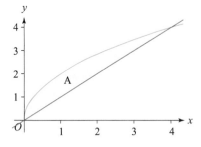

4 The shaded region is bounded by the curve with equation $y = x^3 + 8$ and the coordinate axes.

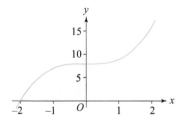

Find the volume of the solid formed when the shaded region is rotated π radians around

a The x-axis, **b** The y-axis.

5 A region is bounded by the curve with equation $x = \dfrac{16}{y}$, the lines $y = x$ and $y = 7$, and the y-axis. Calculate the volume of revolution when this region is rotated 360° around the y-axis.

6 The shaded region shown is bounded by the curve $y = x^2 - 1$, the line $y = -2x$ and the y-axis. Calculate the volume of the solid formed when the region is rotated 2π radians around the y-axis.

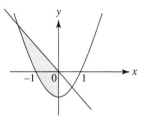

7 The region bounded by the curve with equation $y = 3 - \sqrt{x}$ and the positive x- and y-axes is rotated k radians around the y-axis.

The solid formed has a volume 27π cubic units.

Calculate the value of k in terms of π.

8 The shaded region is bounded by the curve with equation $y = \sqrt{\dfrac{x}{2}}$, the y-axis and the line $y = 6(1 - x)$

a Calculate the area of the shaded region.

Give your answers to 3 significant figures.

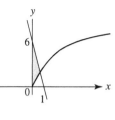

The shaded region is rotated 360° around the y-axis.

b Calculate the volume of revolution. Give your answer to 3 significant figures.

Chapter summary

- The area enclosed between the curve with equation $y = f(x)$, the x-axis and the lines $x = a$ and $x = b$ is given by $\int_a^b y \, dx$
- The volume of the solid formed by rotating the curve $y = f(x)$ between $x = a$ and $x = b$ a full turn around the x-axis is given by $V = \int_a^b \pi y^2 \, dx$
- The volume of the solid formed by rotating the curve $x = f(y)$ between $y = a$ and $y = b$ a full turn around the y-axis is given by $V = \int_a^b \pi x^2 \, dy$

Check and review

You should now be able to...	Try Questions
Find the area enclosed by a curve and lines.	1, 5
Calculate the volume of revolution when rotated around the x-axis.	1–3, 7
Calculate the volume of revolution when rotated around the y-axis.	4–7
Calculate more complicated volumes of revolution by adding or subtracting volumes.	8–11

1 The region R is enclosed by the curve with equation $y = 3 - \dfrac{1}{x^2}$, the x-axis and the line $x = 2$

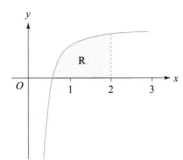

a Show that the area of R is $\dfrac{13 - 4\sqrt{3}}{2}$ square units.

The region R is rotated 360° around the x-axis.

b Calculate the volume of the solid formed.

2 Calculate the volume of the solid formed when the line with equation $y = 1 - 3x$ between $x = 2$ and $x = 5$ is rotated 2π radians around the x-axis.

3 Calculate the volume of the solid formed when the curve with equation $y = 5\sqrt{x}$ between $x = 1$ and $x = 3$ is rotated 180° around the x-axis.

4 Show that the volume of revolution when $x = 4y^2$ between $y = 0$ and $y = \dfrac{1}{2}$ is rotated 2π radians around the y-axis is $\dfrac{\pi}{10}$ cubic units.

5 The shaded region is bounded by the curve with equation $x = y^4 + 1$, $y \geq 0$, the coordinate axes and the line with equation $y = 2$

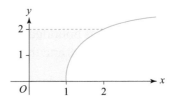

a Show that the area of the shaded region is $\dfrac{42}{5}$ square units.

The region is rotated 360° around the y-axis.

b Calculate the volume of the solid formed.

6 Find the volume of revolution when the area bounded by the x-axis, the line $x = 1$ and the curve $y = 8x^{\frac{3}{4}}$ is rotated 360° around

   a The x-axis,   **b** The y-axis.

7 The region A bounded by the curve with equation $y = 2 - \sqrt{x}$, the line $y = 2 - \dfrac{x}{2}$ and the y-axis is rotated 360° around the x-axis.

   a Sketch a graph showing the region A

   b Find the volume of revolution when region A is rotated 2π radians around the x-axis.

   c Find the volume of revolution when region A is rotated 2π radians around the y-axis.

8 The shaded region is bounded by the curve with equation $y = 5x - 4x^2$, the line $y = 1$ and the x-axis.

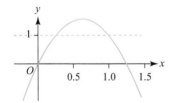

   a Calculate the area of the shaded region.

   b Calculate the volume of the solid formed when the shaded region is rotated 180° around the x-axis.

9 The shaded region shown is bounded by the curve $y = 5 - x^2$, the line $x = 1$ and the coordinate axes.

Calculate the volume of the solid formed by rotating the region 180° around the y-axis.

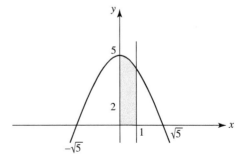

10 Region R is bounded by the curve with equation $y = 3 - \sqrt{x}$, the y-axis and the line $y = 2x$

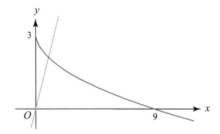

Calculate the volume of revolution when R is rotated 2π radians around the y-axis.

Give your answer in terms of π

11 The area enclosed by the curve $y = x^2 - 3$, above the line $y = -2x$ and below the x-axis is rotated 360° around the x-axis. Calculate the volume of the solid formed.

Investigation

A different method of finding the volume of a solid by considering the revolution of a function, relies on considering the solid to be composed of thin cylindrical shells.

Each cylindrical shell is formed by rotating a vertical strip about the y-axis. The inner surface area of this cylinder is a rectangle with width $2\pi x$ and height y, so it has area $2\pi xy$

This means that the volume, δv, of a cylinder of width δx is given by $\delta v = 2\pi xy\delta x$

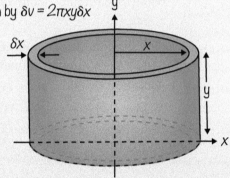

Summing these cylinders between $x = x_1$ and $x = x_2$, and letting δx tend to zero, results in the formula $v = 2\pi \int_{x_2}^{x_2} xy\,dx$

Convince yourself that this formula is correct by using it to find
a The volume of a cylinder of radius a and height h,
b The volume of a cone of radius a and height h

Information

Gabriel's horn is a geometric solid that has a finite volume but an infinite surface area. This is sometimes known as the painter's paradox as although the horn would contain a finite volume of paint you would never have enough paint to paint its outside.

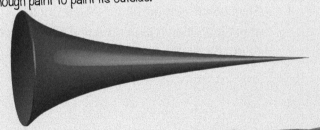

The solid is formed by rotating the function $f(x) = \frac{1}{x}$ about the x-axis for $x > 1$

Have a go

When carrying out an integral with no limit on the upper bound, it is usual practice to take an arbitrary value, say $x = a$, and then consider what happens as a tends to infinity.

Try the method described above to calculate the finite volume of Gabriel's horn.

Challenge

The formula for a **surface area of revolution** that is found by rotating $y = f(x)$ from $x = a$ to $x = b$ about the x-axis is given by

$$A = 2\pi \int_a^b y\sqrt{1 + \left(\frac{dy}{d}\right)^2}\,dx$$

Use this formula to show that Gabriel's horn has an infinite surface area.

1 The shaded region is bounded by the curve
$y = x^2 - 2x$, the x-axis and the line $x = 4$

   a Find the value of α **[2 marks]**

   b Calculate the area of the shaded region.   **[3]**

The shaded area is rotated one full turn around
the x-axis.

   c Find the volume of the solid formed.   **[4]**

2 Find the volume of revolution when the region bounded by the curve $y = 2\sqrt{x}$,
the x-axis and the line $x = 3$ is rotated $360°$ about the x-axis.

Give your answer in terms of π **[4]**

3 The region R is bounded by the curve $y = \dfrac{2}{x} + x^3$, the x-axis and the lines $x = 1$ and
$x = 2$. Find the volume of revolution when R is rotated $360°$ around the x-axis.   **[4]**

4 Find the volume of revolution when the region bounded by the curve $y = x^2(5 - x)$ is
rotated $180°$ around the x-axis. Give your answer to 3 significant figures.   **[5]**

5 The region bounded by the x-axis and the curve with equation $x = 2y^3 + 4y^2$
is rotated 2π radians around the y-axis. Find the volume of the solid formed.   **[6]**

6 The curve with equation $y = x - k\sqrt{x}$ between $x = 0$ and $x = 1$ is rotated by
2π radians around the x-axis and the volume of revolution is $\dfrac{11}{15}\pi$.
Calculate the possible values of k **[5]**

7 The region R is bounded by the curve $y = (x - 2)^3$,
the line $y = 1$ and the coordinate axes.

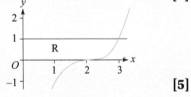

Calculate the volume of revolution when R is rotated
$360°$ around the y-axis.   **[5]**

8 The region A is bounded by the curve $y = 3 - x^2$,
the line $y = 2x$ and the y-axis.

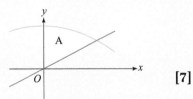

Calculate the volume of revolution when A is
rotated 2π radians around the y-axis.   **[7]**

9 The shaded region is bounded by the curve $y = 4x^4$
and the line $y = 1$

Calculate the exact volume of revolution when the
region is rotated $360°$ around the x-axis.   **[6]**

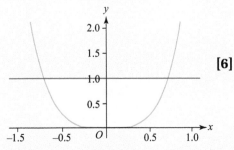

10 The region R is bounded by the curve with equation $x = \dfrac{y-1}{\sqrt{y}}$, and the lines $y = 1$ and $y = 4$

   a   Calculate the exact area of R   **[4]**

   b   Calculate the exact volume of the solid formed when R is rotated 2π radians around the y-axis.   **[5]**

11 The shaded region shown is part of an ellipse with equation $\dfrac{x^2}{9} + \dfrac{y^2}{4} = 1$

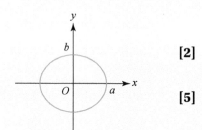

   a   Write the values of a and b in the diagram.   **[2]**

   b   Calculate the volume when the shaded area is rotated π radians about the x-axis.   **[5]**

   c   Calculate the volume when the shaded area is rotated π radians about the y-axis.   **[5]**

12 Part of the curve $y = \dfrac{1}{30}(x^3 - 20x^2 + 100x + 90)$ between $x = 0$ and $x = 13$ is rotated $360°$ to form the shape of a vase.

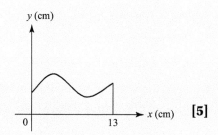

Given that each unit on the coordinate axes represents 1 cm, calculate the capacity of the vase. Give your answer in litres, to 2 significant figures.   **[5]**

13 The shaded region is bounded by the circle with equation $x^2 + (y-1)^2 = 2$ and the x-axis.

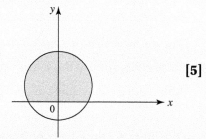

Calculate the exact volume of the solid formed when the shaded region is rotated $180°$ around the y-axis.   **[5]**

14 Region R is bounded by the curve with equation $y = x^2 - 9$, the y-axis and the line $y = x + 3$

   a   Calculate the area of R.   **[7]**

   b   Calculate the volume of revolution when R is rotated $180°$ around the y-axis.   **[8]**

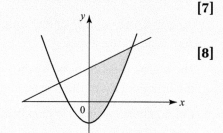

Matrices

4

Matrices have many different applications. One area of their use that has grown tremendously in recent years is in computer graphics, for example in the video gaming industry. The mathematics of matrices is used by programmers to make objects move with realism in three dimensions. **Transformation matrices** are used to translate, rotate and scale objects as well as provide realistic shading or rendering, shift viewpoints and transform colours. As well as in video gaming, these techniques are used in the film industry where accurate coding and artistic talent allow us to merge reality and graphics in ways that make it difficult to know what is real and what is virtual.

Other sectors that require accurate visualisations also rely on the mathematics of matrices. For example they can be used in developing software to support design work in architecture, interior design and the automotive industry.

Orientation

What you need to know	What you will learn	What this leads to
KS4 • Transforming objects by reflection, rotation and enlargement.	• To identify the order of a matrix. • To add, subtract and multiply matrices by a scalar or a conformable matrix. • To apply linear transformations given as matrices, describe transformations given as matrices and write linear transformations as a matrix.	**FP2 Ch8 Further Matrices** • Eigenvalues and eigenvectors. • The Cayley-Hamilton theorem. • Matrix diagonilisation.
Maths Ch1 • Solving simultaneous equations.		**D1 Ch2 Graphs and networks** • Matrix formulation of Prim's algorithm
Maths Ch6 • Using vectors.	• To find invariant points and lines of linear transformations. • To calculate determinants and inverses of matrices. • To use matrices to solve systems of linear equations.	**D2 Ch6 Transportation and allocation problems** • Allocation problems **D2 Ch7 Game theory** • Zero sum games

Properties and arithmetic

Fluency and skills

See Maths Ch6

For a reminder of vectors.

Matrices are a way of representing information in a form that can be manipulated mathematically. You will have already used vectors, which are a particular sort of matrix.

Key point

A **matrix** with n rows and m columns has **order** $n \times m$

For example, the matrix $\begin{pmatrix} 5 & 3 & -1 \\ -4 & 0 & 5 \end{pmatrix}$ is said to have order 2×3

The matrix $\begin{pmatrix} 2 \\ 7 \\ -5 \end{pmatrix}$ order 3×1

Computers use matrices to carry out an operation, such as addition or subtraction, and to carry it out on multiple numbers simultaneously. This is particularly used in the processing of computer graphics.

Key point

If two matrices have the same order then they can be added or subtracted by adding or subtracting their corresponding **elements**.

You can think of putting the two matrices on top of each other and adding or subtracting the elements that match.

Key point

To multiply a matrix by a constant, you should multiply each of its elements by that constant.

Example 1

You are given that $\mathbf{A} = \begin{pmatrix} 2 & 5 \\ 0 & 4 \\ -1 & -3 \end{pmatrix}$, $\mathbf{B} = \begin{pmatrix} 2 & -3 \\ 0 & 5 \end{pmatrix}$ and $\mathbf{C} = \begin{pmatrix} 6 & 15 \\ 0 & 12 \\ -3 & -9 \end{pmatrix}$. Find, if possible,

a $\mathbf{A} + \mathbf{B}$ b $\mathbf{A} + \mathbf{C}$ c $2\mathbf{B}$ d $\mathbf{C} - 3\mathbf{A}$

a   Not possible as **A** and **B** are not of the same order.

b $\mathbf{A} + \mathbf{C} = \begin{pmatrix} 2 & 5 \\ 0 & 4 \\ -1 & -3 \end{pmatrix} + \begin{pmatrix} 6 & 15 \\ 0 & 12 \\ -3 & -9 \end{pmatrix}$

$= \begin{pmatrix} 2+6 & 5+15 \\ 0+0 & 4+12 \\ -1+-3 & -3+-9 \end{pmatrix}$ — Add corresponding elements of **A** and **C**

$= \begin{pmatrix} 8 & 20 \\ 0 & 16 \\ -4 & -12 \end{pmatrix}$

(Continued on the next page)

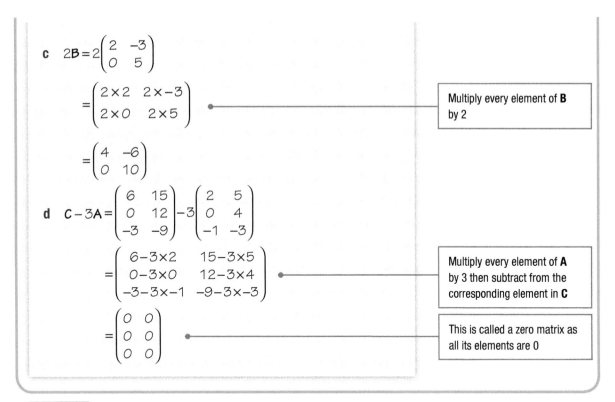

c $2\mathbf{B} = 2\begin{pmatrix} 2 & -3 \\ 0 & 5 \end{pmatrix}$

$= \begin{pmatrix} 2 \times 2 & 2 \times -3 \\ 2 \times 0 & 2 \times 5 \end{pmatrix}$ ——— Multiply every element of **B** by 2

$= \begin{pmatrix} 4 & -6 \\ 0 & 10 \end{pmatrix}$

d $\mathbf{C} - 3\mathbf{A} = \begin{pmatrix} 6 & 15 \\ 0 & 12 \\ -3 & -9 \end{pmatrix} - 3\begin{pmatrix} 2 & 5 \\ 0 & 4 \\ -1 & -3 \end{pmatrix}$

$= \begin{pmatrix} 6-3\times 2 & 15-3\times 5 \\ 0-3\times 0 & 12-3\times 4 \\ -3-3\times -1 & -9-3\times -3 \end{pmatrix}$ ——— Multiply every element of **A** by 3 then subtract from the corresponding element in **C**

$= \begin{pmatrix} 0 & 0 \\ 0 & 0 \\ 0 & 0 \end{pmatrix}$ ——— This is called a zero matrix as all its elements are 0

Key point

The **zero matrix**, **0**, is a matrix, of any order, with all elements equal to zero.

Matrices can only be multiplied together if the number of columns in the first matrix is the same as the number of rows in the second matrix.

Key point

If two matrices can be multiplied, then it is said that they are **conformable for multiplication.**

When multiplying matrices, to find the element in the nth row and mth column, you must multiply the first term of the nth row in the first matrix by the first term of the mth column in the second matrix, then the second term of the nth row by the second term of the mth column and so on, then add these terms together.

Key point

The product of an $n \times m$ matrix and an $m \times p$ matrix has **order** $n \times p$

To find the first term in a matrix multiplication, look at the first row of the first matrix and the first column of the second matrix.

$$\begin{pmatrix} a & d \\ b & e \\ c & f \end{pmatrix}\begin{pmatrix} w & y \\ x & z \end{pmatrix} = \begin{pmatrix} aw+dx & \ldots \\ \ldots & \ldots \\ \ldots & \ldots \end{pmatrix}$$

Then consider the first row and second column.

$$\begin{pmatrix} a & d \\ b & e \\ c & f \end{pmatrix}\begin{pmatrix} w & y \\ x & z \end{pmatrix} = \begin{pmatrix} aw+dx & ay+dz \\ \ldots & \ldots \\ \ldots & \ldots \end{pmatrix}$$

Then move on to using the second row of the first matrix. Continue in this way until the last term which is found by considering the final row of the first matrix and the final column of the second.

$$\begin{pmatrix} a & d \\ b & e \\ c & f \end{pmatrix}\begin{pmatrix} w & y \\ x & z \end{pmatrix} = \begin{pmatrix} aw+dx & ay+dz \\ bw+bx & by+ez \\ cw+fx & cy+fz \end{pmatrix}$$

Example 2

If $A = \begin{pmatrix} 3 & 0 \\ -1 & 2 \\ 7 & -4 \end{pmatrix}$, $B = \begin{pmatrix} 5 & 1 \\ -3 & 0 \end{pmatrix}$ find, if possible, the products

a AB **b** BA **c** B^2

a $AB = \begin{pmatrix} 3 & 0 \\ -1 & 2 \\ 7 & -4 \end{pmatrix} \begin{pmatrix} 5 & 1 \\ -3 & 0 \end{pmatrix}$

$= \begin{pmatrix} (3 \times 5) + (0 \times -3) & (3 \times 1) + (0 \times 0) \\ (-1 \times 5) + (2 \times -3) & (-1 \times 1) + (2 \times 0) \\ (7 \times 5) + (-4 \times -3) & (7 \times 1) + (-4 \times 0) \end{pmatrix}$

$= \begin{pmatrix} 15 & 3 \\ -11 & -1 \\ 47 & 7 \end{pmatrix}$

To find the term in the 3rd row and the 2nd column: multiply the first term from the 3rd row of matrix **A** by the first term from the 2nd column of matrix **B**. Then multiply the second term from the 3rd row of matrix **A** by the second term in the 2nd column of matrix **B**. Finally, add these values together.

Notice how the product of a 3×2 matrix and a 2×2 matrix is a 3×2 matrix.

b Not possible as **B** has 2 columns but **A** has 3 rows.

c $B^2 = \begin{pmatrix} 5 & 1 \\ -3 & 0 \end{pmatrix} \begin{pmatrix} 5 & 1 \\ -3 & 0 \end{pmatrix}$

B^2 means $B \times B$
It does not mean square each element of the matrix.

$= \begin{pmatrix} (5 \times 5) + (1 \times -3) & (5 \times 1) + (1 \times 0) \\ (-3 \times 5) + (0 \times -3) & (-3 \times 1) + (0 \times 0) \end{pmatrix}$

$= \begin{pmatrix} 22 & 5 \\ -15 & -3 \end{pmatrix}$

Key point

It is only possible to find A^2 if **A** is a **square matrix**, that is, it has the same number of rows as columns.

Calculator

Try it on your calculator

Some calculators can be used to add, subtract and multiply matrices.

MatA × MatB

$\begin{bmatrix} 15 & 3 \\ -11 & -1 \\ 47 & 7 \end{bmatrix}$

Activity

Find out how to work out $\begin{pmatrix} 3 & 0 \\ -1 & 2 \\ 7 & -4 \end{pmatrix} \begin{pmatrix} 5 & 1 \\ -3 & 0 \end{pmatrix}$ on *your* calculator.

1 a Write down the order of these matrices.

i $\begin{pmatrix} 5 & -3 \\ 1 & 0 \\ 9 & 11 \end{pmatrix}$ ii $\begin{pmatrix} 9 \\ 3 \\ -2 \end{pmatrix}$

iii $\begin{pmatrix} 3 & -7 \\ -2 & 0 \end{pmatrix}$ iv $\begin{pmatrix} -3 & 0 & 5 & 8 \\ 2 & -1 & 4 & 0 \end{pmatrix}$

b Which of the matrices in part **a** is a square matrix?

2 Calculate

a $\begin{pmatrix} 5 & -1 & 2 \\ 0 & 6 & 4 \end{pmatrix} + \begin{pmatrix} -1 & 4 & 0 \\ 5 & -7 & 2 \end{pmatrix}$

b $\begin{pmatrix} 4 & -2 \\ 8 & -5 \end{pmatrix} - \begin{pmatrix} 8 & 4 \\ 7 & -2 \end{pmatrix}$

c $4\begin{pmatrix} -1 & 0 & 4 \\ 2 & 5 & -3 \\ 8 & -6 & 0 \end{pmatrix}$

d $\begin{pmatrix} -6 & 14 \\ 3 & 9 \\ 2 & -4 \end{pmatrix} + \dfrac{1}{2}\begin{pmatrix} 8 & -4 \\ 6 & 3 \\ -2 & 0 \end{pmatrix}$

3 If $\mathbf{A} = \begin{pmatrix} 9 & -4 \\ 0 & -2 \end{pmatrix}$, $\mathbf{B} = \begin{pmatrix} 8 & 4 \\ 2 & -6 \\ -2 & 0 \end{pmatrix}$,

$\mathbf{C} = \begin{pmatrix} -5 & -2 \\ 7 & 0 \\ 3 & 1 \end{pmatrix}$, $\mathbf{D} = \begin{pmatrix} 0 & 1 \\ -5 & 4 \end{pmatrix}$

a Calculate if possible or, if not, explain why.

i $\mathbf{A} + \mathbf{D}$ ii $\mathbf{A} + \mathbf{B}$

iii $\mathbf{B} - \mathbf{C}$ iv $3\mathbf{B}$

b Show that

i $5\mathbf{A} - \mathbf{D} = \begin{pmatrix} 45 & -21 \\ 5 & -14 \end{pmatrix}$

ii $2\mathbf{C} + 7\mathbf{B} = 2\begin{pmatrix} 23 & 12 \\ 14 & -21 \\ -4 & 1 \end{pmatrix}$

4 Calculate these matrix products.

a $\begin{pmatrix} -3 & 0 \\ 1 & 4 \end{pmatrix}\begin{pmatrix} 2 & 5 & -1 & 6 \\ -3 & 1 & 0 & -4 \end{pmatrix}$

b $\begin{pmatrix} -2 & 1 \\ 0 & 9 \\ -5 & 0 \end{pmatrix}\begin{pmatrix} -6 & 3 \\ 4 & 8 \end{pmatrix}$

c $\begin{pmatrix} 5 & -3 & 0 \\ 2 & -4 & 1 \end{pmatrix}\begin{pmatrix} -3 \\ -5 \\ -1 \end{pmatrix}$

d $(9 \ -2)\begin{pmatrix} 0 & 1 & -2 \\ 4 & 7 & -4 \end{pmatrix}$

e $\begin{pmatrix} 4 \\ 3 \end{pmatrix}(-1 \ 2 \ 5)$

f $\begin{pmatrix} 1 & 3 & -1 \\ -2 & 0 & 2 \\ 0 & -3 & 1 \end{pmatrix}^2$

5 Show that $\begin{pmatrix} 3 & 2 & 0 \\ -1 & 0 & -2 \end{pmatrix}\begin{pmatrix} 4 & 1 \\ 0 & 3 \\ -3 & 0 \end{pmatrix} = \begin{pmatrix} 12 & 9 \\ 2 & -1 \end{pmatrix}$

6 If $\mathbf{A} = \begin{pmatrix} 3 \\ 7 \end{pmatrix}$, $\mathbf{B} = \begin{pmatrix} 0 & 2 \\ -1 & 3 \end{pmatrix}$,

$\mathbf{C} = \begin{pmatrix} 2 & 9 & 4 \\ -3 & 0 & -5 \end{pmatrix}$, $\mathbf{D} = \begin{pmatrix} -1 \\ 5 \\ 2 \end{pmatrix}$,

find, if possible, or if the calculation is not possible, explain why.

a $\mathbf{AB}$ **b** $\mathbf{CD}$ **c** $\mathbf{BCD}$

d $\mathbf{CB}$ **e** $\mathbf{B}^2\mathbf{A}$ **f** $\mathbf{C}^2$

7 Calculate these matrix products, simplifying each element where possible.

a $\begin{pmatrix} 2a & 3 \\ 1 & -1 \end{pmatrix}\begin{pmatrix} 2 & 4 & 3 \\ a & 2 & -a \end{pmatrix}$

b $\begin{pmatrix} a \\ 2 \\ -a \end{pmatrix}(3 -a)$

c $(-1 \ a \ 0)\begin{pmatrix} 3a & a \\ 4 & 2 \\ -1 & 5 \end{pmatrix}$

d $\begin{pmatrix} a & 3 \\ -2 & a \end{pmatrix}^2$

8 Given $\mathbf{A} = \begin{pmatrix} 2 & 5 \\ -1 & 2 \end{pmatrix}$, show that

$\mathbf{A}^3 = \begin{pmatrix} -22 & k \\ -7 & -22 \end{pmatrix}$, stating the value of k

Reasoning and problem-solving

Strategy 1

To solve problems involving matrix arithmetic

① If two matrices are of equal order then equate corresponding elements.

② Solve equations simultaneously to find the values of unknowns.

③ Use subscript notation for the elements of general matrices, for example, a_1, a_2 etc.

You can solve problems involving unknown elements using basic algebra.

Example 3

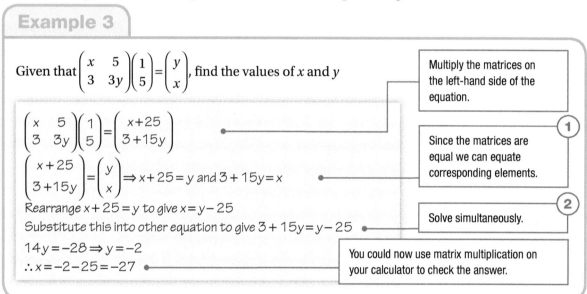

Given that $\begin{pmatrix} x & 5 \\ 3 & 3y \end{pmatrix}\begin{pmatrix} 1 \\ 5 \end{pmatrix} = \begin{pmatrix} y \\ x \end{pmatrix}$, find the values of x and y

Multiply the matrices on the left-hand side of the equation.

$\begin{pmatrix} x & 5 \\ 3 & 3y \end{pmatrix}\begin{pmatrix} 1 \\ 5 \end{pmatrix} = \begin{pmatrix} x+25 \\ 3+15y \end{pmatrix}$

①

$\begin{pmatrix} x+25 \\ 3+15y \end{pmatrix} = \begin{pmatrix} y \\ x \end{pmatrix} \Rightarrow x+25 = y$ and $3+15y = x$

Since the matrices are equal we can equate corresponding elements.

②

Rearrange $x+25 = y$ to give $x = y-25$
Substitute this into other equation to give $3+15y = y-25$

Solve simultaneously.

$14y = -28 \Rightarrow y = -2$

You could now use matrix multiplication on your calculator to check the answer.

$\therefore x = -2-25 = -27$

Alternatively, you may be given a system of linear equations and have to write it in matrix form.

You will be shown later how to use matrices to solve 2×2 systems of equations.

Example 4

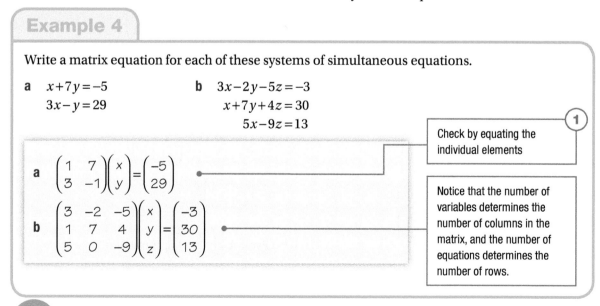

Write a matrix equation for each of these systems of simultaneous equations.

a $\quad x+7y = -5$
$\quad 3x-y = 29$

b $\quad 3x-2y-5z = -3$
$\quad x+7y+4z = 30$
$\quad 5x-9z = 13$

①

Check by equating the individual elements

a $\begin{pmatrix} 1 & 7 \\ 3 & -1 \end{pmatrix}\begin{pmatrix} x \\ y \end{pmatrix} = \begin{pmatrix} -5 \\ 29 \end{pmatrix}$

b $\begin{pmatrix} 3 & -2 & -5 \\ 1 & 7 & 4 \\ 5 & 0 & -9 \end{pmatrix}\begin{pmatrix} x \\ y \\ z \end{pmatrix} = \begin{pmatrix} -3 \\ 30 \\ 13 \end{pmatrix}$

Notice that the number of variables determines the number of columns in the matrix, and the number of equations determines the number of rows.

Matrix addition is both **associative**, so $A + (B + C) = (A + B) + C$, and **commutative**, so $A + B = B + A$

Matrix multiplication is associative, that is, $(AB)C = A(BC)$, but it is not commutative since, in general, $AB \neq BA$

Matrix multiplication is also **distributive**, that is, $A(B + C) = AB + AC$

Example 5

Prove the associative property for matrix multiplication of 2×2 matrices.

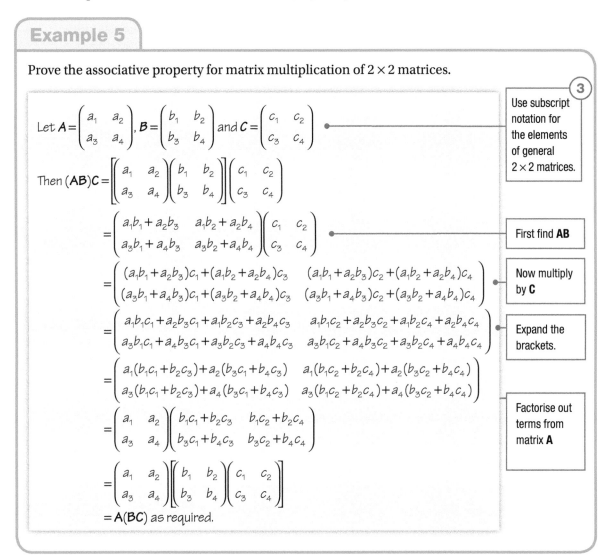

Let $A = \begin{pmatrix} a_1 & a_2 \\ a_3 & a_4 \end{pmatrix}$, $B = \begin{pmatrix} b_1 & b_2 \\ b_3 & b_4 \end{pmatrix}$ and $C = \begin{pmatrix} c_1 & c_2 \\ c_3 & c_4 \end{pmatrix}$

> ③ Use subscript notation for the elements of general 2×2 matrices.

Then $(AB)C = \left[\begin{pmatrix} a_1 & a_2 \\ a_3 & a_4 \end{pmatrix} \begin{pmatrix} b_1 & b_2 \\ b_3 & b_4 \end{pmatrix} \right] \begin{pmatrix} c_1 & c_2 \\ c_3 & c_4 \end{pmatrix}$

$= \begin{pmatrix} a_1 b_1 + a_2 b_3 & a_1 b_2 + a_2 b_4 \\ a_3 b_1 + a_4 b_3 & a_3 b_2 + a_4 b_4 \end{pmatrix} \begin{pmatrix} c_1 & c_2 \\ c_3 & c_4 \end{pmatrix}$

> First find **AB**

$= \begin{pmatrix} (a_1 b_1 + a_2 b_3)c_1 + (a_1 b_2 + a_2 b_4)c_3 & (a_1 b_1 + a_2 b_3)c_2 + (a_1 b_2 + a_2 b_4)c_4 \\ (a_3 b_1 + a_4 b_3)c_1 + (a_3 b_2 + a_4 b_4)c_3 & (a_3 b_1 + a_4 b_3)c_2 + (a_3 b_2 + a_4 b_4)c_4 \end{pmatrix}$

> Now multiply by **C**

$= \begin{pmatrix} a_1 b_1 c_1 + a_2 b_3 c_1 + a_1 b_2 c_3 + a_2 b_4 c_3 & a_1 b_1 c_2 + a_2 b_3 c_2 + a_1 b_2 c_4 + a_2 b_4 c_4 \\ a_3 b_1 c_1 + a_4 b_3 c_1 + a_3 b_2 c_3 + a_4 b_4 c_3 & a_3 b_1 c_2 + a_4 b_3 c_2 + a_3 b_2 c_4 + a_4 b_4 c_4 \end{pmatrix}$

> Expand the brackets.

$= \begin{pmatrix} a_1(b_1 c_1 + b_2 c_3) + a_2(b_3 c_1 + b_4 c_3) & a_1(b_1 c_2 + b_2 c_4) + a_2(b_3 c_2 + b_4 c_4) \\ a_3(b_1 c_1 + b_2 c_3) + a_4(b_3 c_1 + b_4 c_3) & a_3(b_1 c_2 + b_2 c_4) + a_4(b_3 c_2 + b_4 c_4) \end{pmatrix}$

$= \begin{pmatrix} a_1 & a_2 \\ a_3 & a_4 \end{pmatrix} \begin{pmatrix} b_1 c_1 + b_2 c_3 & b_1 c_2 + b_2 c_4 \\ b_3 c_1 + b_4 c_3 & b_3 c_2 + b_4 c_4 \end{pmatrix}$

> Factorise out terms from matrix **A**

$= \begin{pmatrix} a_1 & a_2 \\ a_3 & a_4 \end{pmatrix} \left[\begin{pmatrix} b_1 & b_2 \\ b_3 & b_4 \end{pmatrix} \begin{pmatrix} c_1 & c_2 \\ c_3 & c_4 \end{pmatrix} \right]$

$= A(BC)$ as required.

You can apply the method of proof by induction to matrices.

Strategy 2

To solve induction problems involving matrices

① Check the base case.

② Assume the statement is true for $n = k$

③ Use $A^{k+1} = A \times A^k$ to show statement is true for $n = k + 1$

④ Write a conclusion.

> **See Ch 2.3**
> For a reminder of proof by induction.

Example 6

Prove by induction that $\begin{pmatrix} 1 & 0 \\ 2 & 1 \end{pmatrix}^n = \begin{pmatrix} 1 & 0 \\ 2n & 1 \end{pmatrix}$ for all positive integers n

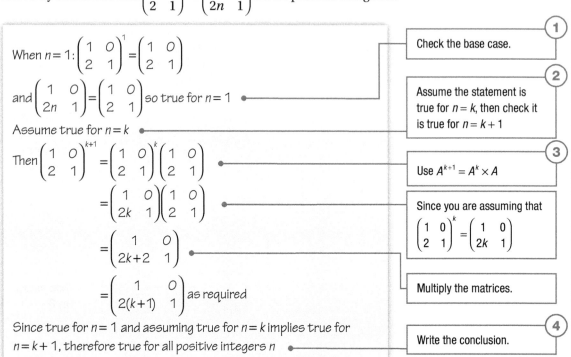

When $n = 1$: $\begin{pmatrix} 1 & 0 \\ 2 & 1 \end{pmatrix}^1 = \begin{pmatrix} 1 & 0 \\ 2 & 1 \end{pmatrix}$

and $\begin{pmatrix} 1 & 0 \\ 2n & 1 \end{pmatrix} = \begin{pmatrix} 1 & 0 \\ 2 & 1 \end{pmatrix}$ so true for $n = 1$

Assume true for $n = k$

Then $\begin{pmatrix} 1 & 0 \\ 2 & 1 \end{pmatrix}^{k+1} = \begin{pmatrix} 1 & 0 \\ 2 & 1 \end{pmatrix}^k \begin{pmatrix} 1 & 0 \\ 2 & 1 \end{pmatrix}$

$= \begin{pmatrix} 1 & 0 \\ 2k & 1 \end{pmatrix} \begin{pmatrix} 1 & 0 \\ 2 & 1 \end{pmatrix}$

$= \begin{pmatrix} 1 & 0 \\ 2k+2 & 1 \end{pmatrix}$

$= \begin{pmatrix} 1 & 0 \\ 2(k+1) & 1 \end{pmatrix}$ as required

Since true for $n = 1$ and assuming true for $n = k$ implies true for $n = k + 1$, therefore true for all positive integers n

1 Check the base case.

2 Assume the statement is true for $n = k$, then check it is true for $n = k + 1$

3 Use $A^{k+1} = A^k \times A$

Since you are assuming that $\begin{pmatrix} 1 & 0 \\ 2 & 1 \end{pmatrix}^k = \begin{pmatrix} 1 & 0 \\ 2k & 1 \end{pmatrix}$

Multiply the matrices.

4 Write the conclusion.

Strategy 3

To solve problems involving tables by using matrices:

(**1**) Convert from tabular form into a matrix of suitable order.

(**2**) Identify any relevant vectors.

(**3**) Multiply the matrix by the vector to perform the necessary data analysis.

(**4**) Write a conclusion, if required.

Example 7

The table shows the probabilities of a spring day being rainy in four UK cities.

	March	April	May
London	0.32	0.3	0.29
Edinburgh	0.38	0.33	0.37
Cardiff	0.42	0.37	0.36
Belfast	0.45	0.38	0.38

a Write a matrix, **A**, to represent the table of information and a vector, **x**, to represent the number of days in each month.

b Show how you can use your matrices to calculate the total number of rainy days expected in each city over the three months.

(Continued on the next page)

a $A = \begin{pmatrix} 0.32 & 0.3 & 0.29 \\ 0.38 & 0.33 & 0.37 \\ 0.42 & 0.37 & 0.36 \\ 0.45 & 0.38 & 0.38 \end{pmatrix}$ $x = \begin{pmatrix} 31 \\ 30 \\ 31 \end{pmatrix}$

The columns of the matrix represent the months and the rows represent the cities.

Note that March and May have 31 days but April has 30

b $\begin{pmatrix} 0.32 & 0.3 & 0.29 \\ 0.38 & 0.33 & 0.37 \\ 0.42 & 0.37 & 0.36 \\ 0.45 & 0.38 & 0.38 \end{pmatrix} \begin{pmatrix} 31 \\ 30 \\ 31 \end{pmatrix} = \begin{pmatrix} 0.32 \times 31 + 0.3 \times 30 + 0.29 \times 31 \\ 0.38 \times 31 + 0.33 \times 30 + 0.37 \times 31 \\ 0.42 \times 31 + 0.37 \times 30 + 0.36 \times 31 \\ 0.45 \times 31 + 0.38 \times 30 + 0.38 \times 31 \end{pmatrix}$

Multiply the vector by the matrix.

$= \begin{pmatrix} 27.91 \\ 33.15 \\ 35.28 \\ 37.13 \end{pmatrix}$

Write a conclusion. Give answers to the nearest day.

So we expect a total of 28 rainy days in London, 33 in Edinburgh, 35 in Cardiff and 37 in Belfast.

Exercise 4.1B Reasoning and problem-solving

1 Find the values of a, b, c and d such that
$$\begin{pmatrix} 1 & a \\ 3 & 7 \end{pmatrix} + \begin{pmatrix} -2 & 4 \\ b & -5 \end{pmatrix} = \begin{pmatrix} c & 1 \\ -2 & d \end{pmatrix}$$

2 Find the values of a, b and c such that
$$\begin{pmatrix} 3 & a \\ 1 & -2 \end{pmatrix} \begin{pmatrix} 5 & b \\ 4 & 1 \end{pmatrix} = \begin{pmatrix} 3 & 9 \\ c & 2 \end{pmatrix}$$

3 Find the values of x and y in each case.

a $\begin{pmatrix} x & 3 \\ 2 & 0 \end{pmatrix} \begin{pmatrix} 2 & 5 \\ y & -y \end{pmatrix} = \begin{pmatrix} 10 & -17 \\ 4 & 10 \end{pmatrix}$

b $\begin{pmatrix} x & -1 \\ y & 2 \end{pmatrix}^2 = \mathbf{0}$

4 Solve each of these equations to find the possible values of x

a $\begin{pmatrix} 3 & x & -2 \\ 4 & 0 & 0 \end{pmatrix} \begin{pmatrix} 1 & 1 \\ 5x & 0 \\ x & 5 \end{pmatrix} = \begin{pmatrix} 6 & -7 \\ 4 & 4 \end{pmatrix}$

b $(7 \quad x \quad 4) \begin{pmatrix} x & 3 & 0 \\ x & 0 & 2 \\ 5 & -1 & x \end{pmatrix} = (10 \quad 17 \quad 6x)$

c $\begin{pmatrix} 5 & x \\ -x & -5 \end{pmatrix}^2 = 16 \begin{pmatrix} 1 & 0 \\ 0 & 1 \end{pmatrix}$

d $\begin{pmatrix} 0 & x \\ x & 1 \end{pmatrix}^3 = \begin{pmatrix} 4 & -10 \\ -10 & 9 \end{pmatrix}$

5 Calculate the values of a, b and c in this matrix equation.
$$\begin{pmatrix} -1 & 2 & 5 \\ 4 & 1 & 0 \\ 0 & 3 & 0 \end{pmatrix} \begin{pmatrix} a \\ b \\ c \end{pmatrix} = \begin{pmatrix} -3 \\ 25 \\ -9 \end{pmatrix}$$

6 Calculate the values of x, y and z in this matrix equation.
$$\begin{pmatrix} 2 & 0 & 0 \\ 0 & 4 & 6 \\ -1 & 3 & 1 \end{pmatrix} \begin{pmatrix} x \\ y \\ z \end{pmatrix} = \begin{pmatrix} 6 \\ 10 \\ 8 \end{pmatrix}$$

7 Calculate the values of a, b and c in this matrix equation.
$$\begin{pmatrix} 1 & 3 & 2 \\ 5 & 0 & 0 \\ 7 & -2 & 1 \end{pmatrix} \begin{pmatrix} a \\ b \\ c \end{pmatrix} = \begin{pmatrix} 5 \\ -5 \\ 3 \end{pmatrix}$$

8 Write a matrix equation for each of these systems of simultaneous equations.

a $2x + 5y = 6$
$6x - y = 3$

b $x + y - z = -4$
$2x + y + z = 4$
$3x + 2y + 2z = 10$

9 Prove by induction that $\begin{pmatrix} 1 & 5 \\ 0 & 1 \end{pmatrix}^n = \begin{pmatrix} 1 & 5n \\ 0 & 1 \end{pmatrix}$ for all positive integers n

10 Prove by induction that
$\begin{pmatrix} 5 & 4 \\ 0 & 1 \end{pmatrix}^n = \begin{pmatrix} 5^n & 5^n - 1 \\ 0 & 1 \end{pmatrix}$ for
all positive integers n

11 Prove by induction that
$\begin{pmatrix} -2 & -1 \\ 9 & 4 \end{pmatrix}^n = \begin{pmatrix} 1 - 3n & -n \\ 9n & 3n + 1 \end{pmatrix}$ for
all positive integers n

12 Prove by induction that
$\begin{pmatrix} 1 & 4 \\ 0 & 2 \end{pmatrix}^n = \begin{pmatrix} 1 & 4(2^n - 1) \\ 0 & 2^n \end{pmatrix}$ for
all positive integers n

13 Prove that $\mathbf{A} + (\mathbf{B} + \mathbf{C}) = (\mathbf{A} + \mathbf{B}) + \mathbf{C}$ for any 3×2 matrices $\mathbf{A}$, $\mathbf{B}$ and $\mathbf{C}$

14 Prove that matrix addition is commutative for any 2×2 matrices.

15 Prove by counter-example that matrix multiplication is not commutative.

16 Prove that $\mathbf{A}(\mathbf{B} + \mathbf{C}) = \mathbf{AB} + \mathbf{AC}$ for any 2×2 matrices $\mathbf{A}$, $\mathbf{B}$ and $\mathbf{C}$

17 Prove that $\mathbf{A}\begin{pmatrix} 1 & 0 \\ 0 & 1 \end{pmatrix} = \begin{pmatrix} 1 & 0 \\ 0 & 1 \end{pmatrix}\mathbf{A}$ for any 2×2 matrix $\mathbf{A}$

18 Prove that $\mathbf{B}\begin{pmatrix} 1 & 0 & 0 \\ 0 & 1 & 0 \\ 0 & 0 & 1 \end{pmatrix} = \begin{pmatrix} 1 & 0 & 0 \\ 0 & 1 & 0 \\ 0 & 0 & 1 \end{pmatrix}\mathbf{B}$ for any 3×3 matrix $\mathbf{B}$

19 The profit (in £1000) made by three employees for a company over the four quarters of a year is shown in the table.

	Q1	Q2	Q3	Q4
Employee A	8	14	15	17
Employee B	6	11	7	9
Employee C	9	18	19	12

The company plans to pay in bonuses 5% of profit from Q1, 2% from Q2, 3% from Q3 and 1% from Q4

a Write a matrix to represent the table of information and a vector to represent the percentage paid in bonuses.

b Show how you can use your matrices to calculate the amount of bonus paid to each of the employees.

c Write down the total amount of bonuses paid to all three employees.

20 The stationery requirements of a group of Maths teachers is given in the table.

	Red Pens	Pencils	Rulers	Board Pens	Paper Clips
Teacher 1	5	20	7	6	50
Teacher 2	4	15	15	8	100
Teacher 3	12	4	2	30	50

Red pens cost 12p each, pencils cost 8p, rulers 18p, board pens 30p and paper clips 1p

a Show how you can use matrices to calculate the expenditure on each type of stationery by each teacher.

b Write down the total cost of all the stationery required.

21 Given that
$\begin{pmatrix} a & b \\ c & d \end{pmatrix}\begin{pmatrix} e & f \\ g & h \end{pmatrix} = \begin{pmatrix} 1 & 0 \\ 0 & 1 \end{pmatrix}$,

a Verify that $e = \dfrac{d}{ad - bc}$ and $g = -\dfrac{c}{ad - bc}$

b Find similar expressions for f and h in terms of a, b, c and d

22 Prove by induction that
$\begin{pmatrix} 0 & 1 \\ 2 & 0 \end{pmatrix}^{2n} = 2^n \begin{pmatrix} 1 & 0 \\ 0 & 1 \end{pmatrix}$ for all $n \in \mathbb{N}$

Fluency and skills

Matrices can be used to represent certain transformations such as some rotations, reflections and enlargements. Transformations that can described in this way are known as **linear**.

In order to find the image of a vector $\begin{pmatrix} x \\ y \end{pmatrix}$ under a transformation $\mathbf{T} = \begin{pmatrix} a & b \\ c & d \end{pmatrix}$, we

pre-multiply the vector by the matrix, so $\begin{pmatrix} x' \\ y' \end{pmatrix} = \begin{pmatrix} a & b \\ c & d \end{pmatrix}\begin{pmatrix} x \\ y \end{pmatrix}$

A point (x, y) can also be represented by the vector $\begin{pmatrix} x \\ y \end{pmatrix}$

To find the image of a point $A(1, -3)$ under the transformation

$T = \begin{pmatrix} 1 & 0 \\ 0 & -1 \end{pmatrix}$, multiply the vector $\begin{pmatrix} 1 \\ 3 \end{pmatrix}$ by T

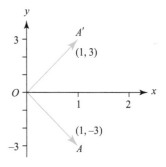

$\begin{pmatrix} x' \\ y' \end{pmatrix} = \begin{pmatrix} 1 & 0 \\ 0 & -1 \end{pmatrix}\begin{pmatrix} 1 \\ -3 \end{pmatrix} = \begin{pmatrix} 1 \\ 3 \end{pmatrix}$

This is a reflection in the x-axis.

The matrix $\begin{pmatrix} 1 & 0 \\ 0 & -1 \end{pmatrix}$ represents a reflection in the x-axis.

The matrix $\begin{pmatrix} -1 & 0 \\ 0 & 1 \end{pmatrix}$ represents a reflection in the y-axis.

The matrix $\begin{pmatrix} 0 & 1 \\ 1 & 0 \end{pmatrix}$ represents a reflection in the line $y = x$

The matrix $\begin{pmatrix} 0 & -1 \\ -1 & 0 \end{pmatrix}$ represents a reflection in the line $y = -x$

You can find matrices that represent particular transformations by considering the effect the transformation will have on a pair of vectors.

Example 1

Find the 2×2 matrix that represents a rotation of $90°$ anticlockwise about the origin.

Let the matrix be given by $\begin{pmatrix} a & b \\ c & d \end{pmatrix}$

Choose two points and their images after the rotation.

$(1, 0)$ moves to $(0, 1)$ and $(0, 1)$ moves to $(-1, 0)$

So $\begin{pmatrix} a & b \\ c & d \end{pmatrix}\begin{pmatrix} 1 \\ 0 \end{pmatrix} = \begin{pmatrix} 0 \\ 1 \end{pmatrix}$ and $\begin{pmatrix} a & b \\ c & d \end{pmatrix}\begin{pmatrix} 0 \\ 1 \end{pmatrix} = \begin{pmatrix} -1 \\ 0 \end{pmatrix}$

So from the first equation: $a = 0, c = 1$

From the second equation: $b = -1$ and $d = 0$

Therefore the transformation matrix is $\begin{pmatrix} 0 & -1 \\ 1 & 0 \end{pmatrix}$

We chose $\begin{pmatrix} 1 \\ 0 \end{pmatrix}$ and $\begin{pmatrix} 0 \\ 1 \end{pmatrix}$ as these are simple to visualise and will give unique values for a, b, c and d

Notice that the first column of the matrix is the image of $\begin{pmatrix} 1 \\ 0 \end{pmatrix}$ and the second column is the image of $\begin{pmatrix} 0 \\ 1 \end{pmatrix}$

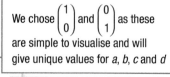

Key point

The matrix $\begin{pmatrix} a & b \\ c & d \end{pmatrix}$ represents the transformation

that maps $\begin{pmatrix} 1 \\ 0 \end{pmatrix}$ to $\begin{pmatrix} a \\ c \end{pmatrix}$ and $\begin{pmatrix} 0 \\ 1 \end{pmatrix}$ to $\begin{pmatrix} b \\ d \end{pmatrix}$

It follows from this fact that the matrix $\begin{pmatrix} 1 & 0 \\ 0 & 1 \end{pmatrix}$ maps all points to themselves.

In number multiplication, the identity is 1 since $1 \times a = a \times 1 = a$

Key point

The matrix $\mathbf{I} = \begin{pmatrix} 1 & 0 \\ 0 & 1 \end{pmatrix}$ is known as the **identity matrix**

and has the property that $\mathbf{AI} = \mathbf{IA} = \mathbf{A}$ for all 2×2 matrices

This can be generalised for any square matrix, in particular $\mathbf{I}_3 = \begin{pmatrix} 1 & 0 & 0 \\ 0 & 1 & 0 \\ 0 & 0 & 1 \end{pmatrix}$

To find the 2×2 matrix that represents a stretch of scale factor 3 parallel to the x-axis, look at what happens to the vectors $\begin{pmatrix} 1 \\ 0 \end{pmatrix}$ and $\begin{pmatrix} 0 \\ 1 \end{pmatrix}$

A stretch 'parallel to the x-axis' can be thought of a stretch 'along the x-axis' so only the x-coordinates are affected.

The image of $\begin{pmatrix} 1 \\ 0 \end{pmatrix}$ under this transformation is $\begin{pmatrix} 3 \\ 0 \end{pmatrix}$. The image of $\begin{pmatrix} 0 \\ 1 \end{pmatrix}$ under this

transformation is $\begin{pmatrix} 0 \\ 1 \end{pmatrix}$

Therefore the transformation matrix is $\begin{pmatrix} 3 & 0 \\ 0 & 1 \end{pmatrix}$

Example 2

Transformations can be applied to several points at once by forming a matrix from their position vectors.

Find the image of the points $(2, -1)$, $(-4, 5)$ and $(-3, 0)$ under the transformation described by the matrix $\begin{pmatrix} 2 & 0 \\ 0 & 2 \end{pmatrix}$. Describe the transformation geometrically.

$$\begin{pmatrix} 2 & 0 \\ 0 & 2 \end{pmatrix}\begin{pmatrix} 2 & -4 & -3 \\ -1 & 5 & 0 \end{pmatrix} = \begin{pmatrix} 4 & -8 & -6 \\ -2 & 10 & 0 \end{pmatrix}$$

Ensure you always pre-multiply by the transformation matrix.

This is an enlargement of scale factor 2, centre the origin.

To find the matrix that represents a 60° rotation anticlockwise about the origin, look at what happens to the vectors $\begin{pmatrix} 1 \\ 0 \end{pmatrix}$ and $\begin{pmatrix} 0 \\ 1 \end{pmatrix}$

See Maths Ch3.1
For a reminder of exact values of trig functions.

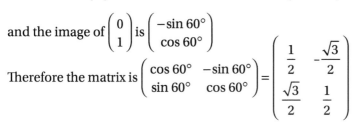

The image of $\begin{pmatrix} 1 \\ 0 \end{pmatrix}$ under this transformation is $\begin{pmatrix} \cos 60° \\ \sin 60° \end{pmatrix}$

and the image of $\begin{pmatrix} 0 \\ 1 \end{pmatrix}$ is $\begin{pmatrix} -\sin 60° \\ \cos 60° \end{pmatrix}$

Therefore the matrix is $\begin{pmatrix} \cos 60° & -\sin 60° \\ \sin 60° & \cos 60° \end{pmatrix} = \begin{pmatrix} \dfrac{1}{2} & -\dfrac{\sqrt{3}}{2} \\ \dfrac{\sqrt{3}}{2} & \dfrac{1}{2} \end{pmatrix}$

This result can be generalised for any angle.

Transformations can also be combined and you can use the product of the matrices to represent the combined transformation.

If the transformation represented by matrix **A** is followed by the transformation represented by matrix **B**, then **BA** is the combined transformation.

Notice that if you wish to apply transformation **A** to a vector first, you should multiply by **A** then by **B** hence **BA**

Example 3

Find the single matrix that represents a rotation of 270° anticlockwise followed by a reflection in the y-axis.

Rotation of 270°: $\begin{pmatrix} 1 \\ 0 \end{pmatrix}$ is transformed to $\begin{pmatrix} 0 \\ -1 \end{pmatrix}$ and $\begin{pmatrix} 0 \\ 1 \end{pmatrix}$ is transformed to $\begin{pmatrix} 1 \\ 0 \end{pmatrix}$

Therefore $\mathbf{A} = \begin{pmatrix} 0 & 1 \\ -1 & 0 \end{pmatrix}$

Reflection in the y-axis: $\begin{pmatrix} 1 \\ 0 \end{pmatrix}$ is transformed to $\begin{pmatrix} -1 \\ 0 \end{pmatrix}$ and $\begin{pmatrix} 0 \\ 1 \end{pmatrix}$ is transformed to $\begin{pmatrix} 0 \\ 1 \end{pmatrix}$

Therefore $\mathbf{B} = \begin{pmatrix} -1 & 0 \\ 0 & 1 \end{pmatrix}$

So combined transformation = **BA**

| The order of the matrices is important. |

$$= \begin{pmatrix} -1 & 0 \\ 0 & 1 \end{pmatrix}\begin{pmatrix} 0 & 1 \\ -1 & 0 \end{pmatrix}$$

$$= \begin{pmatrix} 0 & -1 \\ -1 & 0 \end{pmatrix}$$

| This is a reflection in the line $y = -x$ |

You may recall learning about 3D coordinates in GCSE Maths. Points in 3D space have 3D position vectors, and you will learn more about these in your Further Maths course.

See Ch 5.1

For vectors in Further Maths.

You can use 3×3 matrices to represent transformations in 3D.

To reflect in the plane $x = 0$, the 'mirror' is a plane through the y-axis and extended in the positive and negative z-directions.

The image of the vector $\begin{pmatrix} 1 \\ 0 \\ 0 \end{pmatrix}$ will be $\begin{pmatrix} -1 \\ 0 \\ 0 \end{pmatrix}$

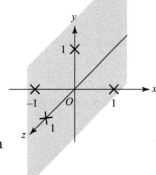

The vectors $\begin{pmatrix} 0 \\ 1 \\ 0 \end{pmatrix}$ and $\begin{pmatrix} 0 \\ 0 \\ 1 \end{pmatrix}$ are in the 'mirror' plane so are unchanged when reflected in x

Therefore the transformation matrix is $\begin{pmatrix} -1 & 0 & 0 \\ 0 & 1 & 0 \\ 0 & 0 & 1 \end{pmatrix}$

The matrices for reflection in the planes $y = 0$ and $z = 0$ can be found using the same method.

The 3D reflection matrices you need to use are as follows.

Reflection in $x = 0$: $\begin{pmatrix} -1 & 0 & 0 \\ 0 & 1 & 0 \\ 0 & 0 & 1 \end{pmatrix}$

Reflection in $y = 0$: $\begin{pmatrix} 1 & 0 & 0 \\ 0 & -1 & 0 \\ 0 & 0 & 1 \end{pmatrix}$

Reflection in $z = 0$: $\begin{pmatrix} 1 & 0 & 0 \\ 0 & 1 & 0 \\ 0 & 0 & -1 \end{pmatrix}$

You also need to know about 3D rotation around one of the coordinate axes.

This diagram illustrates rotating the point $(1, 0, 0)$ by angle θ anticlockwise around the z-axis. Notice how the z-coordinate of the point is unchanged so this is actually the same as rotating around the origin in the 2D case.

Therefore you can represent anticlockwise rotation around the z-axis by the matrix

Notice how this is the 2×2 matrix for anticlockwise rotation around the origin.

$\begin{pmatrix} \cos\theta & -\sin\theta & 0 \\ \sin\theta & \cos\theta & 0 \\ 0 & 0 & 1 \end{pmatrix}$

Consider an anticlockwise rotation around the x-axis or the y-axis in a similar way.

The 3D rotation matrices you need to use are as follows.

Rotation around the x-axis: $\begin{pmatrix} 1 & 0 & 0 \\ 0 & \cos\theta & -\sin\theta \\ 0 & \sin\theta & \cos\theta \end{pmatrix}$

Rotation around the y-axis: $\begin{pmatrix} \cos\theta & 0 & \sin\theta \\ 0 & 1 & 0 \\ -\sin\theta & 0 & \cos\theta \end{pmatrix}$

Rotation around the z-axis: $\begin{pmatrix} \cos\theta & -\sin\theta & 0 \\ \sin\theta & \cos\theta & 0 \\ 0 & 0 & 1 \end{pmatrix}$

Example 4

Find the 3×3 matrix, **T**, that represents a rotation of $45°$ anticlockwise around the x-axis and find the image of the point $P(1, 4, -2)$ under **T**

$$T = \begin{pmatrix} 1 & 0 & 0 \\ 0 & \cos45° & -\sin45° \\ 0 & \sin45° & \cos45° \end{pmatrix}$$

$$= \begin{pmatrix} 1 & 0 & 0 \\ 0 & \dfrac{\sqrt{2}}{2} & -\dfrac{\sqrt{2}}{2} \\ 0 & \dfrac{\sqrt{2}}{2} & \dfrac{\sqrt{2}}{2} \end{pmatrix}$$

$$P' = \begin{pmatrix} 1 & 0 & 0 \\ 0 & \dfrac{\sqrt{2}}{2} & -\dfrac{\sqrt{2}}{2} \\ 0 & \dfrac{\sqrt{2}}{2} & \dfrac{\sqrt{2}}{2} \end{pmatrix} \begin{pmatrix} 1 \\ 4 \\ -2 \end{pmatrix}$$

Pre-multiply the vector representing point P by the matrix **T**

$$= \begin{pmatrix} 1 \\ 2\sqrt{2}+\sqrt{2} \\ 2\sqrt{2}-\sqrt{2} \end{pmatrix}$$

$$= \begin{pmatrix} 1 \\ 3\sqrt{2} \\ \sqrt{2} \end{pmatrix}$$

So the actual point is $(1, 3\sqrt{2}, \sqrt{2})$

Since you are transforming a point you should give the final answer as coordinates.

Exercise 4.2A Fluency and skills

1 The matrix **B** represents a stretch of scale factor 5 parallel to the x-axis and a stretch of scale factor -2 parallel to the y-axis.

   a Use a diagram to show the transformation of the vectors $\begin{pmatrix} 0 \\ 1 \end{pmatrix}$ and $\begin{pmatrix} 1 \\ 0 \end{pmatrix}$ under matrix **B**

   b Write down matrix **B**

2 The matrix **A** represents a rotation of $180°$

   a Use a diagram to show the transformation of the vector $\begin{pmatrix} 1 \\ 0 \end{pmatrix}$ under this transformation.

   b Write down the image of the vector $\begin{pmatrix} 0 \\ 1 \end{pmatrix}$ under this transformation.

   c Write down the matrix **A**

3 Find the 2×2 matrix which represents each of these transformations.

   a Reflection in the line $y = x$

   b Rotation of 30° anticlockwise about the origin.

   c Enlargement of scale factor 5, centre the origin.

   d Stretch parallel to the y-axis of scale factor 4

4 A triangle has vertices $(0, -2)$, $(2, 3)$ and $(-4, 3)$

   a Find the image of these points under the transformation represented by the matrix

$$T = \begin{pmatrix} 2 & 0 \\ 0 & 1 \end{pmatrix}$$

   b Describe geometrically the transformation represented by matrix T.

5 The matrix $\dfrac{\sqrt{2}}{2}\begin{pmatrix} 1 & 1 \\ -1 & 1 \end{pmatrix}$ represents an anticlockwise rotation of θ degrees, centre the origin. Find the value of θ

6 A transformation is given by $A = \begin{pmatrix} 1 & 0 \\ 0 & 0.5 \end{pmatrix}$

   a Find the image of the vectors $\begin{pmatrix} 6 \\ -2 \end{pmatrix}$ and $\begin{pmatrix} -1 \\ 4 \end{pmatrix}$ under the transformation represented by matrix A

   b Describe geometrically the transformation represented by matrix A

7 Write the transformation represented by each of these matrices.

   a $\begin{pmatrix} 3 & 0 \\ 0 & 3 \end{pmatrix}$ **b** $\begin{pmatrix} 1 & 0 \\ 0 & 2 \end{pmatrix}$

   c $\dfrac{\sqrt{2}}{2}\begin{pmatrix} -1 & -1 \\ 1 & -1 \end{pmatrix}$ **d** $\begin{pmatrix} 0 & 1 \\ -1 & 0 \end{pmatrix}$

   e $\begin{pmatrix} -1 & 0 \\ 0 & 1 \end{pmatrix}$ **f** $\begin{pmatrix} \cos 20° & \sin 20° \\ -\sin 20° & \cos 20° \end{pmatrix}$

8 Given the transformation matrices

$$A = \begin{pmatrix} 1 & 1 \\ 0 & 1 \end{pmatrix} \text{ and } B = \begin{pmatrix} 0 & -1 \\ -1 & 1 \end{pmatrix},$$

   a Find the matrix representing

     i Transformation A followed by transformation B,

     ii Transformation B followed by transformation A

   b Comment on your answers to part **a**.

9 Give the transformation matrices

$$A = \begin{pmatrix} 1 & 0 \\ 0 & -1 \end{pmatrix}, B = \begin{pmatrix} 0 & 1 \\ -1 & 0 \end{pmatrix}$$

   a Describe the transformations A and B geometrically.

   b Find the matrix representing

     i Transformation A followed by B,

     ii Transformation B followed by A.

   c Describe both transformations in part **b** geometrically as a single transformation.

10 Find the single 2×2 matrix that represents each of these combinations of transformations.

   a A rotation of 45° anticlockwise about the origin followed by a reflection in the y-axis.

   b A reflection in the line $y = -x$ followed by an enlargement of scale factor 2 about the origin.

   c A stretch parallel to the x-axis of scale factor -2 followed by a clockwise rotation of 90° about the origin.

11 A square has vertices at $(2, 5)$, $(2, 1)$, $(-3, 1)$ and $(-3, 5)$

The square is rotated 225° anticlockwise about the origin and then stretched by a scale factor of -2 in the y-direction.

Find the vertices of the image of the square following these transformations.

12 **a** Write a matrix A that represents a rotation of 135° anticlockwise about the origin.

   b Show that $A^2 = \begin{pmatrix} 0 & 1 \\ -1 & 0 \end{pmatrix}$

   c Show that A^2 represents a clockwise rotation of 90° about the origin.

d Describe the transformation represented by $\mathbf{A}^3$

13 The rotation represented by the 2×2 matrix $\mathbf{A}$ is such that $\mathbf{A}^2 = \mathbf{I}$ and $\mathbf{A} \neq \mathbf{I}$

    a Write down a possible matrix $\mathbf{A}$ and describe the transformation fully.

The rotation represented by the 2×2 matrix $\mathbf{B}$ is such that $\mathbf{B}^3 = \mathbf{I}$ and $\mathbf{B} \neq \mathbf{I}$

    b Write down a possible matrix $\mathbf{B}$ and describe the transformation fully.

14 Matrix $\mathbf{P}$ represents a reflection in the x-axis and matrix $\mathbf{Q}$ represents a stretch of scale factor 3 parallel to the y-axis.

    a Find the single matrix that represents transformation $\mathbf{P}$ followed by transformation $\mathbf{Q}$

    b Describe geometrically the single transformation that could replace this combination of transformations.

15 The matrix $\mathbf{M}$ represents an enlargement around the origin of scale factor k, $k > 0$, followed by an anticlockwise rotation of θ, $0 \leq \theta \leq 180°$ around the origin.

    a Write $\mathbf{M}$ in terms of k and θ

The image of the point $(1, -2)$ under $\mathbf{M}$ is $(2, 6)$

    b Find the values of k and θ

16 A transformation is represented by the matrix $\mathbf{T} = \begin{pmatrix} 1 & 0 & 0 \\ 0 & -1 & 0 \\ 0 & 0 & 1 \end{pmatrix}$

    a Find the image of the points $(4, -1, 0)$ and $(2, 5, 3)$ under $\mathbf{T}$

    b Describe this transformation geometrically.

17 Find the 3×3 matrix that represents each of these transformations.

    a A rotation of $30°$ anticlockwise around the x-axis.

    b A reflection in $z = 0$

    c A rotation of $45°$ anticlockwise around the y-axis.

18 The matrix $\begin{pmatrix} 1 & 0 & 0 \\ 0 & \dfrac{1}{2} & \dfrac{\sqrt{3}}{2} \\ 0 & -\dfrac{\sqrt{3}}{2} & \dfrac{1}{2} \end{pmatrix}$ represents an anticlockwise rotation of θ degrees around one of the coordinate axes.

    a Find the value of θ and state which axis the rotation is around.

    b Find the image of the vector $3\mathbf{i} + 4\mathbf{j} - \mathbf{k}$ under this transformation.

19 The matrix $\dfrac{1}{2}\begin{pmatrix} 1 & \sqrt{3} & 0 \\ -\sqrt{3} & 1 & 0 \\ 0 & 0 & 1 \end{pmatrix}$ represents an anticlockwise rotation of θ around one of the coordinate axes. Find the value of θ and state which axis the rotation is around.

20 The point $A(3, 7, -2)$ is transformed by a transformation matrix $\mathbf{M}$ to the point $A'(3, 7, 2)$

    a Describe this transformation geometrically.

    b Write down the transformation matrix $\mathbf{M}$

21 Describe the transformation represented by each of these matrices.

    a $\begin{pmatrix} 1 & 0 & 0 \\ 0 & 0 & -1 \\ 0 & 1 & 0 \end{pmatrix}$ **b** $\begin{pmatrix} -1 & 0 & 0 \\ 0 & 1 & 0 \\ 0 & 0 & 1 \end{pmatrix}$

    c $\begin{pmatrix} -1 & 0 & 0 \\ 0 & -1 & 0 \\ 0 & 0 & 1 \end{pmatrix}$ **d** $\begin{pmatrix} -\dfrac{1}{2} & 0 & \dfrac{\sqrt{3}}{2} \\ 0 & 1 & 0 \\ -\dfrac{\sqrt{3}}{2} & 0 & -\dfrac{1}{2} \end{pmatrix}$

22 The 3×3 matrix $\mathbf{P}$ represents a clockwise rotation of $135°$ around the y-axis.

    a Find $\mathbf{P}$ in its simplest form.

    b Find the image of the point $(\sqrt{2}, 0, 1)$ under the transformation represented by $\mathbf{P}$

23 The 3×3 matrix $\mathbf{T}$ represents an anticlockwise rotation of $330°$ around the z-axis.

    a Find $\mathbf{T}$ in its simplest form.

    b Find the image of the point $(\sqrt{3}, 0, 1)$ under the transformation $\mathbf{T}$

Reasoning and problem-solving

A point which is unaffected by a transformation is known as an invariant point.

Key point

Given a transformation matrix **T** and a position vector **x**, if **Tx** = **x** then **x** represents an **invariant point**.

You can be asked to find invariant points for a specific transformation.

Strategy

To find invariant points and lines

(1) Write vectors in a general form.

(2) Pre-multiply a vector by a transformation matrix to form a system of simultaneous equations.

(3) Solve simultaneous equations to find either an invariant point or a line of invariant points.

(4) Equate coefficients to find equations of invariant lines.

Example 5

Find any invariant points under the transformation given by $\begin{pmatrix} 3 & 1 \\ -1 & 2 \end{pmatrix}$

Let $\begin{pmatrix} x \\ y \end{pmatrix}$ be the invariant point. ● ——— Write a general vector to represent the invariant point ①

Then $\begin{pmatrix} 3 & 1 \\ -1 & 2 \end{pmatrix}\begin{pmatrix} x \\ y \end{pmatrix} = \begin{pmatrix} x \\ y \end{pmatrix}$

$\therefore 3x + y = x \Rightarrow 2x + y = 0$ ● ——— Form equations. ②

and $-x + 2y = y \Rightarrow y = x$

Therefore $3x = 0 \Rightarrow x = 0$ and $y = 0$. The invariant point is $(0, 0)$ ● ——— In this case only the point $(0, 0)$ is invariant. ③

Key point

For any linear transformation, $(0, 0)$ is an invariant point.

For some transformation, all points satisfying a certain property will be invariant, in which case you can have an invariant line or lines. There are two ways in which a line can be invariant.

For example, consider a reflection in the line $y = x$

All points on the line will be unaffected by the transformation.

Therefore $y = x$ is a line of invariant points.

Key point

If every point on a line is mapped to itself under a transformation then it is known as a **line of invariant points**.

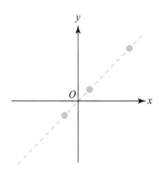

Find the invariant points under the transformation given by $\begin{pmatrix} 2 & 3 \\ -1 & -2 \end{pmatrix}$

Let $\begin{pmatrix} x \\ y \end{pmatrix}$ be an invariant point. •——————

> **1** Write a general vector to represent the invariant point

Then $\begin{pmatrix} 2 & 3 \\ -1 & -2 \end{pmatrix}\begin{pmatrix} x \\ y \end{pmatrix} = \begin{pmatrix} x \\ y \end{pmatrix}$

> **2** Form equations.

$\therefore 2x + 3y = x \Rightarrow x = -3y$ and $-x - 2y = y \Rightarrow x = -3y$

So there is a line of invariant points given by $x = -3y$ •—————

> **3** If the two equations simplify to the same thing then there is a line of invariant points.

In some cases, every point on a line will map to another point on the same line.

For example, consider a rotation of 180° around the origin. The point $(0, 0)$ is invariant but if you take any line through the origin then every other point on that line will map to a different point on the same line. So any line through the origin will be an invariant line.

Key point

If every point on a line is mapped to another point on the same line then it is known as an **invariant line**. Lines of invariant points are a subset of invariant lines.

To find the equations of invariant lines use $y = mx + c$ to rewrite your general vector and $\begin{pmatrix} x' \\ y' \end{pmatrix}$ for the image of that vector. You should then use $y' = mx' + c$ to find the possible values of m and c

Example 7

Find the equations of the invariant lines of the transformation given by $\begin{pmatrix} 1 & 1 \\ 0 & 3 \end{pmatrix}$

$\begin{pmatrix} 1 & 1 \\ 0 & 3 \end{pmatrix}\begin{pmatrix} x \\ mx+c \end{pmatrix} = \begin{pmatrix} x' \\ y' \end{pmatrix}$ •————

> **1** Write the vector in a general form in terms of x

$x + mx + c = x'$

$3(mx + c) = y'$ •————

> **2** Form simultaneous equations.

So $3(mx + c) = m(x + mx + c) + c$ •————

> Using $y' = mx' + c$

$(3m - m - m^2)x + 3c - mc - c = 0$ •————

$(2m - m^2)x + 2c - mc = 0$

> Collect terms in x

$2m - m^2 = 0 \Rightarrow m = 0$ or 2 •————

$2c - mc = 0$ •————

> **4** Equate coefficients of x

Therefore when $m = 0$, $c = 0$

and when $m = 2$ you have $2c - 2c = 0$ so c can be any value.

> **4** Equate constant terms.

Therefore the invariant lines are $y = 0$ and $y = 2x + c$ (for any c). •————

> State the invariant lines.

1   Give the equation of the line of invariant points under the transformation given by each of these matrices.

a $\begin{pmatrix} 1 & -2 \\ 0 & 3 \end{pmatrix}$ b $\begin{pmatrix} 5 & 2 \\ 4 & 3 \end{pmatrix}$

2   Show that the origin is the only invariant point under the transformation given by each of these matrices.

a $\begin{pmatrix} 3 & -2 \\ 2 & 3 \end{pmatrix}$ b $\begin{pmatrix} 2 & 0 & 1 \\ 0 & 3 & -2 \\ 1 & 0 & -4 \end{pmatrix}$

3   In each of these cases, decide whether or not the order the transformations are applied affects the final image. Justify your answers.

a   A reflection in the y-axis and a stretch parallel to the y-axis.

b   A rotation about the origin and an enlargement with centre the origin.

c   A reflection in the line $y = x$ and a stretch along the x-axis.

4   Describe the combination of transformations that will be represented by each of these transformation matrices.

a $\begin{pmatrix} 3 & 0 \\ 0 & -2 \end{pmatrix}$ b $\begin{pmatrix} 0 & 4 \\ 4 & 0 \end{pmatrix}$

5   Find the equations of the invariant lines under each of these transformations.

a $\begin{pmatrix} 2 & 0 \\ 1 & -3 \end{pmatrix}$ b $\begin{pmatrix} 2 & -1 \\ -3 & 0 \end{pmatrix}$

6   a   Suggest two different transformations that have an invariant line given by the equation $x = 0$

    b   Which of your transformations in part **a** has a line of invariant points given by $x = 0$?

7   The invariant lines under transformation **A** are given by the equations $x + y = c$ where c can be any value. Describe a possible transformation **A** geometrically.

8   The invariant lines under transformation **B** are given by the equations $y = kx$ where k can be any value. Describe transformation **B** geometrically.

9   Find the equations of the invariant lines under the transformation given by each of these matrices.

a $\begin{pmatrix} 1 & 2 \\ 2 & -1 \end{pmatrix}$ b $\begin{pmatrix} 3 & 0 \\ 0 & 3 \end{pmatrix}$ c $\begin{pmatrix} -\dfrac{3}{5} & \dfrac{4}{5} \\ \dfrac{4}{5} & \dfrac{3}{5} \end{pmatrix}$

10 a   For each of these transformations, find either the invariant point or the equations of the invariant lines as appropriate.

    i   Reflection in the x-axis.

    ii   Rotation of 90° around the origin.

    iii Stretch of scale factor 2 parallel to the x-axis.

    iv   Reflection in the line $y = -x$

    b   For each of the invariant lines in part **a**, explain whether or not it is a line of invariant points.

11 A transformation is represented by the matrix $\mathbf{T} = \begin{pmatrix} -2 & 3 \\ -3 & 4 \end{pmatrix}$

    a   Find the line of invariant points under transformation **T**

    b   Show that all lines of the form $y = x + c$ are invariant lines of the transformation **T**

12 The matrix representing transformation **A** followed by transformation **B** is given by $\begin{pmatrix} -1 & 0 \\ 0 & 1 \end{pmatrix}$. Find the matrix that represents the transformation **B** followed by the transformation **A**

13 Give either the invariant point or the equation of a plane of invariant points for each of these transformations.

a $\begin{pmatrix} -1 & 0 & 0 \\ 0 & 1 & 0 \\ 0 & 0 & 1 \end{pmatrix}$ b $\begin{pmatrix} 1 & 0 & 0 \\ 0 & 0 & 1 \\ 0 & 1 & 0 \end{pmatrix}$

c $\begin{pmatrix} 0 & 1 & 0 \\ 0 & -1 & 0 \\ 0 & 0 & 1 \end{pmatrix}$ d $\begin{pmatrix} 1 & 0 & 0 \\ 1 & 0 & 0 \\ 0 & 0 & 1 \end{pmatrix}$

Fluency and skills

When you are working with square matrices you can calculate a value known as the **determinant**. The determinant has several uses. Most significantly, it enables you to find the inverse of a matrix.

> **Key point**
>
> If $\mathbf{A} = \begin{pmatrix} a & b \\ c & d \end{pmatrix}$ then the **determinant** of $\mathbf{A}$ is $ad - bc$ and is denoted $|\mathbf{A}|$ or $\det(\mathbf{A})$

To find the determinant of the matrix $\mathbf{A} = \begin{pmatrix} 5 & 2 \\ -1 & -3 \end{pmatrix}$ use the formula $\det(\mathbf{A}) = ad - bc$

$$\det(A) = (5 \times -3) - (2 \times -1)$$
$$= 15 - -2$$
$$= -13$$

Calculator

Try it on your calculator

You can use some calculators to find the determinant.

det(MatA)

26

Activity

Find out how to work out $\det\begin{pmatrix} 2 & -8 \\ 4 & -3 \end{pmatrix}$ on *your* calculator.

> **Key point**
>
> If $\det(\mathbf{A}) = 0$ then $\mathbf{A}$ is called a **singular matrix**.

> **Example 1**
>
> Find the value of k for which the matrix $\begin{pmatrix} 3 & -1 \\ -2 & k \end{pmatrix}$ is singular.
>
> $\det\begin{pmatrix} 3 & -1 \\ -2 & k \end{pmatrix} = (3 \times k) - (-1 \times -2)$ | The determinant is given by $ad - bc$
>
> $\therefore$ if matrix is singular then $3k - 2 = 0$ | A singular matrix has a determinant of 0
>
> $\Rightarrow k = \dfrac{2}{3}$

Consider a square with vertices at $(0, 0)$, $(1, 0)$, $(0, 1)$ and $(1, 1)$.

Under the transformation $\mathbf{T} = \begin{pmatrix} a & 0 \\ 0 & a \end{pmatrix}$ the square is enlarged by

scale factor a centre the origin. So the vertices of the image are $(0, 0)$, $(a, 0)$, $(0, a)$ and (a, a). The area of the original square is 1 and the area of the image is a^2, which is also the determinant of $\mathbf{T}$. This result extends to all linear transformations.

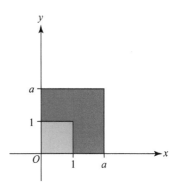

Key point

For a transformation represented by matrix $\mathbf{T}$,
area of image = area of original $\times |\det(\mathbf{T})|$

Example 2

The triangle with vertices $(3, -1)$, $(4, 2)$ and $(0, 2)$ is stretched by scale factor 2 parallel to the x-axis and stretched by scale factor -3 parallel to the y-axis. Find the area of the image under this transformation.

The transformation is given by $\mathbf{T} = \begin{pmatrix} 2 & 0 \\ 0 & -3 \end{pmatrix}$

$\det(\mathbf{T}) = 2 \times -3 = -6$

The area of the original triangle is $\dfrac{1}{2} \times 4 \times 3 = 6$

Therefore area of image is $6 \times |-6| = 36$

> A sketch will help with the area.

> Using area of original $\times |\det(\mathbf{T})|$

Notice that in the example above, the determinant is negative because the orientation of the triangle changes. In other words the triangle has been flipped as well as enlarged.

Key point

If $\det(\mathbf{T}) > 0$ then the transformation represented by $\mathbf{T}$ preserves the orientation of the original shape.

In order to find the determinant of a 3×3 matrix, you need to find the three **minors** of any row of elements in your matrix. Given a 3×3 matrix, the minor of an element is the determinant of the 2×2 matrix remaining when the row and column of the element are crossed out. This means a matrix has nine minors, one for each element in the matrix.

For example, to find the minor of a in the matrix $\begin{pmatrix} a & b & c \\ d & e & f \\ g & h & i \end{pmatrix}$, you cross out the row and the column

involving a then find the determinant of the matrix $\begin{pmatrix} e & f \\ h & i \end{pmatrix}$ that remains. Therefore, the minor of a is $ei - fh$

Key point

$$\det \begin{pmatrix} a & b & c \\ d & e & f \\ g & h & i \end{pmatrix} = a \begin{vmatrix} e & f \\ h & i \end{vmatrix} - b \begin{vmatrix} d & f \\ g & i \end{vmatrix} + c \begin{vmatrix} d & e \\ g & h \end{vmatrix}$$

Example 3

Find the determinant of the matrix $\begin{pmatrix} 3 & 2 & -1 \\ 0 & 4 & -2 \\ -3 & 1 & 5 \end{pmatrix}$

$$\det \begin{pmatrix} 3 & 2 & -1 \\ 0 & 4 & -2 \\ -3 & 1 & 5 \end{pmatrix} = 3 \begin{vmatrix} 4 & -2 \\ 1 & 5 \end{vmatrix} - 2 \begin{vmatrix} 0 & -2 \\ -3 & 5 \end{vmatrix} - 1 \begin{vmatrix} 0 & 4 \\ -3 & 1 \end{vmatrix}$$

Use the formula
$$a \begin{vmatrix} e & f \\ h & i \end{vmatrix} - b \begin{vmatrix} d & f \\ g & i \end{vmatrix} + c \begin{vmatrix} d & e \\ g & h \end{vmatrix}$$

$$= 3(20--2) - 2(0-6) - 1(0--12)$$

Take care with negative signs.

$$= 66 + 12 - 12$$
$$= 66$$

Recall that $\mathbf{I} = \begin{pmatrix} 1 & 0 \\ 0 & 1 \end{pmatrix}$ is the identity matrix.

The inverse of a matrix, $\mathbf{A}$, is $\mathbf{A}^{-1}$ where $\mathbf{AA}^{-1} = \mathbf{A}^{-1}\mathbf{A} = \mathbf{I}$

Key point

If $\mathbf{A} = \begin{pmatrix} a & b \\ c & d \end{pmatrix}$ then the inverse matrix is given by $\mathbf{A}^{-1} = \dfrac{1}{\det(\mathbf{A})} \begin{pmatrix} d & -b \\ -c & a \end{pmatrix}$

Notice how this definition will not work for a matrix which is singular. Singular matrices do not have an inverse. Therefore, only non-singular matrices will have an inverse.

Example 4

Find the inverse of the matrix $\mathbf{B} = \begin{pmatrix} 2 & 5 \\ -1 & 4 \end{pmatrix}$

$$\det(\mathbf{B}) = (2 \times 4) - (5 \times -1)$$

First calculate the determinant using $ad - bc$

$$= 8 - -5$$
$$= 13$$

$$\mathbf{B}^{-1} = \frac{1}{13} \begin{pmatrix} 4 & -5 \\ 1 & 2 \end{pmatrix}$$

Use $\mathbf{B}^{-1} = \dfrac{1}{\det(\mathbf{B})} \begin{pmatrix} d & -b \\ -c & a \end{pmatrix}$

Checking answer:

$$\mathbf{BB}^{-1} = \frac{1}{13} \begin{pmatrix} 2 & 5 \\ -1 & 4 \end{pmatrix} \begin{pmatrix} 4 & -5 \\ 1 & 2 \end{pmatrix}$$

You can check your answer by verifying that $\mathbf{BB}^{-1} = \mathbf{I}$

$$= \frac{1}{13} \begin{pmatrix} 13 & 0 \\ 0 & 13 \end{pmatrix}$$

$$= \begin{pmatrix} 1 & 0 \\ 0 & 1 \end{pmatrix}$$

$$= \mathbf{I}$$

Try it on your calculator

Some calculators can be used to find the inverse of a matrix.

MatA^{-1}

$$\begin{bmatrix} -4 & 3 \\ -3 & 2 \end{bmatrix}$$

Activity

Find out how to work out the inverse of $\begin{pmatrix} 2 & -3 \\ 3 & -4 \end{pmatrix}$ on *your* calculator.

Key point

A matrix **A** is **self-inverse** if $\mathbf{A}^{-1} = \mathbf{A}$

To find the inverse of a 3×3 matrix, **P**

1 First find the **matrix of minors**, **M**. To do this, replace each element in **A** by its minor.

For $\mathbf{P} = \begin{pmatrix} a & b & c \\ d & e & f \\ g & h & i \end{pmatrix}$, the matrix of minors is $\mathbf{M} = \begin{pmatrix} A & B & C \\ D & E & F \\ G & H & I \end{pmatrix}$ where A is the minor of a etc.

2 **Transpose** the matrix (swap rows and columns), and change the sign of alternating elements to

give $\begin{pmatrix} A & -D & E \\ -B & E & -H \\ C & -F & I \end{pmatrix}$

3 Divide this new matrix by the determinant.

Key point

The inverse of **P** is $\mathbf{P}^{-1} = \dfrac{1}{\det(\mathbf{P})} \begin{pmatrix} A & -D & G \\ -B & E & -H \\ C & -F & I \end{pmatrix}$

Example 5

This could be worked out on a calculator but you do need to know the method.

Find the inverse of $\mathbf{A} = \begin{pmatrix} 2 & 1 & -3 \\ 3 & 1 & -2 \\ 0 & 2 & -1 \end{pmatrix}$

$\mathbf{M} = \begin{pmatrix} (1\times-1)-(-2\times2) & (3\times-1)-(-2\times0) & (3\times2)-(1\times0) \\ (1\times-1)-(-3\times2) & (2\times-1)-(-3\times0) & (2\times2)-(1\times0) \\ (1\times-2)-(-3\times1) & (2\times-2)-(-3\times3) & (2\times1)-(1\times3) \end{pmatrix}$

Find the minor of every element.

$= \begin{pmatrix} 3 & -3 & 6 \\ 5 & -2 & 4 \\ 1 & 5 & -1 \end{pmatrix}$

$\det(\mathbf{A}) = (2\times3)-(1\times-3)+(-3\times6) = -9$

Use the minors from the top row of the matrix as

$\det(\mathbf{A}) = 2\begin{vmatrix} 1 & -2 \\ 2 & -1 \end{vmatrix} - 1\begin{vmatrix} 3 & -2 \\ 0 & -1 \end{vmatrix} - 3\begin{vmatrix} 3 & 1 \\ 0 & 2 \end{vmatrix}$

Therefore $\mathbf{A}^{-1} = -\dfrac{1}{9}\begin{pmatrix} 3 & -5 & 1 \\ 3 & -2 & -5 \\ 6 & -4 & -1 \end{pmatrix}$

Use $\mathbf{A}^{-1} = \dfrac{1}{\det(\mathbf{A})}\begin{pmatrix} A & -D & G \\ -B & E & -H \\ C & -F & I \end{pmatrix}$

1 Show that $\det\begin{pmatrix} -7 & 3 \\ 5 & -4 \end{pmatrix} = 13$

2 Calculate the determinants of each of these matrices.

a $\begin{pmatrix} 2 & 4 \\ 5 & 1 \end{pmatrix}$ **b** $\begin{pmatrix} -3 & 2 \\ 0 & 1 \end{pmatrix}$

c $\begin{pmatrix} 4a & 3 \\ -a & -2 \end{pmatrix}$ **d** $\begin{pmatrix} 3\sqrt{2} & -\sqrt{2} \\ -2\sqrt{2} & \sqrt{2} \end{pmatrix}$

3 Show that $\det\begin{pmatrix} a+b & 2a \\ 2b & a+b \end{pmatrix} = (a-b)^2$

4 Find the possible values of k given that
$$\det\begin{pmatrix} 7 & k \\ -k & 2-k \end{pmatrix} = 2$$

5 Decide whether each of these matrices is singular or non-singular. You must show your working.

a $\begin{pmatrix} 3 & 1 \\ -6 & 2 \end{pmatrix}$ **b** $\begin{pmatrix} 2 & 3 \\ 4 & 6 \end{pmatrix}$

c $\begin{pmatrix} -a & -2 \\ -2a & 4 \end{pmatrix}$ **d** $\begin{pmatrix} \sqrt{3} & 3 \\ -2\sqrt{3} & -6 \end{pmatrix}$

6 Find the values of x for which each of these matrices is singular.

a $\begin{pmatrix} x & 1 \\ 5 & 2 \end{pmatrix}$ **b** $\begin{pmatrix} x & -3 \\ -1 & x \end{pmatrix}$

c $\begin{pmatrix} 2x & x \\ -3 & x \end{pmatrix}$ **d** $\begin{pmatrix} 4 & -x \\ -x & x \end{pmatrix}$

7 Given that $\mathbf{A} = \begin{pmatrix} y+3 & y+5 \\ -1 & y-1 \end{pmatrix}$, calculate the possible values of y for which the matrix $\mathbf{A}$ has no inverse.

8 Given the matrices $\mathbf{A} = \begin{pmatrix} -1 & 2 \\ 0 & 5 \end{pmatrix}$ and $\mathbf{B} = \begin{pmatrix} 0 & -3 \\ 2 & 4 \end{pmatrix}$

a Calculate the values of

  i $\det(\mathbf{A}) \times \det(\mathbf{B})$ **ii** $\det(\mathbf{A}) + \det(\mathbf{B})$

b Show that

  i $\det(\mathbf{AB}) = -30$ **iii** $\det(\mathbf{A}+\mathbf{B}) = -7$

9 Calculate the determinant of each of these matrices.

a $\begin{pmatrix} 6 & 2 & -1 \\ -2 & 4 & 0 \\ 3 & 5 & 1 \end{pmatrix}$ **b** $\begin{pmatrix} 2a & 3 & a \\ -b & 0 & b \\ 3 & 5 & 1 \end{pmatrix}$

10 Find the possible values of k given that
$$\det\begin{pmatrix} 4+k & -5 & -k \\ 2 & k+1 & 4 \\ -1 & 2 & 3 \end{pmatrix} = 30$$

11 Find the values of x for which each of these matrices is singular.

a $\begin{pmatrix} 1 & 2 & -x \\ x & 4 & -3x \\ 1 & 0 & 2 \end{pmatrix}$ **b** $\begin{pmatrix} 4 & x & 2 \\ -x & 3 & 3x \\ -2 & x & 1 \end{pmatrix}$

12 A rectangle has area 7 square units.

a Find the area of the image of the rectangle under each of these transformations.

  i $\begin{pmatrix} -2 & 3 \\ 1 & -3 \end{pmatrix}$ **ii** $\begin{pmatrix} 2 & 1 \\ 5 & 2 \end{pmatrix}$

b Explain whether or not each of the transformations in part **a** preserves orientation.

13 A triangle is transformed by the matrix
$$\begin{pmatrix} 1 & -3 \\ 4 & -2 \end{pmatrix}$$
The area of the image is 15 square units.

a Calculate the area of the original triangle.

b Explain whether or not the orientation of the triangle has changed.

14 A triangle with area 5 square units is transformed by the matrix $\begin{pmatrix} 1 & 3 & 2 \\ -2 & 1 & 2 \\ -1 & 2 & 0 \end{pmatrix}$

a Calculate the area of the image.

b Explain whether or not the orientation of the triangle has changed.

15 Find the inverse of each of these matrices.

a $\begin{pmatrix} 7 & 2 \\ 4 & 5 \end{pmatrix}$ **b** $\begin{pmatrix} -4 & 3 \\ 8 & -1 \end{pmatrix}$

c $\begin{pmatrix} 2x & 5 \\ -x & -3 \end{pmatrix}$ **d** $\begin{pmatrix} 4\sqrt{5} & -\sqrt{5} \\ 2\sqrt{5} & \sqrt{20} \end{pmatrix}$

16 Show that the inverse of $\begin{pmatrix} 6 & 1 \\ -9 & -1 \end{pmatrix}$ is

$\begin{pmatrix} -\dfrac{1}{3} & -\dfrac{1}{3} \\ 3 & 2 \end{pmatrix}$

17 $\mathbf{A} = \begin{pmatrix} b & 2a \\ b & 3a \end{pmatrix}$

   a Find the inverse of $\mathbf{A}$, writing each element in its simplest form.

   b Show that $\det(\mathbf{A}^{-1}) = \dfrac{1}{\det(\mathbf{A})}$ for the matrix $\mathbf{A}$ given.

18 Show that the matrix $\begin{pmatrix} 3 & -2 \\ 4 & -3 \end{pmatrix}$ is self-inverse.

19 Find the values of a such that the matrix $\begin{pmatrix} a & -3 \\ 2 & -a \end{pmatrix}$ is self-inverse.

20 Find the value of a such that the matrix $\begin{pmatrix} 1 & 3-a \\ a-5 & 2-a \end{pmatrix}$ is self-inverse.

21 $\mathbf{B} = \begin{pmatrix} x+1 & -x \\ x-1 & 2x \end{pmatrix}$

   a Given that the determinant of $\mathbf{B}$ is 10, find the possible values of x

   b Write the inverse of $\mathbf{B}$

22 Find the inverse of each of these matrices.

   a $\begin{pmatrix} -2 & -1 & 0 \\ 4 & 0 & 2 \\ 5 & 1 & 3 \end{pmatrix}$ **b** $\begin{pmatrix} -4a & 2 & 1 \\ a & 0 & 2 \\ 3a & -1 & 1 \end{pmatrix}$

23 Find the inverse of the matrix

   $\mathbf{B} = \begin{pmatrix} 1 & -1 & k \\ k & 0 & 1 \\ 0 & k & 1 \end{pmatrix}$ in terms of k

24 $\mathbf{T} = \begin{pmatrix} 2 & 0 \\ 0 & 2 \end{pmatrix}$

   a Show that $\mathbf{T}^{-1} = \dfrac{1}{2}\mathbf{I}$

   b Describe the transformations represented by $\mathbf{T}$ and $\mathbf{T}^{-1}$ geometrically.

25 A transformation is represented by the matrix $\mathbf{T} = \begin{pmatrix} 2 & 2 \\ -3 & 3 \end{pmatrix}$

The image of the point A under $\mathbf{T}$ is $A'(10, -27)$. Use the inverse of $\mathbf{T}$ to find the coordinates of A

26 **a** Write the matrix representing rotation of $135°$ anticlockwise about the origin.

   b Show that the inverse of this matrix is $\dfrac{1}{\sqrt{2}}\begin{pmatrix} -1 & 1 \\ -1 & -1 \end{pmatrix}$

27 A transformation is represented by the

matrix $\mathbf{T} = \begin{pmatrix} 1 & 0 & 0 \\ 0 & \dfrac{\sqrt{2}}{2} & \dfrac{\sqrt{2}}{2} \\ 0 & -\dfrac{\sqrt{2}}{2} & \dfrac{\sqrt{2}}{2} \end{pmatrix}$

The image of the point A under $\mathbf{T}$ is A' $(1, -\sqrt{2}, \sqrt{2})$. Use the inverse of $\mathbf{T}$ to find the coordinates of point A

28 For each of these transformations, explain whether it is always, sometimes or never self-inverse. For those which are sometimes self-inverse, state when this occurs.

   a Reflection in a coordinate axis,

   b Enlargement centre the origin,

   c Rotation about the origin,

   d Reflection in the line $y = \pm x$,

Reasoning and problem-solving

Matrices can be used to solve systems of linear equations.

Take, for example, the system of equations $ax + by = x'$
$$cx + dy = y'$$

If $\mathbf{M} = \begin{pmatrix} a & b \\ c & d \end{pmatrix}$ then $\mathbf{M}\begin{pmatrix} x \\ y \end{pmatrix} = \begin{pmatrix} a & b \\ c & d \end{pmatrix}\begin{pmatrix} x \\ y \end{pmatrix} = \begin{pmatrix} ax+by \\ cx+dy \end{pmatrix}$ so the equations can be written using matrices

as $\mathbf{M}\begin{pmatrix} x \\ y \end{pmatrix} = \begin{pmatrix} x' \\ y' \end{pmatrix}$

Then you can pre-multiply both sides by $\mathbf{M}^{-1}$ to give $\mathbf{M}^{-1}\mathbf{M}\begin{pmatrix} x \\ y \end{pmatrix} = \mathbf{M}^{-1}\begin{pmatrix} x' \\ y' \end{pmatrix}$

As $\mathbf{M}^{-1}\mathbf{M} = \mathbf{I}$, this can be rewritten as $\begin{pmatrix} x \\ y \end{pmatrix} = \mathbf{M}^{-1}\begin{pmatrix} x' \\ y' \end{pmatrix}$ and you can calculate the values of x and y

Strategy 1

To solve problems involving systems of linear equations

(1) Rewrite a system of linear equations using matrices.

(2) Pre-multiply or post-multiply a matrix by its inverse.

(3) Use the fact that $\mathbf{A}\mathbf{A}^{-1} = \mathbf{A}^{-1}\mathbf{A} = \mathbf{I}$

Example 6

Use matrices to solve the simultaneous equations $2x + 3y = 11$
$$4x - y = -6$$

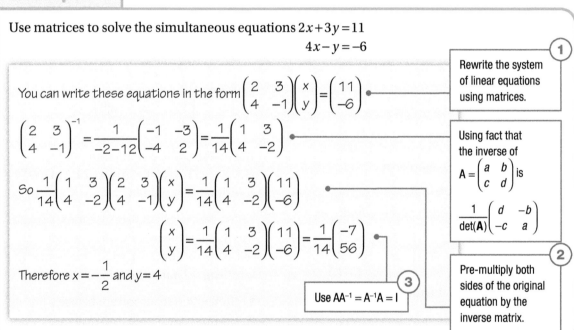

You can write these equations in the form $\begin{pmatrix} 2 & 3 \\ 4 & -1 \end{pmatrix}\begin{pmatrix} x \\ y \end{pmatrix} = \begin{pmatrix} 11 \\ -6 \end{pmatrix}$

(1) Rewrite the system of linear equations using matrices.

$\begin{pmatrix} 2 & 3 \\ 4 & -1 \end{pmatrix}^{-1} = \dfrac{1}{-2-12}\begin{pmatrix} -1 & -3 \\ -4 & 2 \end{pmatrix} = \dfrac{1}{14}\begin{pmatrix} 1 & 3 \\ 4 & -2 \end{pmatrix}$

Using fact that the inverse of $\mathbf{A} = \begin{pmatrix} a & b \\ c & d \end{pmatrix}$ is $\dfrac{1}{\det(\mathbf{A})}\begin{pmatrix} d & -b \\ -c & a \end{pmatrix}$

So $\dfrac{1}{14}\begin{pmatrix} 1 & 3 \\ 4 & -2 \end{pmatrix}\begin{pmatrix} 2 & 3 \\ 4 & -1 \end{pmatrix}\begin{pmatrix} x \\ y \end{pmatrix} = \dfrac{1}{14}\begin{pmatrix} 1 & 3 \\ 4 & -2 \end{pmatrix}\begin{pmatrix} 11 \\ -6 \end{pmatrix}$

$\begin{pmatrix} x \\ y \end{pmatrix} = \dfrac{1}{14}\begin{pmatrix} 1 & 3 \\ 4 & -2 \end{pmatrix}\begin{pmatrix} 11 \\ -6 \end{pmatrix} = \dfrac{1}{14}\begin{pmatrix} -7 \\ 56 \end{pmatrix}$

(2) Pre-multiply both sides of the original equation by the inverse matrix.

Therefore $x = -\dfrac{1}{2}$ and $y = 4$

(3) Use $\mathbf{A}\mathbf{A}^{-1} = \mathbf{A}^{-1}\mathbf{A} = \mathbf{I}$

An equation of the form $ax + by + cz = d$ can represent a plane. If you have three different planes then one of these will apply:

- There are no points that lie on all three planes (their equations are said to be inconsistent). This could be because two or three of the planes are parallel or because the three planes form a triangular prism.

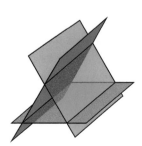

- There are infinitely many solutions as the three planes meet along a line (called a sheaf).

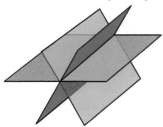

- They intersect at a single point, so there is exactly one solution.

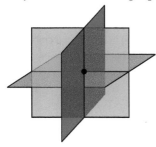

Example 7

Given that there is a unique solution, use matrices to solve this system of equations:

$$x + y + z = 2 \qquad 2x - 3y - z = -1 \qquad 3x - 2y + 2z = 11$$

Write the equations in the form $\begin{pmatrix} 1 & 1 & 1 \\ 2 & -3 & -1 \\ 3 & -2 & 2 \end{pmatrix} \begin{pmatrix} x \\ y \\ z \end{pmatrix} = \begin{pmatrix} 2 \\ -1 \\ 11 \end{pmatrix}$

① Rewrite the system of linear equations using matrices.

$$\det \begin{pmatrix} 1 & 1 & 1 \\ 2 & -3 & -1 \\ 3 & -2 & 2 \end{pmatrix} = 1(-6-2) - 1(4--3) + 1(-4--9) = -10$$

Calculate the determinant.

$$\begin{pmatrix} 1 & 1 & 1 \\ 2 & -3 & -1 \\ 3 & -2 & 2 \end{pmatrix}^{-1} = -\frac{1}{10} \begin{pmatrix} -8 & -4 & 2 \\ -7 & -1 & 3 \\ 5 & 5 & -5 \end{pmatrix}$$

Find the inverse.

$$-\frac{1}{10} \begin{pmatrix} -8 & -4 & 2 \\ -7 & -1 & 3 \\ 5 & 5 & -5 \end{pmatrix} \begin{pmatrix} 1 & 1 & 1 \\ 2 & -3 & -1 \\ 3 & -2 & 2 \end{pmatrix} \begin{pmatrix} x \\ y \\ z \end{pmatrix} = -\frac{1}{10} \begin{pmatrix} -8 & -4 & 2 \\ -7 & -1 & 3 \\ 5 & 5 & -5 \end{pmatrix} \begin{pmatrix} 2 \\ -1 \\ 11 \end{pmatrix}$$

② Pre-multiply both sides of the original equation by the inverse matrix.

$$\begin{pmatrix} x \\ y \\ z \end{pmatrix} = -\frac{1}{10} \begin{pmatrix} -8 & -4 & 2 \\ -7 & -1 & 3 \\ 5 & 5 & -5 \end{pmatrix} \begin{pmatrix} 2 \\ -1 \\ 11 \end{pmatrix}$$

③ Use $AA^{-1} = A^{-1}A = I$

$$= -\frac{1}{10} \begin{pmatrix} 10 \\ 20 \\ -50 \end{pmatrix}$$

So $x = -1$, $y = -2$, $z = 5$

Example 8

Explain why these systems of equations have no solutions.

a $\begin{aligned} 2x-3y-z&=2 \\ 4x-6y-2z&=1 \\ 12x-18y-6z&=3 \end{aligned}\Big\}$ b $\begin{aligned} x+y-z&=5 \\ 2x+2y&=10 \\ x+y+z&=7 \end{aligned}\Big\}$

Also, notice that $12x-18y-6z=3$ is a multiple of $4x-6y-2z=1$ so these represent the same plane.

a   The equation $4x-6y-2z=1$ is a multiple of $2x-3y-z=2$ except for the constant term.

Therefore these represent parallel planes.

Parallel planes do not intersect so there are no solutions.

None of the planes are the same or parallel so we need to start trying to simplify the problem by eliminating a variable.

b   Adding the first and third equations together gives

$2x+2y=12$

However, the second equation is $2x+2y=10$

Therefore the equations are inconsistent and there are no solutions.

The planes form a triangular prism.

The system of equations in Example **9b** can be written in matrix form as

$$\begin{pmatrix} 1 & 1 & -1 \\ 2 & 2 & 0 \\ 1 & 1 & 1 \end{pmatrix}\begin{pmatrix} x \\ y \\ z \end{pmatrix}=\begin{pmatrix} 5 \\ 10 \\ 7 \end{pmatrix}$$

You can attempt to solve these by pre-multiplying both sides of the equation by the inverse of

$\begin{pmatrix} 1 & 1 & -1 \\ 2 & 2 & 0 \\ 1 & 1 & 1 \end{pmatrix}$.

However, for this particular system of equations

$$\det\begin{pmatrix} 1 & 1 & -1 \\ 2 & 2 & 0 \\ 1 & 1 & 1 \end{pmatrix}=1(2-0)-1(2-0)--1(2-2)=0$$

This implies that there is either no solution (as in this case) or an infinite number of solutions.

Strategy 2

To decide on the nature of a system of simultaneous equations

(1) Rewrite the system of linear equations using matrices.

(2) Calculate a determinant – a non-zero determinant implies a unique solution.

(3) Write two of the variables in terms of the third.

(4) Check for inconsistencies.

Example 9

You are given a system of simultaneous equations:

$x + 3y - 7z = 8, \; 2x - 2z = -2, \; x - y + z = -4$

a Decide whether there is a unique solution, an infinite number of solutions or no solutions.

b Describe the geometric significance.

a
$$\begin{pmatrix} 1 & 3 & -7 \\ 2 & 0 & -2 \\ 1 & -1 & 1 \end{pmatrix} \begin{pmatrix} x \\ y \\ z \end{pmatrix} = \begin{pmatrix} 8 \\ -2 \\ -4 \end{pmatrix}$$

> **1** Rewrite the system of linear equations using matrices.

$$\det \begin{pmatrix} 1 & 3 & -7 \\ 2 & 0 & -2 \\ 1 & -1 & 1 \end{pmatrix} = 1(0-2) - 3(2--2) - 7(-2-0)$$
$$= 0$$

> **2** Since the determinant is zero, must either be no solution or an infinite number of solutions.

This implies there is not a unique solution.

Subtracting the third equation from the first equation gives

$4y - 8z = 12 \Rightarrow y = 3 + 2z$

The second equation gives $x = z - 1$

> **3** Here y and x are expressed in terms of z, but you could have chosen any of the three variables to write the other two in terms of.

Verify there are no inconsistencies:

$(z-1) + 3(3+2z) - 7z = 8$

$2(z-1) - 2z = -2$

$(z-1) - (3+2z) + z = -4$

> **4** You can verify there are no inconsistencies by substituting back into the original equations.

So there is an infinite number of solutions. These lie on the line with equations $y = 3 + 2z$, $x = z - 1$

> You will learn more about writing the equation of lines in 3D in Chapter 5

b The planes meet in a line: they form a sheaf.

You can construct matrix proofs by pre-multiplying or post-multiplying both sides of an equation.

Example 10

Prove that $(\mathbf{AB})^{-1} = \mathbf{B}^{-1}\mathbf{A}^{-1}$

> Since matrix multiplication is associative.

$(AB)^{-1}(AB) = I$

$(AB)^{-1}AB = I$

> **2** Post-multiply both sides of the equation by $\mathbf{B}^{-1}$

$(AB)^{-1}ABB^{-1} = IB^{-1}$

$(AB)^{-1}A = B^{-1}$

> **3** Since $\mathbf{BB}^{-1} = \mathbf{I}$ and $\mathbf{IB}^{-1} = \mathbf{B}^{-1}$

$(AB)^{-1}AA^{-1} = B^{-1}A^{-1}$

$(AB)^{-1} = B^{-1}A^{-1}$ as required.

> Since $\mathbf{AA}^{-1} = \mathbf{I}$ **3**

> **2** Post-multiply both sides of the equation by $\mathbf{A}^{-1}$

Key point

If $\mathbf{A}$ and $\mathbf{B}$ are non-singular matrices then $(\mathbf{AB})^{-1} = \mathbf{B}^{-1}\mathbf{A}^{-1}$

1 Use matrices to solve these pairs of simultaneous equations.

a $4x - y = 11$
$2x + 3y = -5$

b $x - 5y = 0$
$2x - 8y = -2$

c $3x + 6y = 12$
$x - 2y = 2$

d $4x + 6y = 1$
$-8x + 3y = 3$

2 Use matrices to find a solution to these systems of linear equations.

a $x + y - 2z = 3$
$2x - 3y + 5z = 4$
$5x + 2y + z = -3$

b $4x + 6y - z = -3$
$2x - 3y = 2$
$8y + 4z = 0$

3 Given that $\begin{pmatrix} a & 3 \\ -2a & -1 \end{pmatrix}\begin{pmatrix} x \\ y \end{pmatrix} = \begin{pmatrix} 5 \\ 10a \end{pmatrix}$,

find expressions in terms of a for x and y
Simplify your answers.

4 You are given the equations $(k+3)x - 2y = k - 1$ and $kx + y = k$

Use matrices to find expressions in terms of k for x and y

5 Given that $\mathbf{A}\begin{pmatrix} -7 & -2 \\ 0 & 1 \end{pmatrix} = \begin{pmatrix} 14 & 8 \\ 7 & 0 \end{pmatrix}$, find the matrix $\mathbf{A}$

6 Given that $\begin{pmatrix} 3 & -5 \\ -1 & 2 \end{pmatrix}\mathbf{B} = \begin{pmatrix} 12 & -1 \\ -4 & 0 \end{pmatrix}$, find the matrix $\mathbf{B}$

7 Given that $\mathbf{A}\begin{pmatrix} 2 & 0 & -3 \\ 0 & 1 & 4 \\ -5 & 2 & -1 \end{pmatrix} = \begin{pmatrix} 14 & -9 & 9 \\ 25 & -10 & 5 \\ 9 & 1 & 7 \end{pmatrix}$,

find the matrix $\mathbf{A}$

8 Given that $\begin{pmatrix} 5 & -1 & 0 \\ 0 & 2 & -2 \\ 3 & 1 & 4 \end{pmatrix}\mathbf{B} = \begin{pmatrix} -32 & 16 & 15 \\ -4 & 8 & 6 \\ 0 & -12 & -3 \end{pmatrix}$,

find the matrix $\mathbf{B}$

9 Given that $\mathbf{P}\begin{pmatrix} 2a & b \\ -a & -b \end{pmatrix} = \begin{pmatrix} a & 2a \\ a & -b \end{pmatrix}$, find an

expression for $\det(\mathbf{P})$ in terms of a and b.
Fully simplify your answer.

10 If $\mathbf{A} = \begin{pmatrix} 3 & 5 \\ 2 & 4 \end{pmatrix}$ and $\mathbf{AB} = \begin{pmatrix} 6 & 8 \\ -2 & 0 \end{pmatrix}$, find $\mathbf{B}$

11 If $\mathbf{B} = \begin{pmatrix} 2 & 3 \\ -3 & -1 \end{pmatrix}$ and $\mathbf{AB} = \begin{pmatrix} 10 & 15 \\ 11 & 13 \end{pmatrix}$, find the matrix $\mathbf{A}$

12 If $\mathbf{A} = \begin{pmatrix} 8 & 0 & 2 \\ 4 & 3 & -1 \\ 5 & 0 & 2 \end{pmatrix}$ and $\mathbf{AB} = \begin{pmatrix} -2 & 28 & 14 \\ 13 & 10 & -5 \\ -2 & 19 & 11 \end{pmatrix}$,

find the matrix $\mathbf{BA}$

13 For each system of equations, show that the planes that the equations represent form a sheaf.

a $x + 2z = 1$
$-2x + y + 4z = 0$
$9x - 2y + 2z = 5$

b $5x - y + z = 0$
$2x + y - 2z = 7$
$36x + 11y - 24z = 91$

14 Find the values of a for which this system of simultaneous equations does not have a unique solution.

$3y + az = 4$
$-x + ay + 2z = 1$
$x + y = 3$

15 Show that each of these systems of linear equations is inconsistent and explain the geometric significance.

a $3x - 2y + z = 7$
$6x - 4y + 2z = 5$
$x + 3y + 2z = 3$

b $-2x + 3y - z = 4$
$6x - 9y + 3z = 7$
$5x + y + z = 3$

16 A family keeps rabbits, hamsters and fish as pets.

They initially have 17 pets in total and one more rabbit than hamsters.

Two of the fish die. They then have 8 more fish than hamsters.

a Write a matrix equation to represent this situation.

b Solve the matrix equation and state how many of each pet the family had initially.

17 A property developer owns 12 homes which have a mix of two, three and four bedrooms.

The total number of bedrooms in all her houses is 36 and she has the same number of two-bedroom homes as four-bedroom homes.

  a Write a matrix equation to represent this situation.

  b Show that there is no unique solution to this problem.

  c How many possible solutions are there? Explain your answer.

18 If $\mathbf{A}^2 = 2\mathbf{I}$, write the 2×2 matrix $\mathbf{A}$

19 Simplify each of these expressions involving the non-singular matrices $\mathbf{A}$ and $\mathbf{B}$

  a $(\mathbf{AB})^{-1}\mathbf{A}$ **b** $\mathbf{B}(\mathbf{A}^{-1}\mathbf{B})^{-1}$

20 If $\mathbf{A} = \mathbf{C}^{-1}\mathbf{BC}$, prove that $\mathbf{B} = \mathbf{CAC}^{-1}$

21 Prove that if $\mathbf{ABA}^{-1} = \mathbf{I}$ then $\mathbf{B} = \mathbf{I}$

22 Prove that $(\mathbf{ABC})^{-1} = \mathbf{C}^{-1}\mathbf{B}^{-1}\mathbf{A}^{-1}$ for non-singular matrices $\mathbf{A}$, $\mathbf{B}$ and $\mathbf{C}$

23 Prove that if $\mathbf{P}$ is self-inverse then $\mathbf{P}^2 = \mathbf{P}^{-1}$

24 Prove that if $\mathbf{PQP} = \mathbf{I}$ for non-singular matrices $\mathbf{P}$ and $\mathbf{Q}$, then $\mathbf{Q} = (\mathbf{P}^{-1})^2$

25 A point P is transformed by the matrix $\mathbf{T} = \begin{pmatrix} 2 & 1 \\ -3 & -4 \end{pmatrix}$ and the coordinates of the image of P are $(9, -11)$

Find the coordinates of P

26 The triangle ABC is stretched by scale factor k parallel to the x-axis then reflected in the line $y = x$. The images of vertices A, B and C are given by $A'(4, -6)$, $B'(0, 0)$ and $C'(4, 2)$

  a Work out the area of the original triangle in terms of k

  b State the range of values of k for which the transformation preserves the orientation of the triangle.

  c Find the coordinates of A, B and C in terms of k

  d If the coordinates of C are $(-1, 4)$, find the value of k

27 Prove that $\det(\mathbf{AB}) = \det(\mathbf{A})\det(\mathbf{B})$ for any non-singular 2×2 matrices $\mathbf{A}$ and $\mathbf{B}$

28 Prove that $\det(\mathbf{A}^{-1}) = [\det(\mathbf{A})]^{-1}$ for any non-singular 2×2 matrix $\mathbf{A}$

29 **a** Write the matrix that represents

  i A rotation of angle θ anticlockwise around the origin,

  ii A rotation of angle θ clockwise around the origin.

  b Hence, deduce the identity $\sin^2\theta + \cos^2\theta \equiv 1$

30 By considering a square with one vertex at the origin, show that, under any linear enlargement, $\mathbf{T}$,

  a The image of the square will be a square.

  b The area of the image will be the area of the original square multiplied by the determinant of $\mathbf{T}$

31 By considering a square with one vertex at the origin, show that, under any stretch, $\mathbf{T}$, in the x- or y-direction (or both),

  a The image of the square will be a rectangle,

  b The area of the image will be the area of the original square multiplied by the determinant of $\mathbf{T}$

Chapter summary

- A matrix with n rows and m columns has order $n \times m$
- Matrices of the same order can be added by adding corresponding elements.
- Matrices can be multiplied by a constant by multiplying each of its elements by that constant.
- Two matrices are conformable for multiplication if the number of columns in the first matrix is the same as the number of rows in the second matrix.
- The product of an $n \times m$ matrix and an $m \times p$ matrix has order $n \times p$
- Under the transformation given by the matrix $\begin{pmatrix} a & b \\ c & d \end{pmatrix}$ the vector $\begin{pmatrix} 1 \\ 0 \end{pmatrix}$ is transformed to $\begin{pmatrix} a \\ c \end{pmatrix}$ and the vector $\begin{pmatrix} 0 \\ 1 \end{pmatrix}$ is transformed to $\begin{pmatrix} b \\ d \end{pmatrix}$
- The matrix representing transformation $\mathbf{A}$ followed by transformation $\mathbf{B}$ is given by the product $\mathbf{BA}$
- The identity matrix is a matrix with ones along the leading diagonal and zeros elsewhere, so in 2D, $\mathbf{I} = \begin{pmatrix} 1 & 0 \\ 0 & 1 \end{pmatrix}$
- The zero matrix is a matrix with 0 as every element, so in 2D, $\mathbf{0} = \begin{pmatrix} 0 & 0 \\ 0 & 0 \end{pmatrix}$
- The invariant points of the matrix $\mathbf{T}$ can be found by solving $\mathbf{T}\begin{pmatrix} x \\ y \end{pmatrix} = \begin{pmatrix} x \\ y \end{pmatrix}$
- The invariant lines of the matrix $\mathbf{T}$ can be found by solving $\mathbf{T}\begin{pmatrix} x \\ mx+c \end{pmatrix} = \begin{pmatrix} x' \\ mx'+c \end{pmatrix}$ to find the values of m and c
- The determinant of a 2×2 matrix is $\det\begin{pmatrix} a & b \\ c & d \end{pmatrix} = ad - bc$
- The determinant of a 3×3 matrix is $\det\begin{pmatrix} a & b & c \\ d & e & f \\ g & h & i \end{pmatrix} = a\begin{vmatrix} e & f \\ h & i \end{vmatrix} - b\begin{vmatrix} d & f \\ g & i \end{vmatrix} + c\begin{vmatrix} d & e \\ g & h \end{vmatrix}$
- A matrix is singular if its determinant is zero.
- Under a transformation $\mathbf{T}$, area of image = area of original $\times |\det(\mathbf{T})|$
- If $\det(\mathbf{T}) > 0$ then the transformation represented by $\mathbf{T}$ preserves orientation.
- If $\mathbf{A} = \begin{pmatrix} a & b \\ c & d \end{pmatrix}$ then its inverse is $\mathbf{A}^{-1} = \dfrac{1}{\det(\mathbf{A})}\begin{pmatrix} d & -b \\ -c & a \end{pmatrix}$
- If $\mathbf{P} = \begin{pmatrix} a & b & c \\ d & e & f \\ g & h & i \end{pmatrix}$ then its inverse is $\mathbf{P}^{-1} = \dfrac{1}{\det(\mathbf{P})}\begin{pmatrix} A & -D & G \\ -B & E & -H \\ C & -F & I \end{pmatrix}$ where A is the minor of a, B is the minor of B and so on.
- For any non-singular matrix $\mathbf{A}$, $\mathbf{AA}^{-1} = \mathbf{A}^{-1}\mathbf{A} = \mathbf{I}$

You should now be able to...	Try Questions
✓ Identify the order of a matrix.	1
✓ Add and subtract matrices of the same order and multiply matrices by a constant.	1, 2
✓ Multiply conformable matrices.	1, 2
✓ Apply a linear transformation given as a matrix to a point or vector.	3, 4
✓ Describe transformations given as matrices geometrically.	3, 4
✓ Write a linear transformation given geometrically as a matrix.	5, 6
✓ Find the matrix that represents a combination of two transformations.	7
✓ Find invariant points and lines.	8, 9
✓ Calculate the determinant of a 2×2 matrix and of a 3 × 3 matrix.	10, 11
✓ Understand what is meant by a singular matrix.	12, 13
✓ Find the inverse of a 2 × 2 matrix and of a 3 × 3 matrix.	14–19
✓ Use matrices to solve systems of linear equations.	20–24
✓ Describe the geometrical significance of solutions.	3, 4, 22–24

1 $\mathbf{A} = \begin{pmatrix} 6 & 2 \\ -4 & 12 \end{pmatrix}$, $\mathbf{B} = \begin{pmatrix} 0 & 2 & -1 \\ 5 & 4 & 0 \end{pmatrix}$, $\mathbf{C} = \begin{pmatrix} -1 & -5 \\ 4 & 0 \\ 0 & 2 \end{pmatrix}$ and $\mathbf{D} = \begin{pmatrix} 8 & -2 & 3 \\ 0 & -1 & 0 \end{pmatrix}$

    a   State the order of each of the matrices.

    b   Find each of these matrices.

        i $\mathbf{B} + \mathbf{D}$ **ii** $3\mathbf{C}$ **iii** $\mathbf{D} - 2\mathbf{B}$ **iv** $\frac{1}{2}\mathbf{A}$

    c   Calculate these matrix products where possible. If it is not possible, explain why not.

        i $\mathbf{AB}$ **ii** $\mathbf{BA}$ **iii** $\mathbf{BC}$ **iv** $\mathbf{CA}$

        v $\mathbf{AC}$ **vi** $\mathbf{CD}$ **vii** $\mathbf{A}^2$ **viii** $\mathbf{C}^2$

2 Given that $\mathbf{A} = \begin{pmatrix} a & 3 \\ a & 2 \end{pmatrix}$, $\mathbf{B} = \begin{pmatrix} -1 & a & 3 \\ 0 & 1 & a \\ 2a & -3 & 0 \end{pmatrix}$ and $\mathbf{C} = \begin{pmatrix} a & -1 \\ 0 & 3 \\ -2 & 2a \end{pmatrix}$,

    find each of these matrices in terms of a

    a $\mathbf{CA}$ **b** $\mathbf{BC}$ **c** $\mathbf{A}^2$ **d** $\mathbf{B}^2$

3 **a** Apply each of these transformations to the point $(-1, 2)$ and describe the effect of the transformation geometrically.

        i $\begin{pmatrix} 0 & 1 \\ 1 & 0 \end{pmatrix}$ **ii** $\begin{pmatrix} 7 & 0 \\ 0 & 1 \end{pmatrix}$ **iii** $\begin{pmatrix} 0 & -1 \\ 1 & 0 \end{pmatrix}$ **iv** $\begin{pmatrix} -\dfrac{\sqrt{3}}{2} & \dfrac{1}{2} \\ \dfrac{1}{2} & -\dfrac{\sqrt{3}}{2} \end{pmatrix}$

    b   A triangle has area 5 square units. Give the area of the image of the triangle under each of the transformations given in part **a**.

    c   Explain which of the transformations in part **a** preserves orientation.

4 Apply each of these transformations to the point $(3, -2, 4)$ and describe the effect of the transformation geometrically.

a $\begin{pmatrix} 1 & 0 & 0 \\ 0 & \dfrac{\sqrt{2}}{2} & -\dfrac{\sqrt{2}}{2} \\ 0 & \dfrac{\sqrt{2}}{2} & \dfrac{\sqrt{2}}{2} \end{pmatrix}$ **b** $\begin{pmatrix} 1 & 0 & 0 \\ 0 & 1 & 0 \\ 0 & 0 & -1 \end{pmatrix}$

5 Write down a 2×2 matrix to represent each of these transformations.

a Reflection in the line $x = 0$

b Rotation of $315°$ anticlockwise around the origin.

c Enlargement of scale factor -3, centre the origin.

6 Write down a 3×3 matrix to represent each of these 3D transformations.

a Rotation of $45°$ anticlockwise around the y-axis.

b Reflection in the plane $y = 0$

7 Write down the single 2×2 matrix that represents each of these transformations.

a Rotation of $270°$ anticlockwise around the origin followed by a stretch of scale factor 2 parallel to the y-axis.

b Reflection in the line $y = -x$ followed by reflection in the x-axis.

c Reflection in the x-axis followed by rotation of $150°$ anticlockwise about the origin.

8 Find the invariant points under each of these transformations.

a $\begin{pmatrix} -1 & 0 \\ 4 & -1 \end{pmatrix}$ **b** $\begin{pmatrix} 2 & -3 \\ -1 & 4 \end{pmatrix}$

9 Find all of the invariant lines under each of these transformations.

a $\begin{pmatrix} 0 & 2 \\ 3 & 1 \end{pmatrix}$ **b** $\dfrac{1}{5}\begin{pmatrix} -4 & 3 \\ 3 & 4 \end{pmatrix}$

10 Work out the determinant of each of these matrices.

a $\begin{pmatrix} 2 & -1 \\ 4 & -3 \end{pmatrix}$ **b** $\begin{pmatrix} 3a & b \\ -2a & b \end{pmatrix}$

11 Work out the determinant of each of these matrices.

a $\begin{pmatrix} 3 & 2 & -1 \\ 4 & 0 & 5 \\ 2 & 1 & -3 \end{pmatrix}$ **b** $\begin{pmatrix} a & -3 & 2a \\ 2b & 4 & b \\ 5 & 0 & -1 \end{pmatrix}$

12 Calculate the values of x for which these matrices are singular.

a $\begin{pmatrix} 3x & 4 \\ 5 & 5 \end{pmatrix}$ **b** $\begin{pmatrix} x & -1 \\ 4 & -2x \end{pmatrix}$

13 Calculate the values of x for which these matrices are singular.

a $\begin{pmatrix} 2 & -1 & 1 \\ x & 3 & 0 \\ -x & 4 & 2 \end{pmatrix}$ **b** $\begin{pmatrix} -x & 3 & 4 \\ 1 & 2 & x \\ 0 & 1 & 2 \end{pmatrix}$

14 Find the inverse, if it exists, of each of these matrices. If it does not exist, explain why not.

a $\begin{pmatrix} 6 & 3 \\ -1 & -2 \end{pmatrix}$ b $\begin{pmatrix} 0 & 1 \\ 3 & -2 \end{pmatrix}$ c $\begin{pmatrix} 3 & -6 \\ -4 & 8 \end{pmatrix}$ d $\begin{pmatrix} 5a & 2b \\ 7a & 3b \end{pmatrix}$

15 Find the inverse, if it exists, of each of these matrices. If it does not exist, explain why not.

a $\begin{pmatrix} 3 & 2 & -1 \\ 4 & 0 & 5 \\ 2 & 1 & -3 \end{pmatrix}$ b $\begin{pmatrix} 6 & -2 & -3 \\ 1 & 0 & -1 \\ 0 & 2 & -3 \end{pmatrix}$

16 Given that $\mathbf{A} = \begin{pmatrix} 5 & 7 \\ 1 & -4 \end{pmatrix}$ and $\mathbf{AB} = \begin{pmatrix} 10 & 22 \\ 2 & -1 \end{pmatrix}$, work out matrix $\mathbf{B}$

17 Given that $\mathbf{C} = \begin{pmatrix} 9 & -2 \\ 5 & 1 \end{pmatrix}$ and $\mathbf{BC} = \begin{pmatrix} -17 & -11 \\ -60 & 7 \end{pmatrix}$, work out matrix $\mathbf{B}$

18 Given that $\mathbf{A} = \begin{pmatrix} 1 & 3 & 0 \\ -2 & 4 & 1 \\ 0 & 2 & -1 \end{pmatrix}$ and $\mathbf{AB} = \begin{pmatrix} 11 & -1 & -11 \\ -1 & 5 & -18 \\ 3 & -3 & -8 \end{pmatrix}$, work out matrix $\mathbf{B}$

19 Given that $\mathbf{C} = \begin{pmatrix} 4 & 1 & -1 \\ -2 & 1 & 0 \\ 3 & -2 & 1 \end{pmatrix}$ and $\mathbf{BC} = \begin{pmatrix} 76 & -4 & -8 \\ -2 & -9 & 5 \\ -2 & -1 & 4 \end{pmatrix}$, work out matrix $\mathbf{B}$

20 Use matrices to solve each pair of simultaneous equations.

a $x + 6y = 7$ b $5x - 7y = 8$
$3x - y = -17$ $10x + 3y = -1$

21 Use matrices to find a solution to each of the systems of simultaneous equations.

a $\left.\begin{array}{l} 3x - 2y + z = 6 \\ x - y + 3z = -23 \\ 5x - 3y + 2z = 5 \end{array}\right\}$ b $\left.\begin{array}{l} 2x + 5y - z = 6 \\ x - y + 6z = -13 \\ 3x + 2y + 4z = -2 \end{array}\right\}$

22 Explain why each of these systems of equations does not have any solutions.

a $\left.\begin{array}{l} 6x - 2y + 4z = 4 \\ 5x - 3y + z = -5 \\ -3x + y - 2z = 7 \end{array}\right\}$ b $\left.\begin{array}{l} -3x - z = 1 \\ -3y + 2z = 2 \\ x + 2y - z = -3 \end{array}\right\}$

23 A system of equations is given by

$3x - y + 3z = 14$
$2y - z = -9$
$3x + y + 2z = 10$

a Show that there is not a unique solution to the system of equations.

b Show that the planes described by the three equations form a triangular prism.

24 A investor buys shares in three different companies: A, B and C

The total amount invested is £12 000 and he invests three times as much in company A as in company B

After one year he has made a total of £555 profit. His shares in companies A and B both increased in value by 10% but his shares in company C lost 5% of their value.

a Represent this situation as a system of linear equations

b Solve these equations to find the amount initially invested in each of the companies.

c Interpret the solution geometrically.

History

The Russian mathematician **Andrey Andreyevich Markov** (1856 – 1922) used matrices in his work on **stochastic processes**. Stochastic processes are collections of random variables. They can be used to show how a system might change over time.

There are now many applications of **Markov processes** including random walks, the Gambler's ruin problem, modelling queues arriving at an airport, exchange rates and the PageRank algorithm.

ICT

Here is an example of a Markov process that helps to predict long term weather probabilities.

In a simplified model the weather on any one day can either be dry or wet.

Assume that the weather today is dry so the probabilities of dry and wet are given by $W_0 = [1 \quad 0]$

Let $P = \begin{bmatrix} 0.75 & 0.25 \\ 0.6 & 0.4 \end{bmatrix}$ where $P = \begin{bmatrix} P(\text{dry tomorrow} | \text{dry today}) & P(\text{wet tomorrow} | \text{dry today}) \\ P(\text{dry tomorrow} | \text{wet today}) & P(\text{wet tomorrow} | \text{wet today}) \end{bmatrix}$

It follows that the weather tomorrow is given by

$W_1 = W_0 P = \begin{bmatrix} 1 & 0 \end{bmatrix} \begin{bmatrix} 0.75 & 0.25 \\ 0.6 & 0.4 \end{bmatrix} = \begin{bmatrix} 0.75 & 0.25 \end{bmatrix}$

The weather the next day would be given by

$W_2 = W_1 P = \begin{bmatrix} 0.75 & 0.25 \end{bmatrix} \begin{bmatrix} 0.75 & 0.25 \\ 0.6 & 0.4 \end{bmatrix} = \begin{bmatrix} 0.7125 & 0.2875 \end{bmatrix}$

Using a spreadsheet, find (to 4 decimal places) the 1×2 matrix that the weather probabilities tend towards.

B5   {=MMULT(B4:C4, D2:E3)}

	A	B	C	D	E	F	G
1	Day	p(dry)	P(wet)	Stochastic matrix			
2				0.75	0.25		
3				0.6	0.4		
4		1	1	0			
5		2	0.75	0.25			
6		3	0.7125	0.2875			
7		4	0.706875	0.293125			
8							

0.75

0.25

0.4

Dry

Wet

0.6

Karl Friedrich Gauß.

Research

Matrices can be used to solve systems of linear equations using a method now known as **Gaussian elimination**. The method was first invented in China before being reinvented in Europe in the 1700s.
It is named after **Carl Friedrich Gauss** after the adoption, by professional computers, of a specialised notation that Gauss devised.

Find out about Gaussian elimination. Try using it to solve the equations
$x + 2y + 3z = 1, \quad 2x - y - z = 3, \quad x + y + z = 3$

1 Given that $\mathbf{M} = \begin{pmatrix} 5 & 2 \\ -3 & 1 \end{pmatrix}$ and $\mathbf{N} = \begin{pmatrix} -4 & 3 \\ 0 & 2 \end{pmatrix}$, calculate

   a $3\mathbf{M}$ **b** $-2\mathbf{N}$ **c** $\mathbf{M} - \mathbf{N}$ **d** $\mathbf{MN}$ **[4 marks]**

2 You are told that $\mathbf{M} = \begin{pmatrix} 1 & 2 \\ 4 & -3 \end{pmatrix}$ and $\mathbf{N} = \begin{pmatrix} 2 & 7 \\ 5 & 1 \end{pmatrix}$

   a Calculate $\mathbf{MN}$ **b** Calculate $\mathbf{NM}$ **c** What does this say about $\mathbf{M}$ and $\mathbf{N}$?   **[3]**

3 Given $\mathbf{M} = \begin{pmatrix} 2 & 7 \\ 9 & 11 \end{pmatrix}$, calculate

   a $\mathbf{MI}$ **b** $\mathbf{0} + 2\mathbf{M}$ **c** $\mathbf{M}^2$ **[3]**

4 You are told that $\begin{pmatrix} a & -1 \\ 2 & 3 \end{pmatrix}\begin{pmatrix} b & 5 \\ 6 & 2 \end{pmatrix} \equiv \begin{pmatrix} -2 & 18 \\ 20 & c \end{pmatrix}$ is an identity. Find a, b and c **[2]**

5 You are told that $\begin{pmatrix} a & 2 & 5 \\ -3 & -1 & -4 \\ -6 & 1 & 2 \end{pmatrix}\begin{pmatrix} \dfrac{2}{35} & \dfrac{1}{35} & \dfrac{-3}{35} \\ \dfrac{30}{35} & \dfrac{50}{35} & \dfrac{25}{35} \\ \dfrac{-9}{35} & \dfrac{-22}{35} & b \end{pmatrix} \equiv \begin{pmatrix} 1 & 0 & 0 \\ 0 & c & 0 \\ 0 & 0 & 1 \end{pmatrix}$ is an identity.

   a Find a, b and c

   b What does this say about the two matrices on the left-hand side?   **[3]**

6 Each of these matrices represents a transformation. Fill in the blanks in each sentence.

   a The matrix $\begin{pmatrix} 6 & 0 \\ 0 & 6 \end{pmatrix}$ describes an _____ with scale factor _____.

   b The matrix $\begin{pmatrix} 0 & -1 \\ 1 & 0 \end{pmatrix}$ describes a rotation about _____ by _____ in the anticlockwise direction.

   c The matrix $\begin{pmatrix} 1 & 0 \\ 0 & -1 \end{pmatrix}$ describes a _____ in _____.   **[6]**

7 The 3×3 matrix $\mathbf{T}$ represents a rotation of 30° clockwise around the y-axis.

   a Find $\mathbf{T}$ in its simplest form.   **[2]**

   b Find the image of the point $(\sqrt{3}, 5, -2)$ **[2]**

8 $\mathbf{M}$ is the matrix which describes a rotation by 135° anticlockwise about the origin. $\mathbf{N}$ is the matrix which describes a rotation by 45° clockwise about the origin.

   a Write matrices $\mathbf{M}$ and $\mathbf{N}$ **b** Calculate $\mathbf{MN}$

   c What effect does $\mathbf{MN}$ have and how does it relate to $\mathbf{M}$ and $\mathbf{N}$?   **[5]**

9 Calculate any invariant points for these matrices.

   a $\begin{pmatrix} 2 & -1 \\ 2 & 3 \end{pmatrix}$ **b** $\begin{pmatrix} 5 & 2 \\ -4 & -1 \end{pmatrix}$ **c** $\begin{pmatrix} 1 & 0 \\ 2 & 7 \end{pmatrix}$ **[6]**

10 Find the invariant line under these transformations.

a Reflection in the y-axis in 2D space.

b Rotation by 180° clockwise about the z-axis in 3D space. **[2]**

11 The matrix $\mathbf{M} = \begin{pmatrix} 4 & 1 \\ 3 & 7 \end{pmatrix}$ has the inverse $\mathbf{M}^{-1} = \dfrac{1}{25}\begin{pmatrix} 7 & -1 \\ -3 & 4 \end{pmatrix}$

Find the solution to the equation $\begin{pmatrix} 4 & 1 \\ 3 & 7 \end{pmatrix}\begin{pmatrix} x \\ y \end{pmatrix} = \begin{pmatrix} 10 \\ 13 \end{pmatrix}$ **[2]**

12 The matrix $\mathbf{M} = \begin{pmatrix} 3 & 10 \\ 7 & 5 \end{pmatrix}$ has the inverse $\mathbf{M}^{-1} = \dfrac{1}{55}\begin{pmatrix} -5 & 10 \\ 7 & -3 \end{pmatrix}$

Find the solution to the equation $\begin{pmatrix} 3 & 10 \\ 7 & 5 \end{pmatrix}\begin{pmatrix} x \\ y \end{pmatrix} = \begin{pmatrix} 42 \\ 43 \end{pmatrix}$ **[2]**

13 The determinant of the matrix $\begin{pmatrix} 3 & 1 \\ 6 & a \end{pmatrix}$ is $3a - 6$

a For what value of a does the matrix $\begin{pmatrix} 3 & 1 \\ 6 & a \end{pmatrix}$ have no inverse?

b For that value of a find any invariant points under $\begin{pmatrix} 3 & 1 \\ 6 & a \end{pmatrix}$ **[4]**

14 Calculate the determinants of these matrices.

a $\begin{pmatrix} 6 & 0 \\ 3 & 2 \end{pmatrix}$ **b** $\begin{pmatrix} 1 & 5 \\ 2 & 10 \end{pmatrix}$ **c** $\begin{pmatrix} 0 & 9 & 2 \\ -1 & 5 & 4 \\ 7 & 0 & -2 \end{pmatrix}$ **[4]**

15 Find the inverses of these matrices.

a $\begin{pmatrix} 9 & 16 \\ 5 & 9 \end{pmatrix}$ **b** $\begin{pmatrix} 4 & -2 \\ 5 & 3 \end{pmatrix}$ **c** $\begin{pmatrix} 2 & -1 & -2 \\ 6 & 1 & 4 \\ -8 & 1 & 0 \end{pmatrix}$ **[8]**

16 Find the values of a for which the matrix $\begin{pmatrix} a & 0 & 2 \\ 2 & -a & 3 \\ 9 & 7 & 1 \end{pmatrix}$ is singular. **[4]**

17 a Find the inverse of the matrix $\begin{pmatrix} 3 & -1 & -1 \\ 1 & 2 & 3 \\ 4 & 1 & -2 \end{pmatrix}$ **[3]**

b Hence or otherwise solve this system of equations. **[2]**

$3x - y - z = 2$
$x + 2y + 3z = 29$
$4z + y - 2z = -17$

18 Give a geometrical description of each of these systems of equations.

a $\left.\begin{array}{r} x + 7y - 2z = 9 \\ 3x + 10y - 2z = 15 \\ 2x - 30y + 12z = -30 \end{array}\right\}$ **b** $\left.\begin{array}{r} 9x - 4y + 2z = 10 \\ 14x - 4y - z = 18 \\ x + 4y - 8z = 20 \end{array}\right\}$ **[8]**

19 a Find the inverses of $\mathbf{A} = \begin{pmatrix} 3 & 1 \\ 5 & 2 \end{pmatrix}$ and $\mathbf{B} = \begin{pmatrix} -2 & 4 \\ 1 & -1 \end{pmatrix}$

b Calculate $\mathbf{B}^{-1}\mathbf{A}^{-1}$

c Find the inverse of $\mathbf{B}^{-1}\mathbf{A}^{-1}$

d How does this relate to $\mathbf{A}$ and $\mathbf{B}$? **[6]**

5 Vectors

Structural and civil engineers use complex mathematics in their development of modern public buildings. To ensure that the building is structurally sound, vectors are used to model the forces that act based on the design of the building and the materials to be used. **Vector calculations** are then carried out to ensure that the building will be safe and sound in a range of situations. First, the framework must be strong enough to support the structure itself and any eventual contents. Secondly, the building will need to withstand extreme weather or even an earth tremor.

This is just one of many uses of vectors – they are used extensively in many other branches of engineering. For example, ideas of vectors are important throughout electrical and electronic engineering, and are therefore crucial to the design of modern communications systems such as mobile phone networks, radio and telecommunications, aircraft control, and so on.

Orientation

What you need to know	What you will learn	What this leads to
KS4 • The equation of a line.	• To write and use the equation of a line in Cartesian and vector form. • To decide whether lines are intersecting, parallel or skew and determine any points of intersection • To calculate the scalar product and use it to find angles and show lines are perpendicular • To write the equation of a plane in Cartesian, vector and scalar product form • To find points of intersection, angles and calculate distances between points, planes and lines.	**FP1 Ch4 Further Vectors** • The vector product. • The scalar triple product. • Lines and planes.
Maths Ch1 • Parallel and perpendicular lines. • Points of intersection. • Solving simultaneous equations.		**FM1 Ch2 Momentum and collisions** • Momentum and impulse in two dimensions.
Maths Ch6 • Position vector and displacements • Magnitude and direction. • i, j form.		

The vector equation of a line

Fluency and skills

See Maths Ch6.1
For a reminder of magnitude and direction of vectors.

You have already used vectors to describe magnitude and direction in 2D. In this section, you will learn how to express the equation of a line in 2D and 3D in terms of vectors, as well as how to convert these to Cartesian form by using parametric equations. In 3D,

the vector $\begin{pmatrix} x \\ y \\ z \end{pmatrix}$ can also be written $x\mathbf{i} + y\mathbf{j} + z\mathbf{k}$ where $\mathbf{i}$, $\mathbf{j}$ and $\mathbf{k}$ are

perpendicular unit vectors.

Vectors can also be used to describe points, lines and planes in 2D or 3D.

Suppose you have a line that passes through the point with position vector $\mathbf{a}$ and is parallel to vector $\mathbf{b}$ then the general point on this line will have position vector $\mathbf{r}$ where $\mathbf{r} = \mathbf{a} + \lambda \mathbf{b}$. You can see this from the triangle in the diagram. λ is a scalar quantity. It can take any value and determines how far along the line you move.

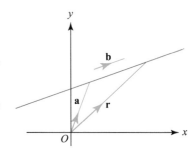

Therefore, in either 2D or 3D:

Key point

$\mathbf{r} = \mathbf{a} + \lambda \mathbf{b}$ is the vector equation of a line parallel to the vector $\mathbf{b}$ and passing through the point with position vector $\mathbf{a}$

Example 1

See Maths Ch6.1
For a reminder of finding parallel vectors.

Work out the vector equation of the line passing through the points with position vectors

$\begin{pmatrix} 1 \\ -3 \\ 2 \end{pmatrix}$ and $\begin{pmatrix} 3 \\ 0 \\ -1 \end{pmatrix}$

First, find a vector in the direction of the line.

$\begin{pmatrix} 3 \\ 0 \\ -1 \end{pmatrix} - \begin{pmatrix} 1 \\ -3 \\ 2 \end{pmatrix} = \begin{pmatrix} 2 \\ 3 \\ -3 \end{pmatrix}$ is parallel to the line

> To find a vector parallel to the line, subtract the two position vectors of points on the line.

Therefore the equation of the line is $\mathbf{r} = \begin{pmatrix} 1 \\ -3 \\ 2 \end{pmatrix} + \lambda \begin{pmatrix} 2 \\ 3 \\ -3 \end{pmatrix}$

> You can use either of the position vectors given.

Or, alternatively, $\mathbf{r} = \begin{pmatrix} 3 \\ 0 \\ -1 \end{pmatrix} + \lambda \begin{pmatrix} 2 \\ 3 \\ -3 \end{pmatrix}$

Suppose you have a line which is parallel to vector $\mathbf{b} = \begin{pmatrix} b_1 \\ b_2 \\ b_3 \end{pmatrix}$ and passes through the point with position vector $\mathbf{a} = \begin{pmatrix} a_1 \\ a_2 \\ a_3 \end{pmatrix}$

Let the general point on this line be $\mathbf{r} = \begin{pmatrix} x \\ y \\ z \end{pmatrix}$. Then the equation of the line is $\begin{pmatrix} x \\ y \\ z \end{pmatrix} = \begin{pmatrix} a_1 \\ a_2 \\ a_3 \end{pmatrix} + \lambda \begin{pmatrix} b_1 \\ b_2 \\ b_3 \end{pmatrix}$

Therefore the line can be written using **parametric equations**: $x = a_1 + \lambda b_1, y = a_2 + \lambda b_2, z = a_3 + \lambda b_3$

These can be rearranged to give $\dfrac{x - a_1}{b_1} = \dfrac{y - a_2}{b_2} = \dfrac{z - a_3}{b_3} (= \lambda)$, which are the **Cartesian equations of the line**.

See Maths
Ch12.3
For a reminder of parametric equations.

Key point

The **Cartesian equations of a line** parallel to the vector $\mathbf{b} = \begin{pmatrix} b_1 \\ b_2 \\ b_3 \end{pmatrix}$ and passing through the point with position vector $\mathbf{a} = \begin{pmatrix} a_1 \\ a_2 \\ a_3 \end{pmatrix}$ are $\dfrac{x - a_1}{b_1} = \dfrac{y - a_2}{b_2} = \dfrac{z - a_3}{b_3}$

Example 2

Give the Cartesian equations of the line with vector equation $\mathbf{r} = \begin{pmatrix} 0 \\ 1 \\ -4 \end{pmatrix} + \lambda \begin{pmatrix} 5 \\ -2 \\ 3 \end{pmatrix}$

You have $\mathbf{a} = \begin{pmatrix} 0 \\ 1 \\ -4 \end{pmatrix}$ and $\mathbf{b} = \begin{pmatrix} 5 \\ -2 \\ 3 \end{pmatrix}$

The parametric equations are $x = 0 + 5\lambda$

$y = 1 - 2\lambda$

$z = -4 + 3\lambda$

So the Cartesian equations are $\dfrac{x - 0}{5} = \dfrac{y - 1}{-2} = \dfrac{z + 4}{3}$

Using $\dfrac{x - a_1}{b_1} = \dfrac{y - a_2}{b_2} = \dfrac{z - a_3}{b_3}$

To determine whether or not a point lies on a line, you can make an equation by using $\mathbf{r} = \mathbf{a} + \lambda \mathbf{b}$ and then substitute the position vector of the given point for $\mathbf{r}$. Then, by considering the components of $\mathbf{i}$, $\mathbf{j}$ and $\mathbf{k}$ separately, you can determine whether or not there is a unique value of λ.

Example 3

Establish whether or not the point $(-5, 5, 6)$ lies on the line $\mathbf{r} = \begin{pmatrix} 3 \\ 1 \\ 0 \end{pmatrix} + \lambda \begin{pmatrix} 4 \\ -2 \\ -3 \end{pmatrix}$

$\begin{pmatrix} -5 \\ 5 \\ 6 \end{pmatrix} = \begin{pmatrix} 3 \\ 1 \\ 0 \end{pmatrix} + \lambda \begin{pmatrix} 4 \\ -2 \\ -3 \end{pmatrix}$

> Form an equation and substitute in the position vector of the point in place of **r**

$\mathbf{i}$ components: $-5 = 3 + 4\lambda \Rightarrow \lambda = -2$

$\mathbf{j}$ components: $5 = 1 - 2\lambda \Rightarrow \lambda = -2$

$\mathbf{k}$ components: $6 = -3\lambda \Rightarrow \lambda = -2$

> Find the value of λ arising from each of the components.

Since you get the same value of λ for each component, $(-5, 5, 6)$ does lie on the line.

> If λ was not the same for all three equations, then you would conclude that the point is not on the line.

Exercise 5.1A Fluency and skills

1 Write a vector equation of the line that is parallel to the vector $\begin{pmatrix} 1 \\ -2 \\ 0 \end{pmatrix}$ and passes through the point with position vector $\begin{pmatrix} 0 \\ 4 \\ 3 \end{pmatrix}$

2 Write a vector equation of the line that is parallel to the vector $5\mathbf{i} - 3\mathbf{j} + \mathbf{k}$ and passes through the point with position vector $7\mathbf{i} - \mathbf{j} + 2\mathbf{k}$

3 Write a vector equation of the line that passes through the points with position vectors $\begin{pmatrix} -7 \\ 2 \\ -5 \end{pmatrix}$ and $\begin{pmatrix} 4 \\ -9 \\ -2 \end{pmatrix}$

4 Write a vector equation of the line that passes through the points with position vectors $2\mathbf{i} + 3\mathbf{k}$ and $2\mathbf{i} + \mathbf{j} - \mathbf{k}$

5 Write Cartesian equations of the line that is parallel to the vector $\begin{pmatrix} 0 \\ 4 \\ -1 \end{pmatrix}$ and passes through the point with position vector $\begin{pmatrix} 5 \\ 2 \\ -7 \end{pmatrix}$

6 Write the Cartesian equations of the line that is parallel to the vector $8\mathbf{i} - 2\mathbf{j} + 3\mathbf{k}$ and passes through the point with position vector $4\mathbf{i} - 2\mathbf{j}$

7 Write the Cartesian equations of the line that passes through the points with position vectors $\begin{pmatrix} 9 \\ 4 \\ -3 \end{pmatrix}$ and $\begin{pmatrix} 1 \\ 0 \\ 5 \end{pmatrix}$

8 Write the Cartesian equations of the line that passes through the points with position vectors $5\mathbf{j} + \mathbf{k}$ and $4\mathbf{i} - \mathbf{j} - \mathbf{k}$

9 Convert each of these vector equations to Cartesian form.

a $\mathbf{r} = \begin{pmatrix} 0 \\ 1 \\ -2 \end{pmatrix} + \lambda \begin{pmatrix} -4 \\ 0 \\ 8 \end{pmatrix}$ **b** $\mathbf{r} = \begin{pmatrix} -3 \\ 0 \\ 7 \end{pmatrix} + \lambda \begin{pmatrix} 5 \\ -1 \\ 3 \end{pmatrix}$

10 Convert each of these Cartesian equations into the form $\mathbf{r} = \mathbf{a} + \lambda \mathbf{b}$

a $\dfrac{x-5}{2} = \dfrac{y-1}{-3} = \dfrac{z+3}{1}$ **b** $\dfrac{x+2}{4} = \dfrac{y}{-2} = \dfrac{5-z}{3}$

11 Establish whether the point $(3, 2, 1)$ lies on each of these lines.

a $\mathbf{r} = \begin{pmatrix} 1 \\ -4 \\ 3 \end{pmatrix} + \lambda \begin{pmatrix} 1 \\ 3 \\ -1 \end{pmatrix}$ **b** $\mathbf{r} = \begin{pmatrix} 5 \\ 2 \\ 0 \end{pmatrix} + \lambda \begin{pmatrix} 2 \\ 0 \\ 1 \end{pmatrix}$

12 Establish whether the point $(-2, 0, 4)$ lies on each of these lines.

a $\dfrac{x-1}{3} = \dfrac{y+3}{-3} = \dfrac{z-0}{4}$ **b** $\dfrac{x+8}{3} = \dfrac{y+2}{1} = \dfrac{z-6}{-1}$

Reasoning and problem-solving

A vector equation of a line is not unique. There are lots of points you can use to set the vector from the origin and you can use any vector which is parallel to the line to give its direction, so there are many correct vector equations for any one line. You need to be able to decide whether or not two equations represent the same line.

Strategy 1

To check whether two equations represent the same line

1. Check whether their direction vectors are equivalent.

2. Check whether they have a point in common.

3. If they have the same direction and a point in common then they must represent the same line.

Example 4

Decide which of these equations represent the same line.

$$L_1 : \mathbf{r} = \begin{pmatrix} 9 \\ 1 \\ -2 \end{pmatrix} + \lambda \begin{pmatrix} -2 \\ -1 \\ 1 \end{pmatrix} \qquad L_2 : \mathbf{r} = \begin{pmatrix} 3 \\ 0 \\ 4 \end{pmatrix} + \lambda \begin{pmatrix} 4 \\ 2 \\ -2 \end{pmatrix} \qquad L_3 : \frac{x-1}{2} = \frac{y+3}{1} = \frac{z-2}{-1}$$

All three lines have the same direction since $\begin{pmatrix} -2 \\ -1 \\ 1 \end{pmatrix} = -\begin{pmatrix} 2 \\ 1 \\ -1 \end{pmatrix}$ and $\begin{pmatrix} 4 \\ 2 \\ -2 \end{pmatrix} = 2\begin{pmatrix} 2 \\ 1 \\ -1 \end{pmatrix}$

(1) If a vector is a multiple of another vector then they are parallel.

Take the point $(1, -3, 2)$ which you know lies on line L_3

Check L_1 : $\begin{pmatrix} 1 \\ -3 \\ 2 \end{pmatrix} = \begin{pmatrix} 9 \\ 1 \\ -2 \end{pmatrix} + \lambda \begin{pmatrix} -2 \\ -1 \\ 1 \end{pmatrix}$

(2) You could alternatively pick a point you know lies on line L_1 or on line L_2

$9 - 2\lambda = 1 \Rightarrow \lambda = 4$

$1 - \lambda = -3 \Rightarrow \lambda = 4$

$-2 + \lambda = 2 \Rightarrow \lambda = 4$

Therefore $(1, -3, 2)$ lies on line L_1

(2) Since the same value of λ is found for each of the components.

So L_1 and L_3 are equations of the same line.

Check L_2 : $\begin{pmatrix} 1 \\ -3 \\ 2 \end{pmatrix} = \begin{pmatrix} 3 \\ 0 \\ 4 \end{pmatrix} + \lambda \begin{pmatrix} 4 \\ 2 \\ -2 \end{pmatrix}$

(3) Since they have the same direction and have a point in common.

$3 + 4\lambda = 1 \Rightarrow \lambda = -\frac{1}{2}$

$2\lambda = -3 \Rightarrow \lambda = -\frac{3}{2}$

$4 - 2\lambda = 2 \Rightarrow \lambda = 1$

So $(1, -3, 2)$ does not lie on L_2

Therefore L_2 does not represent the same line as L_1 and L_3

(3) L_2 is parallel to the other lines but does not have a point in common.

If you have two lines in 2D then they must either intersect or be parallel. However, with two lines in 3D they could also be **skew**, that is, they do not lie on the same plane so do not intersect but are also not parallel.

For example, in the cuboid shown:

- AB and DC are parallel
- AB and AD intersect
- AB and GH are skew (they will never intersect however far they are extended, but they are not parallel)

You already know that lines are parallel if their direction vectors are multiples of each other. To decide whether or not two lines intersect, you must attempt to solve their equations simultaneously and see if they are consistent.

Strategy 2

To establish if lines are parallel, intersecting or skew

(1) Check whether their direction vectors are multiples of each other, in which case they are parallel.

(2) Attempt to solve their equations simultaneously.

(3) If the solution is consistent then they intersect so you can find the point of intersection.

(4) If the solutions leads to an inconsistency then the lines are skew.

Example 5

For each pair of lines, decide whether they are parallel, intersecting or skew.

If they intersect, find their point of intersection.

a $L_1 : \mathbf{r} = 2\mathbf{i} - \mathbf{j} + \lambda(3\mathbf{i} + \mathbf{j} - 2\mathbf{k})$, $L_2 : 3\mathbf{i} + 2\mathbf{j} - \mathbf{k} + \mu(2\mathbf{i} - \mathbf{k})$

b $L_1 : \mathbf{r} = \begin{pmatrix} 1 \\ 0 \\ -8 \end{pmatrix} + \lambda \begin{pmatrix} 2 \\ -2 \\ -3 \end{pmatrix}$, $L_2 : \mathbf{r} = \begin{pmatrix} -4 \\ 7 \\ -1 \end{pmatrix} + \mu \begin{pmatrix} 3 \\ -1 \\ -5 \end{pmatrix}$

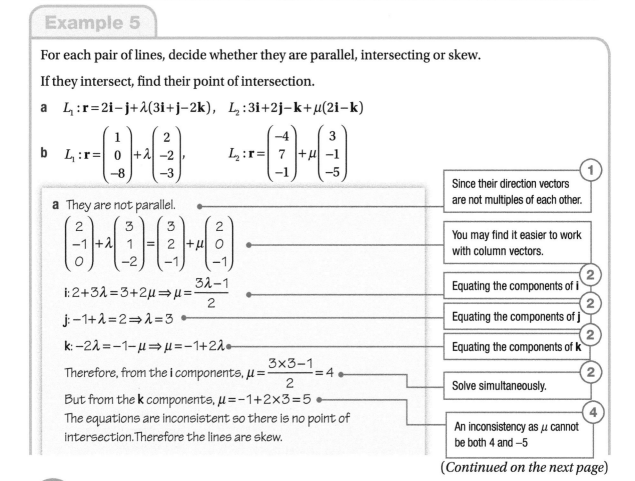

a They are not parallel. ●————— (1) Since their direction vectors are not multiples of each other.

$$\begin{pmatrix} 2 \\ -1 \\ 0 \end{pmatrix} + \lambda \begin{pmatrix} 3 \\ 1 \\ -2 \end{pmatrix} = \begin{pmatrix} 3 \\ 2 \\ -1 \end{pmatrix} + \mu \begin{pmatrix} 2 \\ 0 \\ -1 \end{pmatrix}$$ ●————— You may find it easier to work with column vectors.

$\mathbf{i}: 2 + 3\lambda = 3 + 2\mu \Rightarrow \mu = \dfrac{3\lambda - 1}{2}$ ●————— (2) Equating the components of $\mathbf{i}$

$\mathbf{j}: -1 + \lambda = 2 \Rightarrow \lambda = 3$ ●————— (2) Equating the components of $\mathbf{j}$

$\mathbf{k}: -2\lambda = -1 - \mu \Rightarrow \mu = -1 + 2\lambda$ ●————— (2) Equating the components of $\mathbf{k}$

Therefore, from the $\mathbf{i}$ components, $\mu = \dfrac{3 \times 3 - 1}{2} = 4$ ●————— (2) Solve simultaneously.

But from the $\mathbf{k}$ components, $\mu = -1 + 2 \times 3 = 5$ ●

The equations are inconsistent so there is no point of intersection. Therefore the lines are skew. ————— (4) An inconsistency as μ cannot be both 4 and −5

(Continued on the next page)

Vectors The vector equation of a line

b They are not parallel.

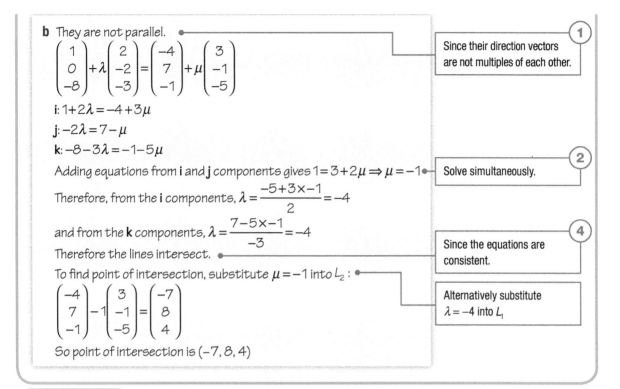

$$\begin{pmatrix} 1 \\ 0 \\ -8 \end{pmatrix} + \lambda \begin{pmatrix} 2 \\ -2 \\ -3 \end{pmatrix} = \begin{pmatrix} -4 \\ 7 \\ -1 \end{pmatrix} + \mu \begin{pmatrix} 3 \\ -1 \\ -5 \end{pmatrix}$$

Since their direction vectors are not multiples of each other.

i: $1 + 2\lambda = -4 + 3\mu$

j: $-2\lambda = 7 - \mu$

k: $-8 - 3\lambda = -1 - 5\mu$

Adding equations from **i** and **j** components gives $1 = 3 + 2\mu \Rightarrow \mu = -1$

Solve simultaneously.

Therefore, from the **i** components, $\lambda = \dfrac{-5 + 3 \times -1}{2} = -4$

and from the **k** components, $\lambda = \dfrac{7 - 5 \times -1}{-3} = -4$

Therefore the lines intersect.

Since the equations are consistent.

To find point of intersection, substitute $\mu = -1$ into L_2:

Alternatively substitute $\lambda = -4$ into L_1

$$\begin{pmatrix} -4 \\ 7 \\ -1 \end{pmatrix} - 1 \begin{pmatrix} 3 \\ -1 \\ -5 \end{pmatrix} = \begin{pmatrix} -7 \\ 8 \\ 4 \end{pmatrix}$$

So point of intersection is $(-7, 8, 4)$

Strategy 3

See Ch 4.2
For a reminder of using matrices for linear transformations.

To apply a matrix transform to line:

① Form a vector for a general point on the line.

② Pre-multiply this vector by the transformation matrix.

③ Write the equation of the image.

Example 6

A line can be transformed by a linear transformation to another line or to a point.

The line with equation $\mathbf{r} = \mathbf{i} - \mathbf{j} + \lambda(\mathbf{i} + 3\mathbf{j})$ is transformed by the matrix $\mathbf{T} = \begin{pmatrix} 1 & -2 \\ 2 & 0 \end{pmatrix}$.
Work out the equation of the image of the line.

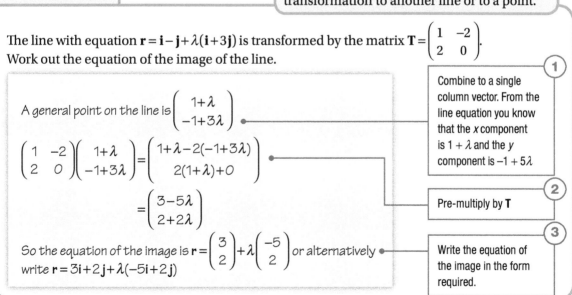

A general point on the line is $\begin{pmatrix} 1 + \lambda \\ -1 + 3\lambda \end{pmatrix}$

Combine to a single column vector. From the line equation you know that the x component is $1 + \lambda$ and the y component is $-1 + 5\lambda$

$$\begin{pmatrix} 1 & -2 \\ 2 & 0 \end{pmatrix} \begin{pmatrix} 1 + \lambda \\ -1 + 3\lambda \end{pmatrix} = \begin{pmatrix} 1 + \lambda - 2(-1 + 3\lambda) \\ 2(1 + \lambda) + 0 \end{pmatrix}$$

$$= \begin{pmatrix} 3 - 5\lambda \\ 2 + 2\lambda \end{pmatrix}$$

Pre-multiply by **T**

So the equation of the image is $\mathbf{r} = \begin{pmatrix} 3 \\ 2 \end{pmatrix} + \lambda \begin{pmatrix} -5 \\ 2 \end{pmatrix}$ or alternatively write $\mathbf{r} = 3\mathbf{i} + 2\mathbf{j} + \lambda(-5\mathbf{i} + 2\mathbf{j})$

Write the equation of the image in the form required.

1 For each pair of equations, decide whether or not they represent the same line. Show your working clearly.

a $\mathbf{r}=\begin{pmatrix}1\\5\end{pmatrix}+\lambda\begin{pmatrix}4\\-3\end{pmatrix}$ and $\mathbf{r}=\begin{pmatrix}2\\3\end{pmatrix}+\mu\begin{pmatrix}-2\\6\end{pmatrix}$

b $\mathbf{r}=\begin{pmatrix}4\\-1\end{pmatrix}+\lambda\begin{pmatrix}1\\-1\end{pmatrix}$ and $\mathbf{r}=\begin{pmatrix}1\\2\end{pmatrix}+\mu\begin{pmatrix}-1\\1\end{pmatrix}$

c $\mathbf{r}=2\mathbf{i}+7\mathbf{j}+\lambda(\mathbf{i}-\mathbf{j})$ and $y=9-x$

2 For each pair of equations, decide whether or not they represent the same line. Show your working clearly.

a $\mathbf{r}=\begin{pmatrix}-3\\3\\3\end{pmatrix}+\lambda\begin{pmatrix}2\\0\\-1\end{pmatrix}$ and $\mathbf{r}=\begin{pmatrix}1\\3\\-1\end{pmatrix}+\mu\begin{pmatrix}-4\\0\\2\end{pmatrix}$

b $\mathbf{r}=\begin{pmatrix}2\\3\\0\end{pmatrix}+\lambda\begin{pmatrix}1\\1\\-1\end{pmatrix}$ and $\mathbf{r}=\begin{pmatrix}2\\5\\-2\end{pmatrix}+\mu\begin{pmatrix}1\\-1\\1\end{pmatrix}$

c $\mathbf{r}=2\mathbf{j}-\mathbf{k}+\lambda(\mathbf{i}+2\mathbf{j}-3\mathbf{k})$ and $\mathbf{r}=3\mathbf{i}+8\mathbf{j}-10\mathbf{k}+\mu(2\mathbf{i}+4\mathbf{j}-6\mathbf{k})$

d $\dfrac{x-2}{1}=\dfrac{y-3}{3}=\dfrac{z+1}{-2}$ and $\dfrac{x+3}{-1}=\dfrac{y+12}{-3}=\dfrac{z-9}{2}$

e $\dfrac{x+5}{-1}=\dfrac{y-0}{2}=\dfrac{z-1}{2}$ and $\mathbf{r}=\begin{pmatrix}1\\-12\\11\end{pmatrix}+\lambda\begin{pmatrix}2\\-4\\-4\end{pmatrix}$

3 Find the point of intersection between these pairs of lines.

a $\mathbf{r}=\begin{pmatrix}-1\\1\end{pmatrix}+\lambda\begin{pmatrix}-2\\3\end{pmatrix}$ and $\mathbf{r}=\begin{pmatrix}1\\7\end{pmatrix}+\mu\begin{pmatrix}4\\-3\end{pmatrix}$

b $\mathbf{r}=-4\mathbf{i}+14\mathbf{j}+\lambda(-\mathbf{i}+2\mathbf{j})$ and $\mathbf{r}=\begin{pmatrix}4\\1\end{pmatrix}+\mu\begin{pmatrix}-2\\3\end{pmatrix}$

c $\mathbf{r}=\begin{pmatrix}0\\2\end{pmatrix}+\lambda\begin{pmatrix}-1\\3\end{pmatrix}$ and $y=2x-3$

4 For each pair of lines, decide whether they are parallel, skew or intersecting. If they are intersecting, find their point of intersection.

a $\mathbf{r}=\begin{pmatrix}7\\-6\\-2\end{pmatrix}+\lambda\begin{pmatrix}3\\0\\-1\end{pmatrix}$ and $\mathbf{r}=\begin{pmatrix}3\\4\\0\end{pmatrix}+\mu\begin{pmatrix}-2\\5\\1\end{pmatrix}$

b $\mathbf{r}=\mathbf{i}+2\mathbf{k}+\lambda(3\mathbf{i}+4\mathbf{j}-2\mathbf{k})$ and $\mathbf{r}=3\mathbf{i}+\mathbf{j}-\mathbf{k}+\mu(-6\mathbf{i}-8\mathbf{j}+4\mathbf{k})$

c $\dfrac{x+5}{-1}=\dfrac{y}{2}=\dfrac{z-1}{2}$ and $\dfrac{x-2}{1}=\dfrac{y+9}{3}=\dfrac{z-5}{-2}$

d $\dfrac{-4-x}{3}=\dfrac{y-2}{-1}=\dfrac{z}{5}$ and $\dfrac{x-6}{2}=\dfrac{y+8}{-2}=\dfrac{5-z}{-1}$

5 Two lines, L_1 and L_2, have equations

$\dfrac{x+10}{5}=\dfrac{y-5}{-1}=\dfrac{z-5}{2}$ and $\mathbf{r}=\begin{pmatrix}3\\3\\5\end{pmatrix}+\lambda\begin{pmatrix}-3\\0\\4\end{pmatrix}$ respectively.

Given that L_1 and L_2 intersect at the point A, calculate the length of $\vec{OA}$

6 Find the image of the line $\mathbf{r}=\begin{pmatrix}1\\3\end{pmatrix}+\lambda\begin{pmatrix}-2\\1\end{pmatrix}$ under the transformation given by the matrix $\begin{pmatrix}2&1\\0&-2\end{pmatrix}$

7 Find the image of the line $\mathbf{r}=\begin{pmatrix}2\\-1\end{pmatrix}+\lambda\begin{pmatrix}2\\1\end{pmatrix}$ under a rotation of 90° anticlockwise about the origin.

8 Find the image of the line $\mathbf{r}=\begin{pmatrix}3\\0\\-1\end{pmatrix}+\lambda\begin{pmatrix}1\\-1\\-2\end{pmatrix}$ under each of these transformations.

a Rotation 90° anticlockwise around the x-axis.

b Rotation 180° around the y-axis.

c Reflection in the plane $z=0$

9 Find the image of the line $\dfrac{x-1}{2}=\dfrac{y+3}{-1}=\dfrac{5-z}{-4}$ after a reflection in the plane $y=0$

See Ch 4.2
For a reminder of writing transformations as matrices.

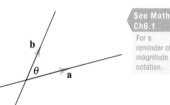

Fluency and skills

A very important concept in the application of vectors is the **scalar product**. As its name suggests, the result of finding the scalar product of two vectors will be a scalar quantity.

Key point

The scalar product $\mathbf{a} \cdot \mathbf{b}$ of vectors $\mathbf{a}$ and $\mathbf{b}$ is defined as $\mathbf{a} \cdot \mathbf{b} = |\mathbf{a}||\mathbf{b}|\cos\theta$ where θ is the angle between the vectors $\mathbf{a}$ and $\mathbf{b}$. The vectors $\mathbf{a}$ and $\mathbf{b}$ are both directed away from (or towards) the angle and $0 \le \theta \le 180°$

See Maths Ch6.1

For a reminder of magnitude notation.

The scalar product is sometimes called the **dot product**. The dot product is a distributive operation.

Example 1

Find the value of

a $\mathbf{i} \cdot \mathbf{i}$ **b** $\mathbf{k} \cdot \mathbf{j}$ **c** $\mathbf{i} \cdot (\mathbf{j} + \mathbf{k})$

$\mathbf{i}$ is the unit vector in the positive x-direction so has magnitude 1

a $\mathbf{i} \cdot \mathbf{i} = 1 \times 1 \times \cos 0°$
  $= 1$

b $\mathbf{k} \cdot \mathbf{j} = 1 \times 1 \times \cos 90°$
  $= 0$

$\mathbf{k}$ and $\mathbf{j}$ are the unit vectors in the positive z-and y-directions respectively so the angle between them is 90°

c $\mathbf{i} \cdot (\mathbf{i} + \mathbf{k}) = \mathbf{i} \cdot \mathbf{i} + \mathbf{i} \cdot \mathbf{k}$
  $= 1 + 1 \times \cos 90°$
  $= 1$

Since the scalar product is distributive.

If vectors $\mathbf{a}$ and $\mathbf{b}$ are perpendicular then $\mathbf{a} \cdot \mathbf{b} = 0$. This is because $\cos 90° = 0$.

Key point

If you have two vectors $\mathbf{a} = a_1\mathbf{i} + a_2\mathbf{j} + a_3\mathbf{k}$ and $\mathbf{b} = b_1\mathbf{i} + b_2\mathbf{j} + b_3\mathbf{k}$, then the scalar product is defined as $\mathbf{a} \cdot \mathbf{b} = a_1b_1 + a_2b_2 + a_3b_3$

Example 2

Given $\mathbf{a} = \begin{pmatrix} 1 \\ -3 \\ 2 \end{pmatrix}$ and $\mathbf{b} = \begin{pmatrix} -4 \\ -1 \\ 5 \end{pmatrix}$, calculate

a The scalar product $\mathbf{a} \cdot \mathbf{b}$,   **b** The angle between vectors $\mathbf{a}$ and $\mathbf{b}$

a $\mathbf{a} \cdot \mathbf{b} = \begin{pmatrix} 1 \\ -3 \\ 2 \end{pmatrix} \cdot \begin{pmatrix} -4 \\ -1 \\ 5 \end{pmatrix}$
  $= (1 \times -4) + (-3 \times -1) + (2 \times 5)$
  $= 9$

Using $\mathbf{a} \cdot \mathbf{b} = a_1b_1 + a_2b_2 + a_3b_3$

(*Continued on the next page*)

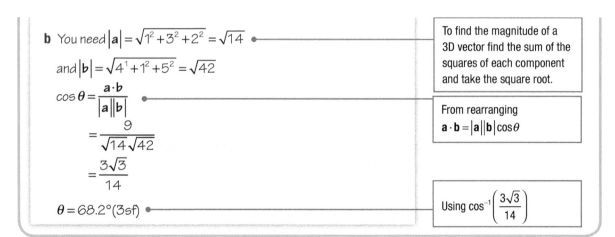

b You need $|\mathbf{a}| = \sqrt{1^2 + 3^2 + 2^2} = \sqrt{14}$

and $|\mathbf{b}| = \sqrt{4^1 + 1^2 + 5^2} = \sqrt{42}$

$\cos\theta = \dfrac{\mathbf{a}\cdot\mathbf{b}}{|\mathbf{a}||\mathbf{b}|}$

$= \dfrac{9}{\sqrt{14}\sqrt{42}}$

$= \dfrac{3\sqrt{3}}{14}$

$\theta = 68.2°(3\text{sf})$

To find the magnitude of a 3D vector find the sum of the squares of each component and take the square root.

From rearranging $\mathbf{a}\cdot\mathbf{b} = |\mathbf{a}||\mathbf{b}|\cos\theta$

Using $\cos^{-1}\left(\dfrac{3\sqrt{3}}{14}\right)$

Calculator

Try it on your calculator

You can use a calculator to find the scalar product

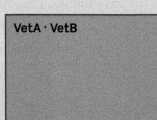

VetA · VetB

-6

Activity

Find out how to work out

$\begin{pmatrix} 2 \\ 3 \\ -1 \end{pmatrix} \cdot \begin{pmatrix} -1 \\ 0 \\ 4 \end{pmatrix}$ on *your* calculator.

Exercise 5.2A Fluency and skills

1 Find the value of

 a $\mathbf{j}\cdot\mathbf{j}$ **b** $\mathbf{i}\cdot\mathbf{k}$ **c** $2\mathbf{i}\cdot\mathbf{j}$ **d** $\mathbf{j}\cdot(\mathbf{i}+\mathbf{j})$
 e $\mathbf{i}\cdot(2\mathbf{i})$ **f** $(\mathbf{i}+\mathbf{j})\cdot(\mathbf{i}-\mathbf{j})$
 g $(\mathbf{j}+\mathbf{k})\cdot(\mathbf{i}+\mathbf{j})$ **h** $(\mathbf{i}+\mathbf{j}+\mathbf{k})\cdot(\mathbf{i}+\mathbf{j}+\mathbf{k})$

2 Calculate these scalar products.

 a $\begin{pmatrix} 2 \\ 4 \\ -1 \end{pmatrix} \cdot \begin{pmatrix} 5 \\ -3 \\ -6 \end{pmatrix}$ **b** $(2\mathbf{i}-\mathbf{j}+3\mathbf{k})\cdot(\mathbf{i}+4\mathbf{j}-\mathbf{k})$

 c $\begin{pmatrix} \sqrt{2} \\ 1 \\ 3 \end{pmatrix} \cdot \begin{pmatrix} 2 \\ 0 \\ -\sqrt{2} \end{pmatrix}$ **d** $(5\mathbf{i}+\mathbf{k})\cdot(2\mathbf{j}-3\mathbf{k})$

3 Find an expression in terms of a for the dot product of $(2a\mathbf{i}-\mathbf{j}+a\mathbf{k})$ and $(3a\mathbf{i}-a\mathbf{j}-5\mathbf{k})$

4 Find an expression in terms of k of the dot product of $\begin{pmatrix} 2 \\ k \\ -3 \end{pmatrix}$ and $\begin{pmatrix} k \\ -1 \\ 2 \end{pmatrix}$

5 Calculate the angle between the vectors

 a $\begin{pmatrix} 5 \\ 1 \\ 3 \end{pmatrix}$ and $\begin{pmatrix} 2 \\ -2 \\ 4 \end{pmatrix}$

 b $(5\mathbf{i}+\mathbf{j}-2\mathbf{k})$ and $(4\mathbf{i}+3\mathbf{j}-\mathbf{k})$

 c $\begin{pmatrix} 2\sqrt{3} \\ 4 \\ \sqrt{3} \end{pmatrix}$ and $\begin{pmatrix} -1 \\ \sqrt{3} \\ 2 \end{pmatrix}$ **d** $(3\mathbf{j}+2\mathbf{k})$ and $(4\mathbf{i}-\mathbf{j})$

6 Calculate the cosine of the acute angle between each pair of vectors.

 Give your answers in terms of a where $a>0$

 a $\begin{pmatrix} a \\ 1 \\ 0 \end{pmatrix}$ and $\begin{pmatrix} 1 \\ 0 \\ a \end{pmatrix}$

 b $3a\mathbf{i}-4a\mathbf{j}+5\mathbf{k}$ and $5a\mathbf{j}+5\mathbf{k}$

7 Show that the vectors $\begin{pmatrix} 6 \\ -4 \end{pmatrix}$ and $\begin{pmatrix} -8 \\ -12 \end{pmatrix}$ are perpendicular.

8 Show that the vectors $\begin{pmatrix} 4 \\ -1 \\ 5 \end{pmatrix}$ and $\begin{pmatrix} -2 \\ -3 \\ 1 \end{pmatrix}$ are perpendicular.

9 Show that the vectors $3\mathbf{i}-5\mathbf{j}-2\mathbf{k}$ and $2\mathbf{i}+4\mathbf{j}-7\mathbf{k}$ are perpendicular.

10 Show that the vectors $2\mathbf{i}+\mathbf{k}$ and $4\mathbf{i}-\mathbf{j}-3\mathbf{k}$ are not perpendicular.

11 Find the value of a for which the vectors $\begin{pmatrix} 3a \\ 2 \\ -2 \end{pmatrix}$ and $\begin{pmatrix} 2 \\ a \\ 4 \end{pmatrix}$ are perpendicular.

12 Find the value of b for which the vectors $(2b\mathbf{i}+\mathbf{j}-\mathbf{k})$ and $(3\mathbf{i}-b\mathbf{j}+10\mathbf{k})$ are perpendicular.

13 Find the values of c for which the vectors $\begin{pmatrix} c \\ 3 \\ -1 \end{pmatrix}$ and $\begin{pmatrix} -c \\ c \\ 2 \end{pmatrix}$ are perpendicular.

Reasoning and problem-solving

Strategy 1

To find the angle between two intersecting lines

① Identify the direction vector of each line, $\mathbf{a}$ and $\mathbf{b}$

② Use $\cos\theta = \dfrac{\mathbf{a}\cdot\mathbf{b}}{\|\mathbf{a}\|\|\mathbf{b}\|}$ to find the acute angle between the lines.

③ Subtract the acute angle from 180° to find the obtuse angle between the lines.

Example 3

Given that L_1 and L_2 have equations $L_1 : \mathbf{r} = \begin{pmatrix} 1 \\ -1 \\ 2 \end{pmatrix} + \lambda \begin{pmatrix} 1 \\ -2 \\ 1 \end{pmatrix}$ and $L_2 : \dfrac{x+4}{2} = \dfrac{y+3}{-1} = \dfrac{z-1}{1}$ and that

they intersect at the point A, find the obtuse angle between L_1 and L_2

The direction vectors are $\begin{pmatrix} 1 \\ -2 \\ 1 \end{pmatrix}$ and $\begin{pmatrix} 2 \\ -1 \\ 1 \end{pmatrix}$ ⟶ Identify direction vectors. ①

$\begin{pmatrix} 1 \\ -2 \\ 1 \end{pmatrix} \cdot \begin{pmatrix} 2 \\ -1 \\ 1 \end{pmatrix} = 2+2+1 = 5$

$\left| \begin{pmatrix} 1 \\ -2 \\ 1 \end{pmatrix} \right| = \sqrt{1^2+2^2+1^2} = \sqrt{6}$

$\left| \begin{pmatrix} 2 \\ -1 \\ 1 \end{pmatrix} \right| = \sqrt{2^2+1^2+1^2} = \sqrt{6}$ ⟶ Find the magnitude of the two direction vectors. ②

$\cos\theta = \dfrac{5}{\sqrt{6}\sqrt{6}}$ ⟶ First find the acute angle. ②

$\theta = 33.6°$

So obtuse angle is $180 - 33.6 = 146.4°$ (1dp) ⟶ Then, subtract from 180° to find the obtuse angle. ③

To prove properties of the scalar product

(1) Use general vectors, for example $\mathbf{a} = a_1\mathbf{i} + a_2\mathbf{j} + a_3\mathbf{k}$

(2) Use the fact that if $\mathbf{a}$, $\mathbf{b}$ are perpendicular vectors then $\mathbf{a} \cdot \mathbf{b} = 0$

(3) Use the fact that $\mathbf{i} \cdot \mathbf{i} = \mathbf{j} \cdot \mathbf{j} = \mathbf{k} \cdot \mathbf{k} = 1$

Example 4

Show that $\mathbf{a} \cdot \mathbf{b} = \mathbf{b} \cdot \mathbf{a}$ for any 3D vectors $\mathbf{a}$ and $\mathbf{b}$

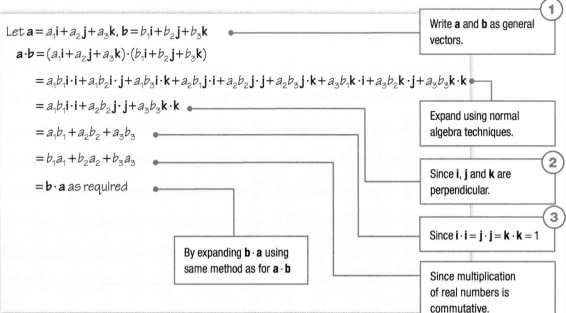

Let $\mathbf{a} = a_1\mathbf{i} + a_2\mathbf{j} + a_3\mathbf{k}$, $\mathbf{b} = b_1\mathbf{i} + b_2\mathbf{j} + b_3\mathbf{k}$ ← **(1)** Write $\mathbf{a}$ and $\mathbf{b}$ as general vectors.

$\mathbf{a} \cdot \mathbf{b} = (a_1\mathbf{i} + a_2\mathbf{j} + a_3\mathbf{k}) \cdot (b_1\mathbf{i} + b_2\mathbf{j} + b_3\mathbf{k})$

$= a_1 b_1 \mathbf{i} \cdot \mathbf{i} + a_1 b_2 \mathbf{i} \cdot \mathbf{j} + a_1 b_3 \mathbf{i} \cdot \mathbf{k} + a_2 b_1 \mathbf{j} \cdot \mathbf{i} + a_2 b_2 \mathbf{j} \cdot \mathbf{j} + a_2 b_3 \mathbf{j} \cdot \mathbf{k} + a_3 b_1 \mathbf{k} \cdot \mathbf{i} + a_3 b_2 \mathbf{k} \cdot \mathbf{j} + a_3 b_3 \mathbf{k} \cdot \mathbf{k}$ ← Expand using normal algebra techniques.

$= a_1 b_1 \mathbf{i} \cdot \mathbf{i} + a_2 b_2 \mathbf{j} \cdot \mathbf{j} + a_3 b_3 \mathbf{k} \cdot \mathbf{k}$ ← **(2)** Since $\mathbf{i}$, $\mathbf{j}$ and $\mathbf{k}$ are perpendicular.

$= a_1 b_1 + a_2 b_2 + a_3 b_3$ ← **(3)** Since $\mathbf{i} \cdot \mathbf{i} = \mathbf{j} \cdot \mathbf{j} = \mathbf{k} \cdot \mathbf{k} = 1$

$= b_1 a_1 + b_2 a_2 + b_3 a_3$ ← Since multiplication of real numbers is commutative.

$= \mathbf{b} \cdot \mathbf{a}$ as required ← By expanding $\mathbf{b} \cdot \mathbf{a}$ using same method as for $\mathbf{a} \cdot \mathbf{b}$

Exercise 5.2B Reasoning and problem-solving

1 Show that the lines $= \begin{pmatrix} 3 \\ -2 \end{pmatrix} + \lambda \begin{pmatrix} -1 \\ 3 \end{pmatrix}$ and
$\mathbf{r} = 2\mathbf{i} - \mathbf{j} + \mu(6\mathbf{i} + 2\mathbf{j})$ are perpendicular.

2 Show that the lines $\mathbf{r} = \begin{pmatrix} 5 \\ -2 \\ 4 \end{pmatrix} + \lambda \begin{pmatrix} 6 \\ 7 \\ -5 \end{pmatrix}$ and
$\mathbf{r} = \begin{pmatrix} 1 \\ 0 \\ -3 \end{pmatrix} + \lambda \begin{pmatrix} 2 \\ -1 \\ 1 \end{pmatrix}$ are perpendicular.

3 Use the scalar product to show that the lines
$\mathbf{r} = \begin{pmatrix} 1 \\ 1 \\ 2 \end{pmatrix} + \lambda \begin{pmatrix} 3 \\ -2 \\ 6 \end{pmatrix}$ and $\dfrac{x-5}{2} = \dfrac{y+7}{6} = \dfrac{z}{1}$ are
perpendicular.

4 Show that the lines with equations
$\mathbf{r} = \begin{pmatrix} 6 \\ 6 \end{pmatrix} + \lambda \begin{pmatrix} 2 \\ 1 \end{pmatrix}$ and $\mathbf{r} = \begin{pmatrix} 4 \\ 0 \end{pmatrix} + \mu \begin{pmatrix} 3 \\ -1 \end{pmatrix}$
intersect and find the acute angle between them.

5 Show that the lines with equations
$\mathbf{r} = \begin{pmatrix} 2 \\ 1 \\ -3 \end{pmatrix} + s \begin{pmatrix} 0 \\ 2 \\ 1 \end{pmatrix}$ and $\mathbf{r} = \begin{pmatrix} 5 \\ 11 \\ 3 \end{pmatrix} + t \begin{pmatrix} 3 \\ -4 \\ -1 \end{pmatrix}$
intersect and find the acute angle between them to the nearest degree.

6 Given that they intersect, find the obtuse angle between the lines with equations
$$\mathbf{r} = \begin{pmatrix} 4 \\ 0 \\ 12 \end{pmatrix} + \lambda \begin{pmatrix} 0 \\ 2 \\ 5 \end{pmatrix} \text{ and } \frac{x-1}{3} = \frac{y+2}{-6} = \frac{z+10}{2}$$

7 Show that the lines with equations
$\mathbf{r} = (2\mathbf{i} - \mathbf{j} + 3\mathbf{k}) + \lambda(\mathbf{i} + \mathbf{j} - \mathbf{k})$ and
$\mathbf{r} = (2\mathbf{i} - 2\mathbf{j} + \mathbf{k}) + \mu(-\mathbf{i} + 3\mathbf{k})$ intersect and find the acute angle between them to 3 significant figures.

8 Find the values of a for which the lines
$$\mathbf{r} = \begin{pmatrix} 3 \\ -1 \\ 2 \end{pmatrix} + \lambda \begin{pmatrix} 2 \\ a \\ 0 \end{pmatrix} \text{ and } \mathbf{r} = \mu \begin{pmatrix} 3 \\ 0 \\ \sqrt{3} \end{pmatrix} \text{ meet at a } 60°$$
angle.

9 Find the values of a for which the lines
$$\mathbf{r} = \begin{pmatrix} 1 \\ 4 \\ 4 \end{pmatrix} + \lambda \begin{pmatrix} -2 \\ a\sqrt{2} \\ 2 \end{pmatrix} \text{ and } \frac{x-3}{1} = \frac{y+2}{0} = \frac{z-1}{-1}$$
meet at a 135° angle.

10 The line with equations $\frac{x}{4} = \frac{y+1}{8} = \frac{z-3}{1}$ and $\frac{x-1}{2} = \frac{y}{6} = \frac{z+2}{11}$ intersect at the point A

  a Find the coordinates of the point A

  b Calculate the exact cosine of the acute angle between the two lines at the point A

11 The acute angle between the vectors $\mathbf{a} = \mathbf{i} - k\mathbf{j}$ and $\mathbf{b} = \mathbf{i} + \mathbf{j}$ is 60°

  Calculate the possible values of k

12 The acute angle between the vectors $\begin{pmatrix} 1 \\ k \\ 2 \end{pmatrix}$ and $\begin{pmatrix} 0 \\ -1 \\ 1 \end{pmatrix}$ is 45°

  Calculate the possible values of k

13 You are given that $\mathbf{a} = 3\mathbf{i} - \mathbf{k}$ and $\mathbf{b} = \mathbf{i} - 2\mathbf{j} + 2\mathbf{k}$

  Calculate the exact area of triangle ABC

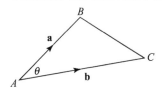

14 The points A, B and C have position vectors
$$\begin{pmatrix} 4 \\ 0 \\ 1 \end{pmatrix}, \begin{pmatrix} 5 \\ -2 \\ 0 \end{pmatrix} \text{ and } \begin{pmatrix} 3 \\ -1 \\ 1 \end{pmatrix} \text{ respectively.}$$

Calculate the exact area of triangle ABC

15 Given that $\mathbf{a} = a_1\mathbf{i} + a_2\mathbf{j} + a_3\mathbf{k}$
and $\mathbf{b} = b_1\mathbf{i} + b_2\mathbf{j} + b_3\mathbf{k}$, show that
$\mathbf{a} \cdot \mathbf{b} = a_1 b_1 + a_2 b_2 + a_3 b_3$

16 Show that the dot product is distributive, that is, $\mathbf{a} \cdot (\mathbf{b} + \mathbf{c}) = \mathbf{a} \cdot \mathbf{b} + \mathbf{a} \cdot \mathbf{c}$ for any 3D vectors $\mathbf{a}$, $\mathbf{b}$ and $\mathbf{c}$

17 a Prove that $\mathbf{a} \cdot \mathbf{a} = |\mathbf{a}|^2$ for any three dimensional vector $\mathbf{a}$

  b Hence prove the cosine rule:
$|\mathbf{a}|^2 = |\mathbf{b}|^2 + |\mathbf{c}|^2 - 2|\mathbf{b}||\mathbf{c}|\cos\theta$ for the triangle shown.

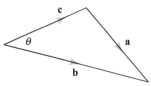

18 a Show that the line $\frac{x+2}{-4} = \frac{y-5}{5} = \frac{z-3}{3}$ cuts the x-axis and find the intercept's coordinates.

  b Calculate the acute angle between the x-axis and the line.

19 Three lines have equations as follows:
L_1 : $\mathbf{r} = 6\mathbf{i} - 3\mathbf{j} + \lambda(\mathbf{i} + \mathbf{k})$, L_2 : $\mathbf{r} = s(3\mathbf{i} - \mathbf{j} + \mathbf{k})$
and L_3 : $\mathbf{r} = t(\mathbf{j} + 2\mathbf{k})$

The lines L_1 and L_2 intersect at the point A, L_1 and L_2 intersect at the point B and L_2 and L_3 intersect at the point C as shown.

Calculate the exact area of triangle ABC

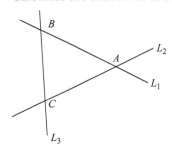

The vector equation of a plane

Fluency and skills

A plane is a 2D surface, extending infinitely far in both directions. You can write the equation of a plane in Cartesian, vector and scalar product form.

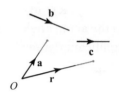

The vector equation of a plane

Suppose you wish to find the equation of the plane containing a point with position vector **a** and two non-parallel vectors **b** and **c** Any other vector on the plane can be described as a multiple of **b** plus a multiple of **c**. Therefore the position vector of any point on the plane is given by $\mathbf{r} = \mathbf{a} + s\mathbf{b} + t\mathbf{c}$ for some values of s and t

Key point

The vector equation of the plane containing a point with position vector **a** and the non-parallel vectors **b** and **c** is $\mathbf{r} = \mathbf{a} + s\mathbf{b} + t\mathbf{c}$

Example 1

Find the equation of the plane containing the points $(1, 4, -2)$, $(0, 3, 2)$ and $(-5, 0, 3)$

$\mathbf{a} = \mathbf{i} + 4\mathbf{j} - 2\mathbf{k}$

> You could use the position vector of any of the three points.

A vector on the plane is $(\mathbf{i} + 4\mathbf{j} - 2\mathbf{k}) - (3\mathbf{j} + 2\mathbf{k}) = \mathbf{i} + \mathbf{j} - 4\mathbf{k}$

Another vector on the plane is
$(\mathbf{i} + 4\mathbf{j} - 2\mathbf{k}) - (-5\mathbf{i} + 3\mathbf{k}) = 6\mathbf{i} + 4\mathbf{j} - 5\mathbf{k}$

So the vector equation of the plane is
$\mathbf{r} = \mathbf{i} + 4\mathbf{j} - 2\mathbf{k} + s(\mathbf{i} + \mathbf{j} - 4\mathbf{k}) + t(6\mathbf{i} + 4\mathbf{j} - 5\mathbf{k})$

Alternatively, you can write this in column vector form:

$$\mathbf{r} = \begin{pmatrix} 1 \\ 4 \\ -2 \end{pmatrix} + s \begin{pmatrix} 1 \\ 1 \\ -4 \end{pmatrix} + t \begin{pmatrix} 6 \\ 4 \\ -5 \end{pmatrix}$$

> You need to find two vectors on the plane. Any combination of the position vectors of the three points can be used as long as they do not give parallel vectors.

The scalar product equation of a plane

Consider a plane containing the point with position vector **a** and with a perpendicular vector **n**. Any vector on this plane will be perpendicular to the vector **n**. Using the definition of the scalar product, any point on the plane with position vector **r** satisfies $(\mathbf{r} - \mathbf{a}) \cdot \mathbf{n} = 0$. Expanding the bracket gives $\mathbf{r} \cdot \mathbf{n} - \mathbf{a} \cdot \mathbf{n} = 0$ which you can rearrange to give $\mathbf{r} \cdot \mathbf{n} = \mathbf{a} \cdot \mathbf{n}$

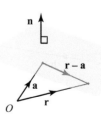

The scalar product equation of the plane perpendicular to the vector **n** and passing through the point with position vector **a** is $\mathbf{r} \cdot \mathbf{n} = p$ where $p = \mathbf{a} \cdot \mathbf{n}$

> A vector perpendicular to a plane is called a **normal** to the plane.

The Cartesian equation of a plane

You can convert into Cartesian form by writing the vector **r** in component form: $x\mathbf{i} + y\mathbf{j} + z\mathbf{k}$

Example 2

> The capital Greek letter Π is often used to denote a plane.

The plane Π passes through the point $(1, 4, -6)$ and is perpendicular to the vector $\mathbf{i} - \mathbf{j} + 3\mathbf{k}$. Find the equation of Π in

a Scalar product form,   **b** Cartesian form.

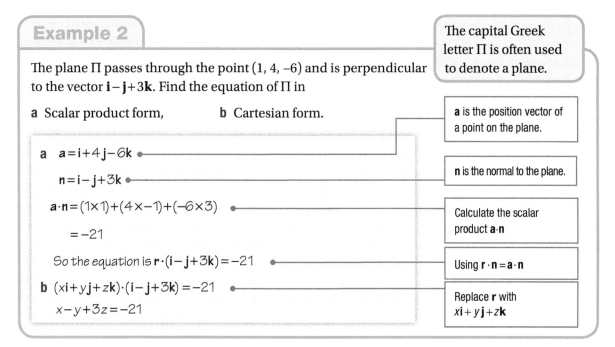

a $\mathbf{a} = \mathbf{i} + 4\mathbf{j} - 6\mathbf{k}$ — *a is the position vector of a point on the plane.*

   $\mathbf{n} = \mathbf{i} - \mathbf{j} + 3\mathbf{k}$ — *n is the normal to the plane.*

   $\mathbf{a} \cdot \mathbf{n} = (1 \times 1) + (4 \times -1) + (-6 \times 3)$ — *Calculate the scalar product $\mathbf{a} \cdot \mathbf{n}$*

   $= -21$

   So the equation is $\mathbf{r} \cdot (\mathbf{i} - \mathbf{j} + 3\mathbf{k}) = -21$ — *Using $\mathbf{r} \cdot \mathbf{n} = \mathbf{a} \cdot \mathbf{n}$*

b $(x\mathbf{i} + y\mathbf{j} + z\mathbf{k}) \cdot (\mathbf{i} - \mathbf{j} + 3\mathbf{k}) = -21$ — *Replace **r** with $x\mathbf{i} + y\mathbf{j} + z\mathbf{k}$*

   $x - y + 3z = -21$

The Cartesian equation of a plane is $ax + by + cz = d$ where the vector $a\mathbf{i} + b\mathbf{j} + c\mathbf{k}$ is perpendicular to the plane.

Points of intersection

You can find the point of intersection between a line and a plane by using the equation of the line to substitute for **r** in the equation of the plane.

Example 3

Find the point of intersection between the line with equation $\mathbf{r} = -4\mathbf{j} + 2\mathbf{k} + \lambda(3\mathbf{i} + \mathbf{j} + \mathbf{k})$ and the plane with equation $\mathbf{r} = \mathbf{i} - 2\mathbf{j} + s(\mathbf{j} + 3\mathbf{k}) + t(2\mathbf{i} + 2\mathbf{j} + \mathbf{k})$

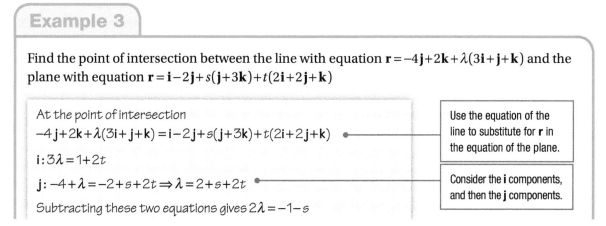

At the point of intersection
$-4\mathbf{j} + 2\mathbf{k} + \lambda(3\mathbf{i} + \mathbf{j} + \mathbf{k}) = \mathbf{i} - 2\mathbf{j} + s(\mathbf{j} + 3\mathbf{k}) + t(2\mathbf{i} + 2\mathbf{j} + \mathbf{k})$ — *Use the equation of the line to substitute for **r** in the equation of the plane.*

$\mathbf{i}: 3\lambda = 1 + 2t$

$\mathbf{j}: -4 + \lambda = -2 + s + 2t \Rightarrow \lambda = 2 + s + 2t$ — *Consider the **i** components, and then the **j** components.*

Subtracting these two equations gives $2\lambda = -1 - s$

(*Continued on the next page*)

which can be rearranged to give $s = -1-2\lambda$

$\mathbf{k} : 2+\lambda = 3s+t$

$\Rightarrow 2+\lambda = 3(-1-2\lambda)+t$

$\Rightarrow t = 5+7\lambda$

Solve this together with $3\lambda = 1+2t$
to give $\lambda = -1$ and $t = -2$

$\Rightarrow s = 1$

$\mathbf{r} = -4\mathbf{j}+2\mathbf{k}-1(3\mathbf{i}+\mathbf{j}+\mathbf{k})$

$= -3\mathbf{i}-5\mathbf{j}+\mathbf{k}$

So the point of intersection is $(-3, -5, 1)$

Now consider $\mathbf{k}$ components.

Substitute $s = -1-2\lambda$

By substituting back into $s = -1-2\lambda$

Substitute $\lambda = -1$ into equation of line.

Could also find by substituting $t = -2$ and $s = 1$ into equation of the plane.

Example 4

Find the point of intersection between the line with equation $\mathbf{r} = \mathbf{i}-\mathbf{k}+t(\mathbf{i}-2\mathbf{j}+3\mathbf{k})$ and the plane with equation $2x - y - 3z = 20$

At the point of intersection $2(1+t)-(-2t)-3(-1+3t) = 20$

$2+2t+2t+3-9t = 20$

$\Rightarrow 5t = -15$

$\Rightarrow t = -3$

So $\mathbf{r} = \mathbf{i}-\mathbf{k}-3(\mathbf{i}-2\mathbf{j}+3\mathbf{k})$

$= -2\mathbf{i}+6\mathbf{j}-10\mathbf{k}$

So the point of intersection is $(-2, 6, -10)$

Substitute $x = 1+t$, $y = -2t$, $z = -1+3t$ from the equation of the line.

Substitute $t = -3$ into equation of line.

Exercise 5.3A Fluency and skills

1   Find, in the form $\mathbf{r} = \mathbf{a}+s\mathbf{b}+t\mathbf{c}$, the equation of the plane that contains the vectors $\mathbf{i}+\mathbf{j}$ and $\mathbf{j}+2\mathbf{k}$ and passes through the point $(1, 5, 2)$

2   Find, in vector form, the equation of the plane that contains the vectors $\begin{pmatrix} 2 \\ 0 \\ -3 \end{pmatrix}$ and $\begin{pmatrix} 5 \\ -2 \\ 4 \end{pmatrix}$ and contains the point $(0, 6, 2)$

3   Find, in vector form, the equation of the plane that contains the points $(3, 1, 0)$, $(2, 4, -2)$ and $(-5, 0, 4)$

4   Find, in vector form, the equation of the plane that contains the points $(5, 1, 1)$, $(-2, 0, 0)$ and $(6, 2, -5)$

5   Find, in vector form, the equation of the plane that contains the points A, B and C with position vectors $\mathbf{a} = \mathbf{i}-2\mathbf{j}$, $\mathbf{b} = 3\mathbf{j}+2\mathbf{k}$ and $\mathbf{c} = -\mathbf{i}+2\mathbf{j}-\mathbf{k}$

6 Find, in the form $\mathbf{r} \cdot \mathbf{n} = p$, the equation of the plane that passes through the point $(3, -1, 4)$ and is perpendicular to the vector $\mathbf{i} + 2\mathbf{k}$

7 Find, in the form $\mathbf{r} \cdot \mathbf{n} = p$, the equation of the plane that passes through the point $(6, 0, -2)$ and has normal $3\mathbf{i} - \mathbf{j} + 4\mathbf{k}$

8 Find the scalar product equation of the plane that passes through the point with position vector $\mathbf{i} + \mathbf{j} + \mathbf{k}$ and is perpendicular to the vector $2\mathbf{j} - 3\mathbf{k}$

9 Find the Cartesian equation of the plane that passes through the point $(0, 3, -3)$ and is perpendicular to the vector $5\mathbf{i} + 4\mathbf{j} - 2\mathbf{k}$

10 Find the Cartesian equation of the plane that passes through the point $(1, -1, 2)$ and is perpendicular to the vector $2\mathbf{i} - 5\mathbf{k}$

11 Find the Cartesian equation of the plane that passes through the point with position vector $3\mathbf{i} + 7\mathbf{k}$ and is perpendicular to the vector $\mathbf{i} - \mathbf{j} + 4\mathbf{k}$

12 The plane Π is perpendicular to the line with equation $\mathbf{r} = \mathbf{i} - \mathbf{j} + t(2\mathbf{i} + \mathbf{j} - 3\mathbf{k})$ and passes through the point $(4, -1, 3)$. Find the equation of Π in

    a Scalar product form,   **b** Cartesian form.

13 The plane Π is perpendicular to the line with equation $\dfrac{x+1}{3} = \dfrac{y-4}{-2} = \dfrac{z}{5}$ and passes through the point $(4, 3, -2)$. Find the equation of Π in

    a Scalar product form,   **b** Cartesian form.

14 Convert each of these equations of planes into Cartesian form.

    a $\mathbf{r} \cdot \begin{pmatrix} 3 \\ 5 \\ -1 \end{pmatrix} = -2$ **b** $\mathbf{r} \cdot (7\mathbf{i} - \mathbf{j} + 8\mathbf{k}) = 3$

15 Convert each of these equations of planes into scalar product form.

    a $9x + 3y - z = 5$ **b** $2x - 7y - 15z + 4 = 0$

16 Show that the line $\mathbf{r} = \begin{pmatrix} -11 \\ -4 \\ 10 \end{pmatrix} + t\begin{pmatrix} 4 \\ -2 \\ 0 \end{pmatrix}$ and the plane $\mathbf{r} = \begin{pmatrix} 3 \\ -5 \\ -10 \end{pmatrix} + \mu\begin{pmatrix} 7 \\ -10 \\ 0 \end{pmatrix} + \lambda\begin{pmatrix} 1 \\ -3 \\ 4 \end{pmatrix}$ intersect and find their point of intersection.

17 Find the point of intersection of the line with equation $\mathbf{r} = -6\mathbf{i} + 2\mathbf{j} + 16\mathbf{k} + \lambda(3\mathbf{i} - \mathbf{j} + \mathbf{k})$ and the plane with equation $\mathbf{r} = -7\mathbf{i} - 9\mathbf{j} + \mathbf{k} + s(5\mathbf{i} + \mathbf{j} + 6\mathbf{k}) + t(-2\mathbf{i} + 4\mathbf{j} + \mathbf{k})$

18 Find the point of intersection of the line with equation $\mathbf{r} = \begin{pmatrix} 1 \\ -2 \\ 5 \end{pmatrix} + t\begin{pmatrix} -5 \\ 1 \\ 2 \end{pmatrix}$ and the plane with equation $\mathbf{r} \cdot \begin{pmatrix} 1 \\ 6 \\ -5 \end{pmatrix} = 0$

19 Find the point of intersection of the line with equation $\dfrac{x}{5} = \dfrac{y+2}{-1} = \dfrac{3-z}{4}$ and the plane with equation $\mathbf{r} \cdot (5\mathbf{i} - \mathbf{j} + 4\mathbf{k}) = -6$

20 Find the point of intersection of the line with equation $\dfrac{x-3}{4} = \dfrac{y-2}{5} = \dfrac{z-1}{-3}$ and the plane with equation $\mathbf{r} = 13\mathbf{i} + 7\mathbf{j} - 6\mathbf{k} + s(\mathbf{i} + 2\mathbf{j} - 5\mathbf{k}) + t(-\mathbf{i} + \mathbf{j} + 2\mathbf{k})$

Reasoning and problem-solving

The vector equation of a plane is not unique, as you can chose any two vectors and any point on the plane to build the equation. In the case of the scalar product equation and the Cartesian equation, all vectors that are perpendicular to the plane are parallel to each other so these forms of the equation will all simplify to a unique equation. You need to be able to tell if two equations represent the same plane.

Key point

A set of vectors are said to be **linearly independent** if one of the points cannot be written as a linear combination of the others.

> For vectors **a**, **b** and **c**, if $\mathbf{a} = s\mathbf{b} + t\mathbf{c}$ for some scalars s and t, then $s\mathbf{b} + t\mathbf{c}$ is a linear combination of **b** and **c**. This would mean that **a**, **b**, and **c** are not linearly independent.

Strategy 1

To check whether two equations represent the same plane

① Select three linearly independent points on one of the planes.

② Substitute to check if they lie on the other plane.

③ Three linearly independent points define a unique plane so if the three points satisfy both equations then they must represent the same plane.

Example 5

Verify whether or not the equations $\mathbf{r}\cdot(\mathbf{i}-6\mathbf{j}+2\mathbf{k})=1$ and $\mathbf{r}=\mathbf{i}+\mathbf{j}+3\mathbf{k}+s(2\mathbf{i}-\mathbf{k})+t(\mathbf{j}+3\mathbf{k})$ represent the same plane.

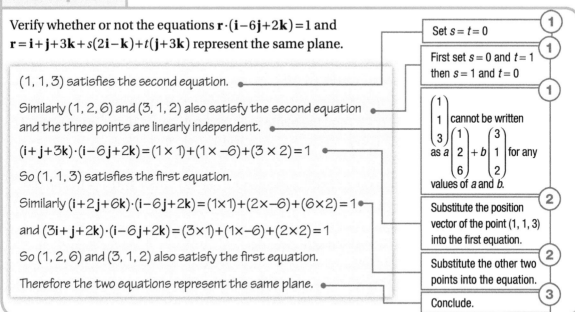

(1, 1, 3) satisfies the second equation.

Similarly (1, 2, 6) and (3, 1, 2) also satisfy the second equation and the three points are linearly independent.

$(\mathbf{i}+\mathbf{j}+3\mathbf{k})\cdot(\mathbf{i}-6\mathbf{j}+2\mathbf{k})=(1\times 1)+(1\times -6)+(3\times 2)=1$

So (1, 1, 3) satisfies the first equation.

Similarly $(\mathbf{i}+2\mathbf{j}+6\mathbf{k})\cdot(\mathbf{i}-6\mathbf{j}+2\mathbf{k})=(1\times1)+(2\times-6)+(6\times2)=1$

and $(3\mathbf{i}+\mathbf{j}+2\mathbf{k})\cdot(\mathbf{i}-6\mathbf{j}+2\mathbf{k})=(3\times1)+(1\times-6)+(2\times2)=1$

So (1, 2, 6) and (3, 1, 2) also satisfy the first equation.

Therefore the two equations represent the same plane.

Side annotations:
- Set $s = t = 0$ ①
- First set $s = 0$ and $t = 1$ then $s = 1$ and $t = 0$ ①
- $\begin{pmatrix}1\\1\\3\end{pmatrix}$ cannot be written as $a\begin{pmatrix}1\\2\\6\end{pmatrix}+b\begin{pmatrix}3\\1\\2\end{pmatrix}$ for any values of a and b. ①
- Substitute the position vector of the point (1, 1, 3) into the first equation. ②
- Substitute the other two points into the equation. ②
- Conclude. ③

Consider the intersection of a plane $\mathbf{r}\cdot\mathbf{n}=p$ and a line $\mathbf{r}=\mathbf{a}+\lambda\mathbf{b}$

You know that the acute angle between the vectors **n** and **b** is given by formula $\cos\alpha = \left|\dfrac{\mathbf{b}\cdot\mathbf{n}}{\|\mathbf{b}\|\|\mathbf{n}\|}\right|$. Therefore, the angle θ is given by $\theta = 90 - \alpha$. Since $\sin(90-\alpha) = \cos\alpha$, this means that $\sin\theta = \left|\dfrac{\mathbf{b}\cdot\mathbf{n}}{\|\mathbf{b}\|\|\mathbf{n}\|}\right|$

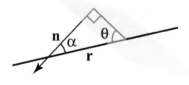

The acute angle, θ, between the plane $\mathbf{r} \cdot \mathbf{n} = p$ and the line

$\mathbf{r} = \mathbf{a} + \lambda\mathbf{b}$ is given by $\sin\theta = \left| \dfrac{\mathbf{b} \cdot \mathbf{n}}{\|\mathbf{b}\|\|\mathbf{n}\|} \right|$

Now consider the intersection of two planes, $\mathbf{r} \cdot \mathbf{n}_1 = p_1$ and $\mathbf{r} \cdot \mathbf{n}_2 = p_2$

To find the angle α between the normals you can use the formula

$\cos\alpha = \dfrac{\mathbf{n}_1 \cdot \mathbf{n}_2}{|\mathbf{n}_1||\mathbf{n}_2|}$

Using the quadrilateral formed from the planes and their normals

you can see that $\theta = 180 - \alpha$ which implies that $\cos\theta = \left| \dfrac{\mathbf{n}_1 \cdot \mathbf{n}_2}{\|\mathbf{n}_1\|\|\mathbf{n}_2\|} \right|$ since

$\cos(180 - \alpha) = \cos\alpha$

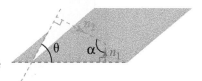

The acute angle, θ, between the planes $\mathbf{r} \cdot \mathbf{n}_1 = p_1$ and $\mathbf{r} \cdot \mathbf{n}_2 = p_2$ is

given by $\cos\theta = \left| \dfrac{\mathbf{n}_1 \cdot \mathbf{n}_2}{\|\mathbf{n}_1\|\|\mathbf{n}_2\|} \right|$

Strategy 2

To calculate the angle between two planes or between a line and a plane

(1) Identify the normal vector(s) to the plane(s).

(2) Identify the direction vector of the line.

(3) Use $\cos\theta = \left| \dfrac{\mathbf{n}_1 \cdot \mathbf{n}_2}{\|\mathbf{n}_1\|\|\mathbf{n}_2\|} \right|$ to find the acute angle between two planes.

(4) Use $\sin\theta = \left| \dfrac{\mathbf{b} \cdot \mathbf{n}}{\|\mathbf{b}\|\|\mathbf{n}\|} \right|$ to find the acute angle between a line $\mathbf{r} = \mathbf{a} + t\mathbf{b}$ and a plane.

(5) Subtract from 180° to give an obtuse angle if necessary.

Example 6

Calculate the obtuse angle between the planes $\mathbf{r} \cdot (13\mathbf{i} - 9\mathbf{j} + 5\mathbf{k}) = 8$ and $\mathbf{r} \cdot (2\mathbf{j} + 3\mathbf{k}) = 4$

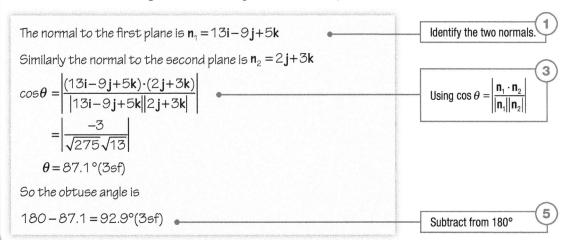

The normal to the first plane is $\mathbf{n}_1 = 13\mathbf{i} - 9\mathbf{j} + 5\mathbf{k}$ — Identify the two normals. (1)

Similarly the normal to the second plane is $\mathbf{n}_2 = 2\mathbf{j} + 3\mathbf{k}$

$\cos\theta = \left| \dfrac{(13\mathbf{i} - 9\mathbf{j} + 5\mathbf{k}) \cdot (2\mathbf{j} + 3\mathbf{k})}{|13\mathbf{i} - 9\mathbf{j} + 5\mathbf{k}||2\mathbf{j} + 3\mathbf{k}|} \right|$ — Using $\cos\theta = \left| \dfrac{\mathbf{n}_1 \cdot \mathbf{n}_2}{\|\mathbf{n}_1\|\|\mathbf{n}_2\|} \right|$ (3)

$= \left| \dfrac{-3}{\sqrt{275}\sqrt{13}} \right|$

$\theta = 87.1°\,(3sf)$

So the obtuse angle is

$180 - 87.1 = 92.9°\,(3sf)$ — Subtract from 180° (5)

Example 7

Calculate the acute angle between the plane $2x+3y-z=1$ and the line $\dfrac{x}{4}=\dfrac{y+1}{-2}=\dfrac{z-2}{7}$

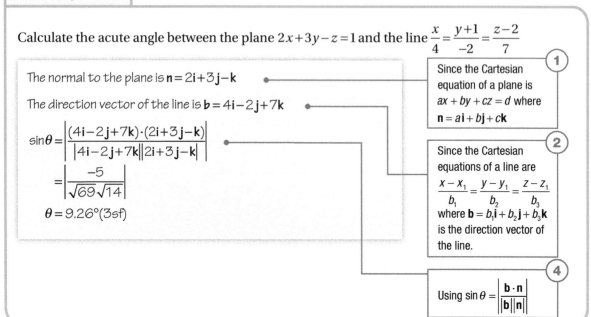

The normal to the plane is $\mathbf{n}=2\mathbf{i}+3\mathbf{j}-\mathbf{k}$

1 Since the Cartesian equation of a plane is $ax + by + cz = d$ where $\mathbf{n} = a\mathbf{i} + b\mathbf{j} + c\mathbf{k}$

The direction vector of the line is $\mathbf{b}=4\mathbf{i}-2\mathbf{j}+7\mathbf{k}$

$$\sin\theta = \left|\frac{(4\mathbf{i}-2\mathbf{j}+7\mathbf{k})\cdot(2\mathbf{i}+3\mathbf{j}-\mathbf{k})}{|4\mathbf{i}-2\mathbf{j}+7\mathbf{k}||2\mathbf{i}+3\mathbf{j}-\mathbf{k}|}\right|$$

$$= \left|\frac{-5}{\sqrt{69}\sqrt{14}}\right|$$

$$\theta = 9.26°\,(3sf)$$

2 Since the Cartesian equations of a line are $\dfrac{x-x_1}{b_1}=\dfrac{y-y_1}{b_2}=\dfrac{z-z_1}{b_3}$ where $\mathbf{b} = b_1\mathbf{i} + b_2\mathbf{j} + b_3\mathbf{k}$ is the direction vector of the line.

4 Using $\sin\theta = \left|\dfrac{\mathbf{b}\cdot\mathbf{n}}{\|\mathbf{b}\|\|\mathbf{n}\|}\right|$

Exercise 5.3B Reasoning and problem-solving

1 On which of these planes does the point $(-7, 10, 0)$ lie? Show your working.

  a $\mathbf{r}=2\mathbf{i}-3\mathbf{j}-7\mathbf{k}+s(-\mathbf{i}+5\mathbf{j}+3\mathbf{k})+t(7\mathbf{i}-3\mathbf{j}-\mathbf{k})$

  b $\mathbf{r}=\begin{pmatrix}2\\0\\8\end{pmatrix}+s\begin{pmatrix}1\\3\\-2\end{pmatrix}+t\begin{pmatrix}-2\\4\\0\end{pmatrix}$

  c $\mathbf{r}\cdot(\mathbf{i}-\mathbf{j}+3\mathbf{k})=17$

  d $2x+3y-8z=16$

2 Find the values of k for which the point $(k, k, -4)$ lies on the plane with equation $\mathbf{r}\cdot\begin{pmatrix}5\\k\\3\end{pmatrix}=24$

3 Find the values of k for which the point $(-6, k, 6)$ lies on the plane with equation

$$\mathbf{r}=\begin{pmatrix}6\\-4\\1\end{pmatrix}+s\begin{pmatrix}1\\3\\-2\end{pmatrix}+t\begin{pmatrix}9\\2\\1\end{pmatrix}$$

4 Verify whether or not each pair of equations represent the same plane.

  a $2x+3y-z-2=0$ and $\mathbf{r}\cdot(2\mathbf{i}+3\mathbf{j}-\mathbf{k})=2$

  b $\mathbf{r}=3\mathbf{i}-\mathbf{k}+s(4\mathbf{i}+\mathbf{j}+2\mathbf{k})+t(5\mathbf{i}+3\mathbf{j})$ and $\mathbf{r}=(\mathbf{i}-4\mathbf{j}+4\mathbf{k})+\lambda(2\mathbf{i}+4\mathbf{j}+5\mathbf{k})+\mu(5\mathbf{i}+3\mathbf{j})$

  c $\mathbf{r}=\begin{pmatrix}1\\0\\-7\end{pmatrix}+s\begin{pmatrix}5\\-2\\4\end{pmatrix}+t\begin{pmatrix}0\\0\\1\end{pmatrix}$ and $\mathbf{r}=\begin{pmatrix}-4\\2\\-7\end{pmatrix}+\lambda\begin{pmatrix}0\\1\\3\end{pmatrix}+\mu\begin{pmatrix}5\\-2\\0\end{pmatrix}$

  d $\mathbf{r}=\begin{pmatrix}-2\\1\\0\end{pmatrix}+s\begin{pmatrix}4\\3\\2\end{pmatrix}+t\begin{pmatrix}-5\\0\\1\end{pmatrix}$ and $\mathbf{r}\cdot\begin{pmatrix}6\\-2\\-1\end{pmatrix}=-14$

  e $\mathbf{r}\cdot(2\mathbf{i}-7\mathbf{j}-3\mathbf{k})=-2$ and $\mathbf{r}=(5\mathbf{i}+4\mathbf{k})+\lambda(3\mathbf{j}-7\mathbf{k})+\mu(2\mathbf{i}+\mathbf{j}-\mathbf{k})$

  f $x+3y+z=-1$ and $\mathbf{r}=(\mathbf{i}-2\mathbf{k})+\lambda(\mathbf{i}-9\mathbf{j}+6\mathbf{k})+\mu(4\mathbf{i}+3\mathbf{j}-3\mathbf{k})$

5 Calculate the acute angle between the line $\mathbf{r} = 2\mathbf{i} - 8\mathbf{j} + \mathbf{k} + t(\mathbf{j} + \mathbf{k})$ and the plane $\mathbf{r} \cdot (6\mathbf{i} - \mathbf{j} + 5\mathbf{k}) = 9$

6 Calculate the obtuse angle between the line $\mathbf{r} = 9\mathbf{i} - 5\mathbf{j} + 3\mathbf{k} + \lambda(\mathbf{i} - \mathbf{j} - 4\mathbf{k})$ and the plane $6x - y + z = 24$

7 Calculate the acute angle between the line $\dfrac{x-1}{5} = \dfrac{y-2}{-2} = \dfrac{z+3}{-2}$ and the plane $2x + 7y - z = 5$

8 Calculate the obtuse angle between the line $\mathbf{r} = \begin{pmatrix} -5 \\ 0 \\ 1 \end{pmatrix} + \lambda \begin{pmatrix} 7 \\ 2 \\ -3 \end{pmatrix}$ and the plane $\mathbf{r} \cdot \begin{pmatrix} -2 \\ 4 \\ -9 \end{pmatrix} = 1$

9 State the value of k for which the line $\mathbf{r} = \begin{pmatrix} -7 \\ 0 \\ 8 \end{pmatrix} + \lambda \begin{pmatrix} 3 \\ k \\ -2 \end{pmatrix}$ is perpendicular to the plane $\mathbf{r} \cdot \begin{pmatrix} -9 \\ 2 \\ 6 \end{pmatrix} = 5$

10 Show that the line $\mathbf{r} = \begin{pmatrix} 1 \\ -3 \\ -2 \end{pmatrix} + \lambda \begin{pmatrix} 0 \\ 4 \\ 6 \end{pmatrix}$ lies on the plane with equation $\mathbf{r} \cdot \begin{pmatrix} 5 \\ 9 \\ -6 \end{pmatrix} = -10$

11 Show that the line $\mathbf{r} = \mathbf{i} - 3\mathbf{k} + \lambda(2\mathbf{j} + 3\mathbf{k})$ lies on the plane with equation $x + 3y - 2z = 7$

12 Calculate the acute angle between the plane $\mathbf{r} \cdot (2\mathbf{i} + \mathbf{j} - 3\mathbf{k}) = 2$ and the plane $\mathbf{r} \cdot (6\mathbf{i} - 4\mathbf{j} - 7\mathbf{k}) = 5$

13 Calculate the acute angle between the plane $\mathbf{r} \cdot (7\mathbf{i} - \mathbf{j} - 5\mathbf{k}) = 9$ and the plane $2x - y + z = 4$

14 Calculate the obtuse angle between the plane $x + y - 7z - 3 = 0$ and the plane $8x - 3y + 4z = 15$

15 State the value of k for which the planes $\mathbf{r} \cdot \begin{pmatrix} 5 \\ 2 \\ k \end{pmatrix} = 1$ and $\mathbf{r} \cdot \begin{pmatrix} 6 \\ -1 \\ 4 \end{pmatrix} = -2$ are perpendicular.

16 The plane Π is perpendicular to the x-axis and passes through the point $(5, 4, -2)$. The line L is parallel to the vector $\mathbf{i} - \mathbf{k}$ and passes through the point $(1, 5, 2)$. Π and L intersect at the point A. Calculate the length $\overrightarrow{OA}$

17 The lines L_1 and L_2 have equations $\mathbf{r} = 2\mathbf{i} + 3\mathbf{j} - \mathbf{k} + \lambda(\mathbf{i} - 6\mathbf{j} + 3\mathbf{k})$ and $\mathbf{r} = 15\mathbf{i} + 23\mathbf{k} + \mu(3\mathbf{i} + 7\mathbf{j} + 4\mathbf{k})$ respectively.

Find the vector equation of the plane that contains L_1 and L_2

18 Find the scalar product equation of the plane that contains the vectors $\begin{pmatrix} 1 \\ 0 \\ -3 \end{pmatrix}$ and $\begin{pmatrix} 2 \\ 6 \\ -1 \end{pmatrix}$ and passes through the point $(3, -1, 0)$

19 Convert this equation of a plane into scalar product form.

$\mathbf{r} = \begin{pmatrix} 0 \\ 3 \\ -2 \end{pmatrix} + \mu \begin{pmatrix} 6 \\ -2 \\ 2 \end{pmatrix} + \lambda \begin{pmatrix} -1 \\ 4 \\ 7 \end{pmatrix}$

20 Find the Cartesian equation of the plane that passes through the points $(3, -5, 1)$, $(2, 4, 5)$ and $(1, 0, 3)$

21 The lines $\mathbf{r} = \begin{pmatrix} 1 \\ 0 \\ -2 \end{pmatrix} + s \begin{pmatrix} 5 \\ -2 \\ -4 \end{pmatrix}$ and $\mathbf{r} = \begin{pmatrix} 16 \\ 10 \\ 1 \end{pmatrix} + t \begin{pmatrix} -1 \\ 6 \\ 6 \end{pmatrix}$ intersect the plane $\mathbf{r} \cdot \begin{pmatrix} 1 \\ -3 \\ 2 \end{pmatrix} = 9$ at the points A and B respectively. Calculate the area of triangle OAB

5.4 Finding distances

Fluency and skills

Suppose you wish to find the shortest distance from a line $\mathbf{r} = \mathbf{a} + \lambda\mathbf{b}$ to a given point, C, with position vector $\mathbf{c}$

The shortest possible distance will be along the line which is perpendicular to the original line. Since $\overrightarrow{DC} = \mathbf{c} - (\mathbf{a} + \lambda\mathbf{b})$, the scalar product $(\mathbf{c} - (\mathbf{a} + \lambda\mathbf{b})) \cdot \mathbf{b} = 0$

This allows you to find the point of intersection between $\overrightarrow{DC}$ and the line $\mathbf{r} = \mathbf{a} + \lambda\mathbf{b}$. You will then be able to find the length of $\overrightarrow{DC}$ and hence the shortest distance from the line to the point.

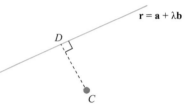

Example 1

Calculate the shortest distance from the point $(3, 3, 1)$ to the line with equation $\mathbf{r} = \begin{pmatrix} 2 \\ 0 \\ -6 \end{pmatrix} + \lambda \begin{pmatrix} -3 \\ 0 \\ -1 \end{pmatrix}$

$\begin{pmatrix} 3 \\ 3 \\ 1 \end{pmatrix} - \left[\begin{pmatrix} 2 \\ 0 \\ -6 \end{pmatrix} + \lambda \begin{pmatrix} -3 \\ 0 \\ -1 \end{pmatrix} \right] = \begin{pmatrix} 1+3\lambda \\ 3 \\ 7+\lambda \end{pmatrix}$

Find a vector that passes through the point and the line.

$\begin{pmatrix} 1+3\lambda \\ 3 \\ 7+\lambda \end{pmatrix} \cdot \begin{pmatrix} -3 \\ 0 \\ -1 \end{pmatrix} = 0$

Since the vectors are perpendicular, their scalar product is zero.

$\Rightarrow -3 - 9\lambda - 7 - \lambda = 0$

$\Rightarrow \lambda = -1$

$\begin{pmatrix} 2 \\ 0 \\ -6 \end{pmatrix} - 1\begin{pmatrix} -3 \\ 0 \\ -1 \end{pmatrix} = \begin{pmatrix} 5 \\ 0 \\ -5 \end{pmatrix}$

This is the position vector of the point of intersection of the line with the perpendicular through $(3, 3, 1)$

$\text{Distance} = \sqrt{(5-3)^2 + (0-3)^2 + (-5-1)^2}$

$= 7$

Find the distance from $(3, 3, 1)$ to the point of intersection.

You already know how to find the point of intersection between two lines, if it exists. If two lines do not intersect, you can instead find the shortest possible distance between them. In the case of parallel lines, this is quite straight forward as the distance between them is constant.

If you have parallel lines with equations $\mathbf{r} = \mathbf{a} + \lambda\mathbf{b}$ and $\mathbf{r} = \mathbf{c} + \mu\mathbf{b}$ then, since the distance between them is always the same, you can

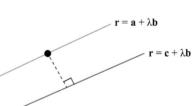

choose any point on either one of the lines (for example, you could choose the point with position vector $\mathbf{a}$ on the line $\mathbf{r} = \mathbf{a} + \lambda\mathbf{b}$), and then use the method above for finding the shortest distance from a point to a line.

You can also find the shortest distance from a point to a plane.

Suppose you have the plane with equation $n_1x + n_2y + n_3z + d = 0$ and the point $P(\alpha, \beta, \gamma)$. The shortest distance from P to the plane will be perpendicular to the plane, therefore you need the length of $\overrightarrow{QP}$, where Q is the point on the perpendicular that intersects the plane.

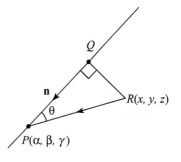

See Maths Ch3.1
For a reminder of trigonometric ratios.

Using the right-angled triangle you can see that $|\overrightarrow{QP}| = |\overrightarrow{RP}|\cos\theta$
$\cos\theta = \left|\dfrac{\overrightarrow{RP}\cdot\mathbf{n}}{|\overrightarrow{RP}||\mathbf{n}|}\right|$, so this becomes $|\overrightarrow{QP}| = |\overrightarrow{RP}|\left|\dfrac{\overrightarrow{RP}\cdot\mathbf{n}}{|\overrightarrow{RP}||\mathbf{n}|}\right|$, so $|\overrightarrow{QP}| = \left|\dfrac{\overrightarrow{RP}\cdot\mathbf{n}}{|\mathbf{n}|}\right|$

Since the equation of the plane is $n_1x + n_2y + n_3z + d = 0$, you know that $\mathbf{n} = n_1\mathbf{i} + n_2\mathbf{j} + n_3\mathbf{k}$

From the diagram, you can see that $\overrightarrow{RP} = (\alpha - x)\mathbf{i} + (\beta - y)\mathbf{j} + (\gamma - z)\mathbf{k}$

So $\overrightarrow{RP}\cdot\mathbf{n} = ((\alpha - x)\mathbf{i} + (\beta - y)\mathbf{j} + (\gamma - z)\mathbf{k})\cdot(n_1\mathbf{i} + n_2\mathbf{j} + n_3\mathbf{k})$

$= n_1(\alpha - x) + n_2(\beta - y) + n_3(\gamma - z)$, which rearranges to

$= (n_1\alpha + n_2\beta + n_3\gamma) - (n_1x + n_2y + n_2z)$

Using the equation of the plane you can replace $n_1x + n_2y + n_2z$ by $-d$, to give $n_1\alpha + n_2\beta + n_3\gamma + d$

Substituting this into $|\overrightarrow{QP}| = \left|\dfrac{\overrightarrow{RP}\cdot\mathbf{n}}{|\mathbf{n}|}\right|$ gives the distance from P to the plane: $\left|\dfrac{n_1\alpha + n_2\beta + n_3\gamma + d}{\sqrt{n_1^2 + n_2^2 + n_3^2}}\right|$

> **Key point**
>
> The perpendicular distance from the point (α, β, γ) to the plane
> $n_1x + n_2y + n_3z + d = 0$ is given by the formula $\left|\dfrac{n_1\alpha + n_2\beta + n_3\gamma + d}{\sqrt{n_1^2 + n_2^2 + n_3^2}}\right|$

Example 2

Find the shortest distance from the plane with equation $\mathbf{r}\cdot\begin{pmatrix}1\\2\\-2\end{pmatrix} = 3$ to the point $(4, 0, 7)$

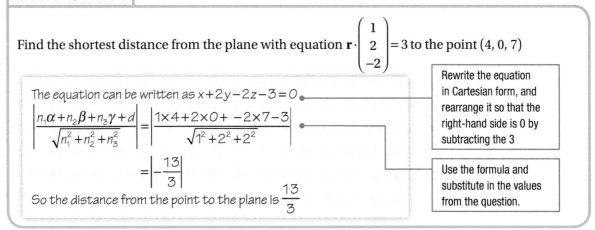

The equation can be written as $x + 2y - 2z - 3 = 0$ ———— Rewrite the equation in Cartesian form, and rearrange it so that the right-hand side is 0 by subtracting the 3

$\left|\dfrac{n_1\alpha + n_2\beta + n_3\gamma + d}{\sqrt{n_1^2 + n_2^2 + n_3^2}}\right| = \left|\dfrac{1\times 4 + 2\times 0 + -2\times 7 - 3}{\sqrt{1^2 + 2^2 + 2^2}}\right|$

Use the formula and substitute in the values from the question.

$= \left|-\dfrac{13}{3}\right|$

So the distance from the point to the plane is $\dfrac{13}{3}$

1   Calculate the shortest distance between the line with equation $\mathbf{r} = 6\mathbf{i} - 4\mathbf{j} + \mathbf{k} + s(2\mathbf{i} - \mathbf{k})$ and the point $(-4, 2, 6)$

2   Find the shortest distance between the line with equation $\mathbf{r} = \begin{pmatrix} 1 \\ 3 \\ -2 \end{pmatrix} + \lambda \begin{pmatrix} 1 \\ -2 \\ 0 \end{pmatrix}$ and the point $(0, 5, 1)$

3   Find the shortest distance between the line with equations $\dfrac{x-3}{2} = \dfrac{y-1}{0} = \dfrac{z+1}{0}$ and the point $(0, 3, 1)$

4   Calculate the shortest distance between the line with equation $\mathbf{r} = \mu(\mathbf{i} + \mathbf{j} - \mathbf{k})$ and the point with position vector $2\mathbf{i} - 3\mathbf{j} + \mathbf{k}$

5   Calculate the shortest distance between these pairs of parallel lines.

   a $\mathbf{r} = 5\mathbf{i} + \mathbf{k} + s(\mathbf{i} - \mathbf{j})$ and $\mathbf{r} = 2\mathbf{i} + \mathbf{j} - 7\mathbf{k} + t(\mathbf{i} - \mathbf{j})$

   b $\mathbf{r} = \begin{pmatrix} 2 \\ 3 \\ 5 \end{pmatrix} + \lambda \begin{pmatrix} 3 \\ 1 \\ -2 \end{pmatrix}$ and $\mathbf{r} = \begin{pmatrix} -6 \\ 1 \\ -1 \end{pmatrix} + \mu \begin{pmatrix} 9 \\ 3 \\ -6 \end{pmatrix}$

   c $x = y + 3 = \dfrac{z-6}{-2}$ and $x + 1 = y - 4 = \dfrac{z-8}{-2}$

d $\mathbf{r} = \begin{pmatrix} 0 \\ 4 \\ -1 \end{pmatrix} + \lambda \begin{pmatrix} -3 \\ 5 \\ 1 \end{pmatrix}$ and $\dfrac{x}{3} = \dfrac{y+3}{-5} = -z - 1$

6   Find the shortest distance between these pairs of points and planes.

   a $(2, 5, 1)$ and $2x - 4y + 4z + 5 = 0$

   b $(3, 0, -4)$ and $3x + y - 2z = 6$

   c $(-1, 5, 4)$ and $\mathbf{r} \cdot (\mathbf{i} - 7\mathbf{j} + 5\mathbf{k}) = -3$

   d $(6, 1, 4)$ and $\mathbf{r} \cdot \begin{pmatrix} 0 \\ -8 \\ 6 \end{pmatrix} = 12$

   e $(-2, -5, -1)$ and $\mathbf{r} \cdot (\mathbf{i} - \mathbf{j} - 3\mathbf{k}) = 1$

7   The plane Π is perpendicular to the y-axis and passes through the point $(1, 2, 0)$

   Calculate the shortest distance from this plane to the point $(3, 4, -1)$

8   Use the formula for the distance between a point and a plane to show that the shortest distance from the origin to the plane $\mathbf{r} \cdot \mathbf{n} = p$ is $\dfrac{p}{|\mathbf{n}|}$

Reasoning and problem-solving

See Maths Ch4.2 For a reminder of differenti- ation.

An alternative approach to finding the minimum distance between a line and a point is to use calculus. Suppose you have the line $\mathbf{r} = \mathbf{a} + t\mathbf{b}$ and a point with position vector $\mathbf{c}$. Then you need to find the minimum value of $x = |\mathbf{a} + t\mathbf{b} - \mathbf{c}|$ which can be found by using $\dfrac{dx}{dt} = 0$

Strategy

To find the minimum distance between a line and a point

①   Find the vector joining the point given to a general point on the line.

②   Find an expression for x, the length of this vector.

③   Use $\dfrac{dx}{dt} = 0$ to calculate the value of t.

④   Substitute the value of t into your expression for the length.

Example 3

Find the minimum distance between the line with equation $r = \mathbf{i}+\mathbf{j}+t(\mathbf{j}-\mathbf{k})$ and the point $(2, 1, -2)$

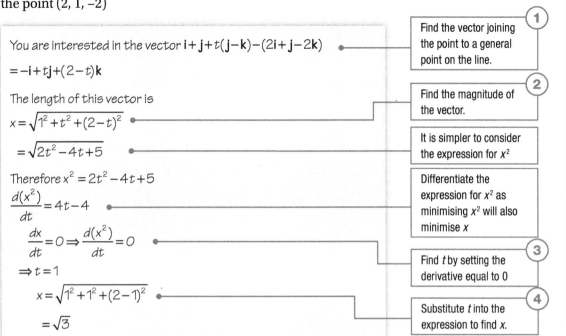

You are interested in the vector $\mathbf{i}+\mathbf{j}+t(\mathbf{j}-\mathbf{k})-(2\mathbf{i}+\mathbf{j}-2\mathbf{k})$

$= -\mathbf{i}+t\mathbf{j}+(2-t)\mathbf{k}$

The length of this vector is

$x = \sqrt{1^2+t^2+(2-t)^2}$

$= \sqrt{2t^2-4t+5}$

Therefore $x^2 = 2t^2-4t+5$

$\dfrac{d(x^2)}{dt} = 4t-4$

$\dfrac{dx}{dt} = 0 \Rightarrow \dfrac{d(x^2)}{dt} = 0$

$\Rightarrow t = 1$

$x = \sqrt{1^2+1^2+(2-1)^2}$

$= \sqrt{3}$

1 Find the vector joining the point to a general point on the line.

2 Find the magnitude of the vector.

It is simpler to consider the expression for x^2

Differentiate the expression for x^2 as minimising x^2 will also minimise x

3 Find t by setting the derivative equal to 0

4 Substitute t into the expression to find x.

A similar process can be used to find the perpendicular distance between a pair of parallel lines.

Exercise 5.4B Reasoning and problem-solving

1 Use calculus to find the shortest distance between the point $(1, 5, -7)$ and the line with equation $\mathbf{r} = \begin{pmatrix} -1 \\ 9 \\ -5 \end{pmatrix} + \lambda \begin{pmatrix} 0 \\ 3 \\ 1 \end{pmatrix}$

2 Use calculus to find the shortest distance between the point $(-4, 0, 2)$ and the line with equation $\dfrac{x+2}{-1} = \dfrac{y+2}{2} = \dfrac{z-1}{1}$

3 Use calculus to find the shortest distance between each pair of lines.

a $x = 3; \dfrac{y+1}{6} = \dfrac{z}{2}$ and $\mathbf{r} = \begin{pmatrix} -1 \\ 1 \\ 5 \end{pmatrix} + \lambda \begin{pmatrix} 0 \\ 3 \\ 1 \end{pmatrix}$

b $\mathbf{r} = \begin{pmatrix} -1 \\ 2 \\ -3 \end{pmatrix} + \lambda \begin{pmatrix} 1 \\ -1 \\ 1 \end{pmatrix}$ and $x+5 = 2-y = z+1$

4 The points A, B, C are defined as $(2, -1, 4)$, $(0, -2, 4)$ and $(1, 0, 5)$ respectively. Calculate the shortest distance from the origin to the plane containing points A, B and C.

5 Which of these lines comes closest to the origin?

$L_1: \mathbf{r} = \begin{pmatrix} 5 \\ -2 \\ -4 \end{pmatrix} + \lambda \begin{pmatrix} 1 \\ 2 \\ 1 \end{pmatrix}$ or $L_2: \mathbf{r} = \begin{pmatrix} -1 \\ 3 \\ 2 \end{pmatrix} + \lambda \begin{pmatrix} 1 \\ 1 \\ 0 \end{pmatrix}$

Fully explain your reasoning.

6 Which of these two points lies closest to the line $x-4 = \dfrac{y+3}{3} = \dfrac{7-z}{2}$?

$A: (4, 1, 6)$ or $B: (6, -2, 6)$

Fully explain your reasoning.

7 The shortest distance between the plane

$\mathbf{r} \cdot \begin{pmatrix} 2 \\ a \\ -1 \end{pmatrix} = 7$ and the point $(-4, 0, -3)$ is 4 units.

Find the possible values of a

8 The line l passes through the points A $(1, 4, -2)$ and $B (0, 2, -7)$

The plane Π has equation $5x-5y+z = 3$

a Find the shortest distance from the plane to the point

  i A **ii** B

b Does the line intersect the plane? Explain your answer.

9 a Find the perpendicular distance between each of these pairs of planes. Give your answer to 3 significant figures.

  i $3x - y + 9z = 15$ and $6x - 2y + 18z - 3 = 0$

  ii $\mathbf{r} \cdot \begin{pmatrix} 4 \\ 1 \\ -2 \end{pmatrix} = 2$ and $\mathbf{r} \cdot \begin{pmatrix} -6 \\ \frac{3}{2} \\ 3 \end{pmatrix} = 5$

  iii $\mathbf{r} \cdot (\mathbf{i} + 3\mathbf{j} - 5\mathbf{k}) = 12$ and $-2x - 6y + 10z + 13 = 0$

b Explain why the planes in part **a** do not meet.

10 The line l_1 has equation $\mathbf{r} = \begin{pmatrix} 2 \\ -3 \\ 1 \end{pmatrix} + \lambda \begin{pmatrix} 3 \\ -5 \\ 2 \end{pmatrix}$

and the line l_2 has equation

$\mathbf{r} = \begin{pmatrix} 9 \\ -10 \\ -3 \end{pmatrix} + \mu \begin{pmatrix} 1 \\ 3 \\ 6 \end{pmatrix}$

a Show that l_1 and l_2 do not intersect.

b Explain the geometric relationship between these two lines.

c Show that the vector $\begin{pmatrix} 18 \\ 8 \\ -7 \end{pmatrix}$ is perpendicular to both lines.

d Hence write a plane containing l_1 and a plane containing l_2

e By finding the distance of each of these planes from the origin, calculate the shortest distance between the two planes. Give your answer to 1 decimal place.

f Hence, give the minimum distance between the lines l_1 and l_2. Explain your answer.

11 The planes Π_1 and Π_2 have equations

$\mathbf{r} = \begin{pmatrix} 5 \\ 3 \\ -2 \end{pmatrix} + s\begin{pmatrix} 4 \\ 2 \\ -1 \end{pmatrix} + t\begin{pmatrix} 0 \\ -5 \\ 1 \end{pmatrix}$ and $\mathbf{r} \cdot \begin{pmatrix} 3 \\ 4 \\ 20 \end{pmatrix} = 12$

a Show that Π_1 and Π_2 are parallel planes.

b Find the shortest distance between Π_1 and Π_2

12 The lines L_1 and L_2 have equations $\mathbf{r} = 8\mathbf{i} - 14\mathbf{j} + 13\mathbf{k} + s(-4\mathbf{i} + 7\mathbf{j} - 6\mathbf{k})$ and $\dfrac{x}{2} = \dfrac{y - 17}{5} = \dfrac{z + 7}{-1}$ respectively. The plane Π contains both L_1 and L_2

a Find the vector equation of the plane.

b Calculate the distance of the plane from the point $(16, 11, -13)$

13 L_1 and L_2 have equations $\mathbf{r} = \mathbf{i} + t\mathbf{j}$ and $\mathbf{r} = \mathbf{j} - \mathbf{k} + s(\mathbf{i} + \mathbf{k})$ respectively and intersect at the point A

The line L_3 is parallel to the vector $\mathbf{i} + \mathbf{j}$ and passes through the point $(2, 1, 0)$

a Calculate the shortest distance from A to the line L_3

b Show that L_3 and L_2 will never meet.

The lines L_3 and L_1 intersect at the point C and the closest point to A on L_3 is called B

c Calculate the area of the triangle ABC

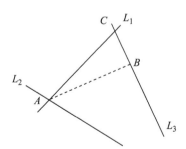

Chapter summary

- The vector equation of a line parallel to the vector **b** and passing through the point with position vector **a** is $\mathbf{r} = \mathbf{a} + \lambda\mathbf{b}$

- The Cartesian equations of a line parallel to the vector $\mathbf{b} = b_1\mathbf{i} + b_2\mathbf{j} + b_3\mathbf{k}$ and passing through the point with position vector $\mathbf{a} = a_1\mathbf{i} + a_2\mathbf{j} + a_3\mathbf{k}$ are $\dfrac{x - a_1}{b_1} = \dfrac{y - a_2}{b_2} = \dfrac{z - a_3}{b_3}$

- Two lines in 3D can be parallel, intersecting or skew.

- The scalar product is defined as $\mathbf{a} \cdot \mathbf{b} = |\mathbf{a}||\mathbf{b}|\cos\theta$ where θ is the angle between the vectors **a** and **b**

- If you have two vectors $\mathbf{a} = a_1\mathbf{i} + a_2\mathbf{j} + a_3\mathbf{k}$ and $\mathbf{b} = b_1\mathbf{i} + b_2\mathbf{j} + b_3\mathbf{k}$, then the scalar product is defined as $\mathbf{a} \cdot \mathbf{b} = a_1b_1 + a_2b_2 + a_3b_3$

- The vector equation of a plane is $\mathbf{r} = \mathbf{a} + s\mathbf{b} + t\mathbf{c}$ where **a** is the position vector of a point on the plane and **b** and **c** are non-parallel vectors on the plane.

- The scalar product equation of a plane is $\mathbf{r} \cdot \mathbf{n} = p$ where **n** is perpendicular to the plane and $p = \mathbf{a} \cdot \mathbf{n}$ for **a** the position vector of a point on the plane.

- The Cartesian equation of a plane is $ax + by + cz = d$ where the vector $a\mathbf{i} + b\mathbf{j} + c\mathbf{k}$ is perpendicular to the plane.

- The acute angle, θ, between the plane $\mathbf{r} \cdot \mathbf{n} = p$ and the line $\mathbf{r} = \mathbf{a} + \lambda\mathbf{b}$ is given by $\sin\theta = \left|\dfrac{\mathbf{b} \cdot \mathbf{n}}{\|\mathbf{b}\|\|\mathbf{n}\|}\right|$

- The acute angle, θ, between the planes $\mathbf{r} \cdot \mathbf{n}_1 = p_1$ and $\mathbf{r} \cdot \mathbf{n}_2 = p_2$ is given by $\cos\theta = \left|\dfrac{\mathbf{n}_1 \cdot \mathbf{n}_2}{\|\mathbf{n}_1\|\|\mathbf{n}_2\|}\right|$

- The perpendicular distance from the point (α, β, γ) to the plane $n_1x + n_2y + n_3z + d = 0$ is given by the formula $\left|\dfrac{n_1\alpha + n_2\beta + n_3\gamma + d}{\sqrt{n_1^2 + n_2^2 + n_3^2}}\right|$

Check and review

You should now be able to...	Try Questions
✓ Write and use the equation of a line in Cartesian and vector form.	1, 2
✓ Decide whether two lines are intersecting, parallel or skew.	3
✓ Find the coordinates of the point of intersection of two lines.	3
✓ Calculate the scalar product of two vectors.	4
✓ Use the scalar product to find the angle between two vectors.	4, 5
✓ Use the scalar product to show that two lines are perpendicular.	6
✓ Write the equation of a plane in the Cartesian, vector and scalar product forms.	7, 8
✓ Calculate the angle between a line and a plane.	9, 10
✓ Calculate the angle between two planes.	11, 12
✓ Find the point of intersection between a line and a plane.	9, 10
✓ Calculate the shortest distance from a point to a line.	13
✓ Calculate the perpendicular distance between parallel lines.	14
✓ Calculate the shortest distance from a point to a plane.	15, 16

1 A line is parallel to the vector $2\mathbf{i}-3\mathbf{j}+\mathbf{k}$ and passes through the point $(1, -1, 4)$. Write the equation of the line in

   a Cartesian form, b Vector form.

2 A line passes through the points $(3, 0, 5)$ and $(-2, 4, 7)$. Write the equation of the line in

   a Cartesian form, b Vector form.

3 For each pair of lines, decide whether they are parallel, skew or intersecting. If they are intersecting, find their point of intersection.

   a $\mathbf{r}=\begin{pmatrix}1\\3\\-2\end{pmatrix}+s\begin{pmatrix}-2\\4\\-1\end{pmatrix}$ and $\mathbf{r}=\begin{pmatrix}-4\\0\\1\end{pmatrix}+t\begin{pmatrix}5\\-3\\3\end{pmatrix}$

   b $\mathbf{r}=4\mathbf{i}-\mathbf{j}+\lambda(2\mathbf{i}+3\mathbf{j}-\mathbf{k})$ and
      $\dfrac{1-x}{10}=\dfrac{1-y}{15}=\dfrac{z+3}{5}$

   c $\dfrac{x-2}{3}=\dfrac{y+1}{2}=\dfrac{z-3}{-1}$ and
      $\dfrac{x-1}{-4}=\dfrac{y+3}{-3}=\dfrac{z-6}{2}$

   d $\mathbf{r}=-5\mathbf{i}+2\mathbf{j}+4\mathbf{k}+\lambda(5\mathbf{i}-3\mathbf{j}+\mathbf{k})$ and
      $\mathbf{r}=12\mathbf{i}-3\mathbf{j}+6\mathbf{k}+\mu(2\mathbf{i}+4\mathbf{j}-\mathbf{k})$

4 For the vectors $\mathbf{a}=2\mathbf{i}-\mathbf{j}+3\mathbf{k}$ and $\mathbf{b}=3\mathbf{i}+4\mathbf{j}-5\mathbf{k}$, calculate

   a $\mathbf{a}\cdot\mathbf{b}$ b the acute angle between $\mathbf{a}$ and $\mathbf{b}$

5 Find the obtuse angle between the line $\mathbf{r}=5\mathbf{i}+3\mathbf{j}+2\mathbf{k}+s(\mathbf{i}+\mathbf{j}-2\mathbf{k})$ and the line $\dfrac{x-1}{2}=\dfrac{y+3}{3}=z$

6 Show that the lines $\dfrac{x-7}{-2}=\dfrac{y+2}{8}=\dfrac{z-3}{-4}$ and
   $\mathbf{r}=\begin{pmatrix}5\\0\\-6\end{pmatrix}+\lambda\begin{pmatrix}-4\\2\\6\end{pmatrix}$ are perpendicular.

7 Write the vector equation of the plane that contains the vectors $-2\mathbf{i}+3\mathbf{k}$ and $\mathbf{i}-4\mathbf{j}-\mathbf{k}$ and passes through the point $(2, 0, -3)$

8 A plane is perpendicular to the vector $7\mathbf{i}-5\mathbf{j}+\mathbf{k}$ and contains the point with position vector $2\mathbf{j}-5\mathbf{k}$. Write the equation of the plane in

   a The form $\mathbf{r}\cdot\mathbf{n}=p$, b Cartesian form.

9 A plane has equation $\mathbf{r}\cdot(\mathbf{i}-2\mathbf{j}+2\mathbf{k})=5$
   The line with equation $\mathbf{r}=\mathbf{i}+t(2\mathbf{i}+3\mathbf{j}-\mathbf{k})$ intersects the plane at the point A

   a Calculate the acute angle between the plane and the line at A

   b Find the coordinates of A

   c Calculate the length OA

10 The line with equation $\dfrac{x-1}{2}=y=\dfrac{z+2}{4}$ intersects the plane $8x+6y-3z-4=0$ at the point A

   a Find the coordinates of A

   b Calculate the obtuse angle between the line and the plane at point A

11 The plane Π_1 has equation $\mathbf{r}\cdot\begin{pmatrix}1\\0\\-4\end{pmatrix}=7$
   and the plane Π_2 has equation $\mathbf{r}\cdot\begin{pmatrix}-3\\1\\2\end{pmatrix}$

   Given that the acute angle between the two planes is θ, find the exact value of $\cos\theta$

12 Calculate the size of the obtuse angle between the planes $x+y-5z=4$ and $2x-y+7z+1=0$

13 Calculate the shortest distance from the point $(5, -2, 4)$ to the line with equation
   $\mathbf{r}=\begin{pmatrix}3\\0\\-1\end{pmatrix}+s\begin{pmatrix}1\\3\\-2\end{pmatrix}$

14 Calculate the shortest distance between the lines $\dfrac{x}{6}=\dfrac{y}{4}=\dfrac{1+z}{2}$ and $\mathbf{r}=\begin{pmatrix}1\\-1\\0\end{pmatrix}+s\begin{pmatrix}-9\\-6\\3\end{pmatrix}$

15 Calculate the shortest distance from the point $(5, 2, 1)$ to the plane $\mathbf{r}\cdot(12\mathbf{i}+7\mathbf{j}-14\mathbf{k})=9$

16 Calculate the shortest distance from the point $(-12, -7, 4)$ to the plane $2x-9y-12z=-2$

Investigation

A **median** of a triangle is the line joining one of the vertices to the midpoint of the opposite side.
The scalar product of vectors can be used to calculate the length of a median.
Consider a triangle defined by the vectors **a**, **b** and **c**,
and the median that cuts side **c** is vector **p**.

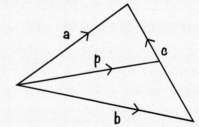

Vector **a** can be expressed as $\mathbf{p} + \dfrac{\mathbf{c}}{2}$ and **b** can be
expressed as $\mathbf{p} - \dfrac{\mathbf{c}}{2}$

Find **a·a** + **b·b** in terms of **p** and **c**

Hence, find an expression for the median in terms of the lengths of the sides of a triangle.
Use it to find the lengths of the medians of a triangle with side lengths 3, 4 and 5

Did you know?

The word vector is derived from the Latin word vector for meaning
"one who carries or conveys", hence its use in medical science to describe
the host organism for a disease.

Have a go

When **R** is a vector representing a constant, resultant force acting on a body which is moved
thorough a displacement represented by the vector **r**, the work done is given by **R·r**
   Using this vector expression, explain why it is a waste of energy to push a heavy object in a
direction near perpendicular to the direction in which you wish.
How could you ensure that more than half a person's efforts are of use?

History

The historical development of vectors relies on many underpinning
ideas in both pure and applied mathematics.
   One important development was that of complex numbers and
their geometrical representation. This led mathematicians to
wonder how to devise a system that allowed analysis of
three-dimensional space.
   Alongside such developments, Newton and others were busy considering
how to understand and analyse forces and motion. In his famous Principia
Mathematica, Newton wrote, "A body, acted on by two forces simultaneously,
will describe the diagonal of a parallelogram in the same time as it would describe the sides by those
forces separately".
   As you now see, such thinking relating to a parallelogram of forces was very close to what we
might now represent by vectors.

Isaac Newton

1 a Write a vector equation of the line that passes through the point with position vector $2\mathbf{i} + 4\mathbf{j} - \mathbf{k}$ and is parallel to the vector $\mathbf{i} + \mathbf{j} - 2\mathbf{k}$ **[1 mark]**

b Show that the point $(-1, 1, 5)$ lies on the line.   **[3]**

2 a Find the vector equation of the line that passes through the points with position vectors $2\mathbf{i} + \mathbf{j}$ and $3\mathbf{i} + \mathbf{k}$ **[2]**

b Show that the point $(8, -5, 7)$ does not lie on the line.   **[3]**

3 Find the vector equation of the line that passes through the point with position vector $\begin{pmatrix} 2 \\ 1 \\ 0 \end{pmatrix}$

and is parallel to the line with equation $\mathbf{r} = \begin{pmatrix} 2 \\ 4 \\ -3 \end{pmatrix} + \lambda \begin{pmatrix} 8 \\ -5 \\ 3 \end{pmatrix}$ **[2]**

4 The position vectors of points A and B are $\begin{pmatrix} 7 \\ 1 \\ -3 \end{pmatrix}$ and $\begin{pmatrix} -2 \\ 8 \\ -1 \end{pmatrix}$ respectively.

a Calculate $\overrightarrow{OA} \cdot \overrightarrow{OB}$ **[2]**

b Find the size, in degrees, of the acute angle AOB.   **[3]**

5 Calculate the cosine of the acute angle between the vectors $\begin{pmatrix} 4 \\ 3 \\ 5 \end{pmatrix}$ and $\begin{pmatrix} -1 \\ -2 \\ 1 \end{pmatrix}$

Give your answer in the form $a\sqrt{3}$ where a is a constant to be found.   **[4]**

6 Show that the vectors $-4\mathbf{i} + \mathbf{k}$ and $2\mathbf{i} - \mathbf{j} + 8\mathbf{k}$ are perpendicular.   **[2]**

7 For these pairs of lines, state whether they are skew, parallel or intersecting.

If they are intersecting find their point of intersection.

a $\mathbf{r} = \begin{pmatrix} 3 \\ -1 \\ 5 \end{pmatrix} + \lambda \begin{pmatrix} 2 \\ 1 \\ -1 \end{pmatrix}$ and $\mathbf{r} = \begin{pmatrix} 13 \\ 1 \\ 6 \end{pmatrix} + \mu \begin{pmatrix} 4 \\ 1 \\ 0 \end{pmatrix}$ **[5]**

b $\mathbf{r} = 3\mathbf{i} + \mathbf{j} - \mathbf{k} + s(\mathbf{i} - 2\mathbf{j})$ and $\mathbf{r} = 5\mathbf{i} + 2\mathbf{j} - 3\mathbf{k} + t(-2\mathbf{i} + 4\mathbf{j})$ **[2]**

c $\mathbf{r} = \mathbf{i} + 2\mathbf{j} + 3\mathbf{k} + \lambda(2\mathbf{i} + \mathbf{j} - \mathbf{k})$ and $\mathbf{r} = 6\mathbf{i} - \mathbf{k} + \mu(3\mathbf{i} + \mathbf{k})$ **[4]**

8 Show that the point $(3, 1, -2)$ lies on each of these planes.

a $\mathbf{r} \cdot \begin{pmatrix} 0 \\ 4 \\ -2 \end{pmatrix} = 8$ **b** $2x + 3y - z = 11$ **[4]**

9 A plane passes through the point with position vector $\mathbf{i} - 5\mathbf{j} + 2\mathbf{k}$ and contains the vectors $-3\mathbf{j} - \mathbf{k}$ and $-2\mathbf{i} + 4\mathbf{j}$

a Find the equation of the plane in vector form.   **[2]**

b Show that the vector $2\mathbf{i} + \mathbf{j} - 3\mathbf{k}$ is perpendicular to the plane.   **[4]**

c Find the equation of the plane in

i Scalar product form,   **ii** Cartesian form.   **[4]**

10 a Find the equation of the plane that passes through the points $(7, 12, -14)$, $(5, 4, -10)$, $(1, 9, -5)$ in the form $\mathbf{r} = \mathbf{a} + s\mathbf{b} + t\mathbf{c}$ **[3]**

b Does the point $(-1, 0, 4)$ lie on the plane? Explain how you know.   **[4]**

11 a Find the Cartesian equation of the line with vector equation

$\mathbf{r} = 2\mathbf{i} - 3\mathbf{k} + s(-\mathbf{i} + 4\mathbf{j} + 3\mathbf{k})$ [2]

b Does this represent the same line as $\dfrac{x-1}{-1} = \dfrac{y-4}{4} = \dfrac{z+6}{-3}$? Explain how you know. [3]

12 The point $(2, 7, -3)$ lies on the line with equation $\dfrac{x-a}{2} = \dfrac{y-b}{-3} = \dfrac{z-c}{1}$

a Write the values of a, b and c [1]

b Find the equation of the line in vector form. [2]

13 Which of these equations, if any, represent the same line?

A: $\dfrac{x-4}{2} = \dfrac{y-7}{-1} = \dfrac{z-2}{5}$ **B:** $\mathbf{r} = \begin{pmatrix} 8 \\ 5 \\ 8 \end{pmatrix} - \lambda \begin{pmatrix} 2 \\ -1 \\ 5 \end{pmatrix}$ **C:** $\mathbf{r} = \begin{pmatrix} 10 \\ 4 \\ 13 \end{pmatrix} + \lambda \begin{pmatrix} 2 \\ -1 \\ 5 \end{pmatrix}$ [3]

14 Calculate the shortest distance from the point $(5, -1, 6)$ to the line with equation

$\mathbf{r} = \begin{pmatrix} 7 \\ -3 \\ 9 \end{pmatrix} + t \begin{pmatrix} 2 \\ 1 \\ -2 \end{pmatrix}$ [4]

15 Calculate the perpendicular distance between the lines with equations

$\mathbf{r} = 3\mathbf{i} - \mathbf{j} + \mu(-2\mathbf{i} + 3\mathbf{j} - \mathbf{k})$ and $\mathbf{r} = 2\mathbf{i} + \mathbf{j} - \mathbf{k} + \lambda(-2\mathbf{i} + 3\mathbf{j} - \mathbf{k})$ [5]

16 The plane Π_1 has equation $\mathbf{r} \cdot (2\mathbf{i} - 3\mathbf{j}) = 4$ and the plane Π_2 has equation $\mathbf{r} \cdot (\mathbf{i} + 2\mathbf{j} - \mathbf{k}) = 7$

Calculate the acute angle between planes Π_1 and Π_2 [3]

17 A plane Π has equation $\mathbf{r} \cdot \begin{pmatrix} -1 \\ 3 \\ -3 \end{pmatrix} = -2$ and a line l has equation $\mathbf{r} = \begin{pmatrix} 5 \\ -7 \\ 1 \end{pmatrix} + t \begin{pmatrix} 0 \\ -4 \\ 2 \end{pmatrix}$

a Calculate the acute angle between the plane Π and the line l [3]

b Find the point of intersection between the plane Π and the line l [4]

18 Find the perpendicular distance from the point $(5, 0, -2)$ to the plane with equation

$2x + y - 3z = 9$ [3]

19 The lines l_1 and l_2 have Cartesian equations $\dfrac{x+1}{2} = \dfrac{y-4}{7} = \dfrac{1+z}{-1}$ and $\mathbf{r} = \begin{pmatrix} -5 \\ 4 \\ 1 \end{pmatrix} + t \begin{pmatrix} -4 \\ -2 \\ 2 \end{pmatrix}$

a Calculate the acute angle between l_1 and l_2 [4]

b Find the point of intersection between l_1 and l_2 [4]

20 Find the equation of the plane through the point $(5, -2, 7)$ that contains the line with

equation $\dfrac{x-2}{4} = \dfrac{y+1}{5} = \dfrac{z-1}{-2}$. Give your answer in the form $\mathbf{r} = \mathbf{a} + \lambda\mathbf{b} + \mu\mathbf{c}$ [6]

21 The line l_1 is given by $\mathbf{r} = 6\mathbf{i} - 3\mathbf{j} + \mathbf{k} + s(-\mathbf{i} + \mathbf{j} - 2\mathbf{k})$ and the line l_2 is given by

$\dfrac{x-4}{-5} = \dfrac{y-2}{2} = \dfrac{z+5}{-8}$. The point A is the intersection of l_1 and l_2 and the point B has coordinates

$(1, -2, 7)$. Calculate the area of triangle OAB [8]

22 Is the point $(-2, 4, 7)$ closer to the line with equation $\mathbf{r} = 5\mathbf{i} + 2\mathbf{k} + \lambda(\mathbf{i} - 2\mathbf{j} + 4\mathbf{k})$ or to the line

parallel to the vector $2\mathbf{i} + \mathbf{j} - \mathbf{k}$ which passes through the origin? [10]

23 The plane Π_1 has equation $\mathbf{r} \cdot \begin{pmatrix} 2 \\ 0 \\ -3 \end{pmatrix} = -11$ and the plane Π_2 has equation $x - 2y - z = 1$ [9]

The line with equation $\dfrac{x+2}{4} = \dfrac{y+1}{-3} = \dfrac{z-4}{1}$ intersects Π_1 at the point A and Π_2 at the point B

Calculate the length of AB

The following mathematical formulae will be provided for you.

Summations

$$\sum_{r=1}^{n} r^2 = \frac{1}{6}n(n+1)(2n+1) \qquad \sum_{r=1}^{n} r^3 = \frac{1}{4}n^2(n+1)^2$$

Matrix transformations

Anticlockwise rotation through θ about O: Reflection in the line $y = (\tan\theta)x$:

$$\begin{pmatrix} \cos\theta & -\sin\theta \\ \sin\theta & \cos\theta \end{pmatrix} \qquad \begin{pmatrix} \cos 2\theta & \sin 2\theta \\ \sin 2\theta & -\cos 2\theta \end{pmatrix}$$

Area of a sector

$$A = \frac{1}{2}\int r^2 \, d\theta \qquad \text{(polar coordinates)}$$

Complex numbers

$$\{r(\cos\theta + i\sin\theta)\}^n = r^n(\cos n\theta + i\sin n\theta)$$

The roots of $z^n = 1$ are given by $z = e^{\frac{2\pi k i}{n}}$ for $k = 0, 1, 2, ..., n-1$

Maclaurin's and Taylor's Series

$$f(x) = f(0) + xf'(0) + \frac{x^2}{2!}f''(0) + ... + \frac{x^r}{r!}f^{(r)}(0) + ...$$

$$e^x = \exp(x) = 1 + x + \frac{x^2}{2!} + ... + \frac{x^r}{r!} + ... \text{ for all } x$$

$$\ln(1+x) = x - \frac{x^2}{2} + \frac{x^3}{3} - ... + (-1)^{r+1}\frac{x^r}{r} + ... \; (-1 < x \le 1)$$

$$\sin x = x - \frac{x^3}{3!} + \frac{x^5}{5!} - ... + (-1)^r \frac{x^{2r+1}}{(2r+1)!} + ... \text{ for all } x$$

$$\cos x = 1 - \frac{x^2}{2!} + \frac{x^4}{4!} - ... + (-1)^r \frac{x^{2r}}{(2r)!} + ... \text{ for all } x$$

$$\arctan x = x - \frac{x^3}{3} + \frac{x^5}{5} - ... + (-1)^r \frac{x^{2r+1}}{2r+1} + ... \; (-1 \le x \le 1)$$

Vectors

Vector products: $\mathbf{a} \times \mathbf{b} = |\mathbf{a}||\mathbf{b}|\sin\theta\,\hat{\mathbf{n}} = \begin{vmatrix} \mathbf{i} & \mathbf{j} & \mathbf{k} \\ a_1 & a_2 & a_3 \\ b_1 & b_2 & b_3 \end{vmatrix} = \begin{pmatrix} a_2 b_3 - a_3 b_2 \\ a_3 b_1 - a_1 b_3 \\ a_1 b_2 - a_2 b_1 \end{pmatrix}$

$$\mathbf{a}\cdot(\mathbf{b}\times\mathbf{c}) = \begin{vmatrix} a_1 & a_2 & a_3 \\ b_1 & b_2 & b_3 \\ c_1 & c_2 & c_3 \end{vmatrix} = \mathbf{b}\cdot(\mathbf{c}\times\mathbf{a}) = \mathbf{c}\cdot(\mathbf{a}\times\mathbf{b})$$

If A is the point with position vector $\mathbf{a} = a_1\mathbf{i} + a_2\mathbf{j} + a_3\mathbf{k}$ and the direction vector $\mathbf{b}$ is given by $\mathbf{b} = b_1\mathbf{i} + b_2\mathbf{j} + b_3\mathbf{k}$, then the straight line through A with direction vector $\mathbf{b}$ has cartesian equation

$$\frac{x-a_1}{b_1} = \frac{y-a_2}{b_2} = \frac{z-a_3}{b_3}(=\lambda)$$

The plane through A with normal vector $\mathbf{n} = n_1\mathbf{i} + n_2\mathbf{j} + n_3\mathbf{k}$ has cartesian equation $n_1x + n_2y + n_3z + d = 0$ where $d = -\mathbf{a.n}$

The plane through non–collinear points A, B and C vector equation

$$\mathbf{r} = \mathbf{a} + \lambda(\mathbf{b} - \mathbf{a}) + \mu(\mathbf{c} - \mathbf{a}) = (1 - \lambda - \mu)\mathbf{a} + \lambda\mathbf{b} + \mu\mathbf{c}$$

The plane through the point with position vector $\mathbf{a}$ and parallel to $\mathbf{b}$ and $\mathbf{c}$ has equation

$$\mathbf{r} = \mathbf{a} + s\mathbf{b} + t\mathbf{c}$$

The perpendicular distance of (α, β, γ) from $n_1x + n_2y + n_3z + d = 0$ is $\dfrac{|n_1\alpha + n_2\beta + n_3\gamma + d|}{\sqrt{n_1^2 + n_2^2 + n_3^2}}$.

Hyperbolic functions

$\cosh^2 x - \sinh^2 x = 1$

$\operatorname{arcosh} x = \ln\left\{x + \sqrt{x^2 - 1}\right\}$ $(x \geq 1)$

$\sinh 2x = 2\sinh x \cosh x$

$\operatorname{arsinh} x = \ln\left\{x + \sqrt{x^2 - 1}\right\}$

$\cosh 2x = \cosh^2 x + \sinh^2 x$

$\operatorname{artanh} x = \dfrac{1}{2}\ln\left(\dfrac{1+x}{1-x}\right)$ $(|x| < 1)$

Differentiation

$f(x)$	$f'(x)$
$\arcsin x$	$\dfrac{1}{\sqrt{1-x^2}}$
$\arccos x$	$-\dfrac{1}{\sqrt{1-x^2}}$
$\arctan x$	$\dfrac{1}{1+x^2}$
$\sinh x$	$\cosh x$
$\cosh x$	$\sinh x$
$\tanh x$	$\operatorname{sech}^2 x$
$\operatorname{arsinh} x$	$\dfrac{1}{\sqrt{1+x^2}}$
$\operatorname{arcosh} x$	$\dfrac{1}{\sqrt{x^2-1}}$
$\operatorname{artanh} x$	$\dfrac{1}{1-x^2}$

Integration (+ constant; $a > 0$ where relevant)

$f(x)$	$\int f(x)\,dx$	
$\sinh x$	$\cosh x$	
$\cosh x$	$\sinh x$	
$\tanh x$	$\ln \cosh x$	
$\dfrac{1}{\sqrt{a^2 - x^2}}$	$\arcsin\left(\dfrac{x}{a}\right)$	$(|x| < a)$
$\dfrac{1}{a^2 + x^2}$	$\dfrac{1}{a}\arctan\left(\dfrac{x}{a}\right)$	
$\dfrac{1}{\sqrt{x^2 - a^2}}$	$\operatorname{arcosh}\left(\dfrac{x}{a}\right),\ \ln\{x + \sqrt{x^2 - a^2}\}$	$(x > a)$
$\dfrac{1}{\sqrt{a^2 + x^2}}$	$\operatorname{arsinh}\left(\dfrac{x}{a}\right),\ \ln\{x + \sqrt{x^2 + a^2}\}$	
$\dfrac{1}{a^2 - x^2}$	$\dfrac{1}{2a}\ln\left|\dfrac{a + x}{a - x}\right| = \dfrac{1}{a}\operatorname{artanh}\left(\dfrac{x}{a}\right)$	$(|x| < a)$
$\dfrac{1}{x^2 - a^2}$	$\dfrac{1}{2a}\ln\left|\dfrac{x - a}{x + a}\right|$	

Mathematical notation
For AS Level Further Maths

You should understand the following notation without need for further explanation.

Set Notation

$\in$	is an element of
$\notin$	is not an element of
$\subseteq$	is a subset of
$\subset$	is a proper subset of
$\{x_1, x_2, \ldots\}$	the set with elements $x_1, x_2, \ldots$
$\{x: \ldots\}$	the set of all x such that ...
$n(A)$	the number of elements in set A
$\varnothing$	the empty set
ε	the universal set
A'	the complement of the set A
$\mathbb{N}$	the set of natural numbers, $\{1, 2, 3, \ldots\}$
$\mathbb{Z}$	the set of integers, $\{0, \pm 1, \pm 2, \pm 3, \ldots\}$
$\mathbb{Z}^+$	the set of positive integers, $\{1, 2, 3, \ldots\}$
$\mathbb{Z}_0^+$	the set of non-negative integers, $\{0, 1, 2, 3, \ldots\}$
$\mathbb{R}$	the set of real numbers
$\mathbb{Q}$	the set of rational numbers, $\left\{\dfrac{p}{q} : p \in \mathbb{Z}, \ q \in \mathbb{Z}^+\right\}$
$\cup$	union
$\cap$	intersection
(x, y)	the ordered pair x, y
$[a, b]$	the closed interval $\{x \in \mathbb{R} : a \le x \le b\}$
$[a, b)$	the interval $\{x \in \mathbb{R} : a \le x < b\}$
$(a, b]$	the interval $\{x \in \mathbb{R} : a < x \le b\}$
(a, b)	the open interval $\{x \in \mathbb{R} : a < x < b\}$
$\mathbb{C}$	the set of complex numbers

Miscellaneous Symbols

$=$	is equal to
$\ne$	is not equal to
$\equiv$	is identical to or is congruent to
$\approx$	is approximately equal to
∞	infinity
$\propto$	is proportional to
$<$	is less than
$\le, \leq$	is less than or equal to, is not greater than
$>$	is greater than
$\ge, \geq$	is greater than or equal to, is not less than
$\therefore$	therefore
$\because$	because
$p \Rightarrow q$	p implies q (if p then q)
$p \Leftarrow q$	p is implied by q (if q then p)
$p \Leftrightarrow q$	p implies and is implied by q (p is equivalent to q)

a	first term for an arithmetic or geometric sequence
l	last term for an arithmetic sequence
d	common difference for an arithmetic sequence
r	common ratio for a geometric sequence
S_n	sum to n terms of a sequence
S_∞	sum to infinity of a sequence

Operations

$a + b$	a plus b
$a - b$	a minus b
$a \times b,\ ab,\ a \cdot b$	a multiplied by b
$a \div b,\ \dfrac{a}{b}$	a divided by b
$\displaystyle\sum_{i=1}^{n} a_i$	$a_1 + a_2 + \ldots + a_n$
$\sqrt{a}$	the non-negative square root of a
$\lvert a \rvert$	the modulus of a
$n!$	n factorial: $n! = n \times (n-1) \times \ldots \times 2 \times 1,\ n \in \mathbb{N};\ 0! = 1$
$\dbinom{n}{r},\ {}^{n}C_r,\ {}_{n}C_r$	the binomial coefficient $\dfrac{n!}{r!(n-r)!}$ for $n,\ r \in \mathbb{Z}_0^{+},\ r \le n$ or $\dfrac{n(n-1)\ldots(n-r+1)}{r!}$ for $n \in \mathbb{Q},\ r \in \mathbb{Z}_0^{+}$

Functions

$f(x)$	the value of the function f at x
$\displaystyle\lim_{x \to a} f(x)$	the limit of $f(x)$ as x tends to a
$f\colon x \mapsto y$	the function f maps the element x to the element y
f^{-1}	the inverse function of the function f
gf	the composite function of f and g which is defined by $gf(x) = g(f(x))$
$\Delta x,\ \delta x$	an increment of x
$\dfrac{dy}{dx}$	the derivative of y with respect to x
$\dfrac{d^n y}{dx^n}$	the nth derivative of y with respect to x
$f'(x) \ldots,\ f^{(n)}(x)$	the first, ..., nth derivatives of $f(x)$ with respect to x
$\dot{x},\ \ddot{x},\ \ldots$	the first, second, ... derivatives of x with respect to t
$\displaystyle\int y\,dx$	the indefinite integral of y with respect to x
$\displaystyle\int_{a}^{b} y\,dx$	the definite integral of y with respect to x between the limits $x = a$ and $x = b$

Exponential and Logarithmic Functions

e	base of natural logarithms
$e^x,\ \exp x$	exponential function of x
$\log_a x$	logarithm to the base a of x
$\ln x,\ \log_e x$	natural logarithm of x

Trigonometric Functions

sin, cos, tan, the trigonometric functions

°	degrees
rad	radians

Complex numbers

i, j	square root of -1
$x + iy$	complex number with real part x and imaginary part y
$r(\cos\theta + i\sin\theta)$	modulus argument form of a complex number with modulus r and argument θ
z	a complex number $z = x + iy = r(\cos\theta + i\sin\theta)$
$\mathrm{Re}(z)$	the real part of z, $\mathrm{Re}(z) = x$
$\mathrm{Im}(z)$	the imaginary part of z, $\mathrm{Im}(z) = y$
$|z|$	the modulus of z, $|z| = \sqrt{x^2 + y^2}$
$\arg(z)$	the argument of z, $\arg(z) = \theta$, $-\pi < \theta < \pi$
z^*	the complex conjugate of z, $x - iy$

Matrices

$\mathbf{M}$	a matrix $\mathbf{M}$
$\mathbf{0}$	zero matrix
$\mathbf{I}$	identity matrix
$\mathbf{M}^{-1}$	the inverse of the matrix $\mathbf{M}$
$\mathbf{M}^{\mathrm{T}}$	the transpose of the matrix $\mathbf{M}$
$\mathbf{Mr}$	Image of the column vector $\mathbf{r}$ under the transformation associated with the matrix $\mathbf{M}$

Vectors

$\mathbf{a}, \underline{a}, \underset{\sim}{a}$	the vector $\mathbf{a}, \underline{a}, \underset{\sim}{a}$
$\overrightarrow{AB}$	the vector represented in magnitude and direction by the directed line segment AB
$\hat{\mathbf{a}}$	a unit vector in the direction of $\mathbf{a}$
$\mathbf{i}, \mathbf{j}, \mathbf{k}$	unit vectors in the directions of the Cartesian coordinate axes
$|\mathbf{a}|, a$	the magnitude of $\mathbf{a}$
$|\overrightarrow{AB}|, AB$	the magnitude of $\overrightarrow{AB}$
$\begin{pmatrix} a \\ b \end{pmatrix}, a\mathbf{i} + b\mathbf{j}$	column vector and corresponding unit vector notation
$\mathbf{r}$	position vector
$\mathbf{s}$	displacement vector
$\mathbf{v}$	velocity vector
$\mathbf{a}$	acceleration vector
$\mathbf{a} \cdot \mathbf{b}$	the scalar product of $\mathbf{a}$ and $\mathbf{b}$

Answers

Chapter 1
Exercise 1.1A

1 **a** $x = \pm 5i$ **b** $x = \pm 11i$

   c $x = \pm 2\sqrt{5}i$ **d** $x = \pm 2\sqrt{2}i$

   e $z = \pm 3i$ **f** $z = \pm 2\sqrt{3}i$

2 **a** $7 - 6i$ **b** $-7 - 10i$

   c $18 - 27i$ **d** $26 + 25i$

   e $-41 - 63i$ **f** $27 - 10i$

3 **a** $7 + 17i$ **b** $39 - 27i$

   c $3 + 8i$ **d** $65 - 72i$

4 **a** $-i$ **b** 1

   c i **d** $-8i$

   e $= 81$ **f** $-112 + 180i$

5 **a** $\dfrac{6}{5} - \dfrac{3}{5}i$ **b** $-\dfrac{5}{13} + \dfrac{1}{13}i$

   c $-\dfrac{2}{5} + \dfrac{11}{5}i$ **d** $-\dfrac{1}{5} - \dfrac{7}{5}i$

   e $(1 - 2\sqrt{2}) + (-2 - \sqrt{2})i$ **f** $-\sqrt{2}$

6 **a** $2 + 4i$ **b** $-11 + 10i$

   c $-\dfrac{5}{17} + \dfrac{14}{17}i$ **d** $-\dfrac{5}{13} - \dfrac{14}{13}i$

Exercise 1.1B

1 $a = \pm 3,\ b = \mp 12$

2 $b = 7,\ a = 34$

3 $4 + i$

4 $x = 2 - i$

5 $z_2 = 8 + 2i,\ z_1 = 3 - 5i$

6 $w = -2 - \dfrac{1}{2}i,\ z = 4 + 7i$

7 $z_2 = -3 - 2i,\ z_1 = \dfrac{1}{2} + i$

8 **a** $w = 3 + 5i$ or $w = -3 - 5i$

   b $w = 1 - 2i$ or $w = -1 + 2i$

   c $w = 5 - 2i$ or $w = -5 + 2i$

9 $-2 + \sqrt{2}i$ and $2 - \sqrt{2}i$

10 $w = -3 \pm i,\ z = 1 \pm 3i$

11 **a** $z = 1 + 7i$ or $z = 1 - 7i$

    b $z = 2 - 3i$ or $z = -2 + 3i$

12 **a** $-7 - 24i$

    b $4282 + 1475i$

    c $44 + 8i$

13 $a = \pm 3,\ b = \pm 96$

Exercise 1.2A

1 **a** $5 + 2i$ **b** $8 - i$

   c $-5i - 6$ **d** $\sqrt{2} + i\sqrt{3}$

   e $\dfrac{1}{3} - 4i$ **f** $-\dfrac{2}{3}i - 5$

2 **a** 85 **b** 18

   c $-4i$ **d** $\dfrac{77}{85} - \dfrac{36}{85}i$

   e $9 - 2i$ **f** $\dfrac{77}{85} + \dfrac{36}{85}i$

3 **a** 8 **b** $-2\sqrt{6}$

   c $2\sqrt{2}i$ **d** $\dfrac{1}{2} - \dfrac{\sqrt{3}}{2}i$

   e 8 **f** 24

4 **a** $x = -\dfrac{5}{2} \pm \dfrac{\sqrt{3}}{2}i$ **b** $x = \dfrac{3}{2} \pm \dfrac{\sqrt{11}}{2}i$

   c $x = -\dfrac{7}{4} \pm \dfrac{\sqrt{7}}{4}i$ **d** $x = \dfrac{5}{3} \pm \dfrac{\sqrt{2}}{3}i$

5 **a** $x^2 - 3x - 28 = 0$ **b** $x^2 - 6x + 34 = 0$

   c $x^2 + 2x + 82 = 0$ **d** $x^2 + 10x + 41 = 0$

6 **a** $\sqrt{3} - i$ **b** $z^2 - 2\sqrt{3}z + 4 = 0$

7 **a** $x^2 - 4x + 5 = 0$ **b** $x^2 - 8x + 25 = 0$

   c $x^2 + 2x + 50 = 0$ **d** $x^2 + 10x + 29 = 0$

   e $x^2 - 2ax + (9 + a^2) = 0$ **f** $x^2 - 10x + (25 + b^2) = 0$

8 **a** $x^3 + x^2 - 7x + 65 = 0$

   b $x^3 - 2x - 4 = 0$

   c $x^3 - 2\sqrt{3}x^2 + 7x = 0$

   d $x^3 + (2\sqrt{2} - 3)x^2 + (3 - 6\sqrt{2})x - 9 = 0$

9 **a** $6 + 2i$

   b $2z^3 - 23z^2 + 68z + 40 = 0$

10 **a** $x^3 + 9x^2 + 25x + 25 = (x + 5)(x^2 + 4x + 5)$

     $f(x) = 0 \Rightarrow x^2 + 4x + 5 = 0$

    b $x = -5,\ -2 \pm i$

11 **a** $1 - 8i$ is also a root giving quadratic factor $x^2 - 2x + 65$

     $x^4 + 4x^3 + 66x^2 + 364x + 845$

     $= (x^2 - 2x + 65)(x^2 + 6x + 13) = 0$

     $\Rightarrow x^2 + 6x + 13 = 0$

    b $x = 1 \pm 8i,\ -3 \pm 2i$

12 **a** $x = -3,\ 1 \pm 3i$ **b** $x = 1,\ 2 \pm 5i$

    c $x = 5,\ -1 \pm 6i$ **d** $x = -2,\ 5 \pm 4i$

13 **a** $x = 1 \pm 2i,\ \pm 2$

    b $x = 2 \pm 5i,\ -\dfrac{1}{2},\ 3$

    c $x = 3 \pm 2i,\ 1 \pm i$

14 **a** $(x - 7i)(x + 7i) = x^2 + 49$

     $x^4 - x^3 + 43x^2 - 49x - 294 = (x^2 + 49)(x^2 - x - 6) = 0$

     So can be written $(x^2 + 49)(x - 3)(x + 2) = 0$

     $(A = 49,\ B = -3,\ C = 2)$

Exercise 1.2B

1 Let $z = a + bi,\ w = c + di,\ a, b, c, d \in \mathbb{R}$

   $(zw)^* = ((a + bi)(c + di))^*$

        $= ((ac - bd) + (bc + ad)i)^*$

        $= (ac - bd) - (bc + ad)i$

        $= (a - bi)(c - di)$

        $= z^*w^*$

   b $(z^*)^* = ((a + bi)^*)^*$

        $= (a - bi)^*$

        $= a + bi$

        $= z$

c $\left(\dfrac{z}{w}\right)^* = \left(\dfrac{a+bi}{c+di}\right)^*$

$= \left(\dfrac{(a+bi)(c-di)}{(c+di)(c-di)}\right)^*$

$= \left(\dfrac{(ac+bd)+(bc-ad)i}{c^2-d^2i^2}\right)^*$

$= \dfrac{ac+bd-(bc-ad)i}{c^2+d^2}$

$\dfrac{z^*}{w^*} = \dfrac{a-bi}{c-di}$

$= \dfrac{(a-bi)(c+di)}{(c-di)(c+di)}$

$= \dfrac{ac+bd-(bc-ad)i}{c^2+d^2}$

$= \left(\dfrac{z}{w}\right)^*$

2 a Let $z = a+bi$, $a, b \in \mathbb{R}$
$z + z^* = (a+bi)+(a-bi)$
$= 2a$ so a real number

b $z - z^* = (a+bi)-(a-bi)$
$= 2bi$ so an imaginary number

c $zz^* = (a+bi)(a-bi)$
$= a^2 - abi + abi - b^2i^2$
$= a^2 + b^2$ so a real number

3 $z = 3 \pm 7i$

4 $w = 2+9i$ or $w = -2+9i$

5 $z = \sqrt{3} - i$

6 $w = 4+2i$

7 a $k = -119$ **b** $x = (7), 1 \pm 4i$

8 $k = -8$
$x = -\dfrac{3}{2} \pm \dfrac{\sqrt{23}}{2}i, 1$

9 $x = 1, 3, \pm i$

10 $x = -3, 7, 1 + \sqrt{2}i, 1 - \sqrt{2}i$

11 a $k = 10$
b $x = -6 \pm i, 3, -1$

12 $A = -10$
$x = 5 \pm i, \pm \sqrt{2}i$

13 a $x^4 - 4x^3 + 24x^2 - 40x + 100 = 0$
b $x^4 + 12x^3 + 62x^2 + 156x + 169 = 0$

14 a $(x+1)(x^2 - 20x + 109)$
b $x = -1, 10 \pm 3i$
c

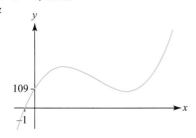

15 a $x = 1 \pm 3i$ (repeated)
b

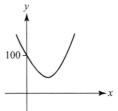

16 a $p(x) = 0$ has a repeated real root and two complex roots
b $x = -5, 2 \pm i$

17 $x^3 - (a+b+c)x^2 + (ab+ac+bc)x - abc = 0$

18 a -5 **b** -12

19 a $-\alpha - \beta - \gamma - \delta$ **b** $\alpha\beta\gamma\delta$

20 a $x = \pm 2\sqrt{2}, x = \pm 2\sqrt{2}i$

b $x = 3, -\dfrac{3}{2} \pm \dfrac{3\sqrt{3}}{2}i$

c $x = -3i, \pm\dfrac{3\sqrt{3}}{2} + \dfrac{3}{2}i$

21 a Order 5 (quantic)
b $x^5 - 6x^4 + 10x^3 - 20x^2 + 9x + 306 = 0$

22 Let $n = 1$, then $(z^n)^* = z^*$
and $(z^*)^n = z^*$ so true for $n = 1$
Assume true for $n = k$ and let $n = k + 1$:

$(z^{k+1})^* = (zz^k)^*$

$= z^*(z^k)^*$ using result from question 1a

$= z^*(z^*)^k$ since assuming true for $n = k$

$= (z^*)^{k+1}$

True for $n = 1$ and true for $n = k$ implies true for $n = k + 1$
therefore true for all positive integers n

Exercise 1.3A

1 $\overrightarrow{OA} = 5+3i$
$\overrightarrow{OB} = -3-6i$
$\overrightarrow{OC} = 6$
$\overrightarrow{OD} = -1+2i$
$\overrightarrow{OE} = -4i$
$\overrightarrow{OF} = -6+5i$

2 $u = -3+5i, v = 2-7i, w = 4i, z = -4+i$

3

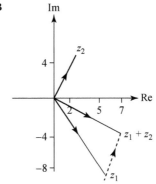

4

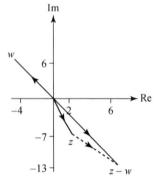

5

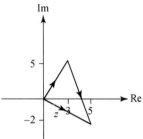

6

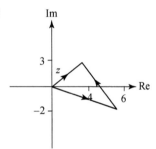

7

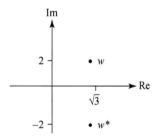

w^* is a reflection of w in the real axis.

8

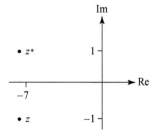

z is a reflection of z^* in the real axis.

9 $x = \pm 4i$

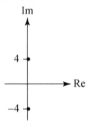

b $x = \pm 4\sqrt{5}\,i$

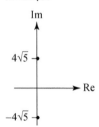

In both cases the points are reflections of each other in the real axis

10 a $z = -1 \pm \sqrt{3}\,i$

b

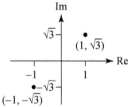

c Reflection in real axis.

11 a $z = 1 \mp 5i$

b

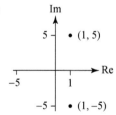

c Reflection in real axis.

12 a

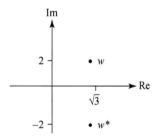

b

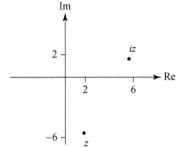

In both cases iz is the image of z rotated $\dfrac{\pi}{2}$ radians (anti–clockwise) about the origin.

13 a

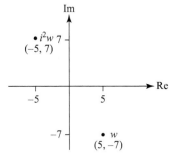

b

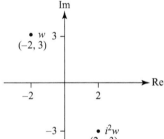

In both cases, i^2w is w rotated π radians around the origin.

Exercise 1.3B

1 a $z_2 = 2 + 5i$

b

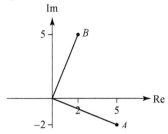

c Gradient of OA is $-\dfrac{2}{5}$, gradient of OB is $\dfrac{5}{2}$

$-\dfrac{2}{5} \times \dfrac{5}{2} = -1$ so OA and OB are perpendicular

$\therefore$ AOB is a right angle

2 a $x = -1 \pm \sqrt{7}i, 1$

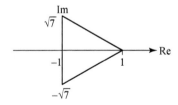

b Isosceles triangle

c $2\sqrt{7}$

3 6 square units

4 a $A = -2, B = 3$

b $x = -\dfrac{3}{2} \pm \dfrac{\sqrt{3}}{2}i, 1 \pm \sqrt{2}i$

c

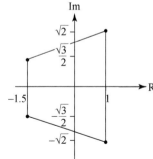

d A trapezium

e 5.70 square units

5 a $(-2)^4 + 8(-2)^3 + 40(-2)^2 + 96(-2) + 80 = 0$

$x^4 + 8x^3 + 40x^2 + 96x + 80 = (x+2)^2(x^2 + 4x + 20) = 0$

b $x = -2 \pm 4i$

c

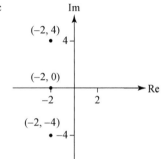

6 a -1 and 2

b 9 square units

7 a $x^3 - x^2 + 9x - 9 = (x - 1)(x^2 + 9) = 0$

$\Rightarrow x = 1, \pm 3i$

$A(1, 0), B(0, 3), C(0, -3)$

$|AB| = \sqrt{1^2 + 3^2} = \sqrt{10}$

$|AC| = \sqrt{1^2 + 3^2} = \sqrt{10}$

$|BC| = \sqrt{0^2 + 6^2} = 6$

So ABC is isosceles

b 3 square units

8 a $A = (5, -2)$

$B = (7, 3)$

$C = (2, 5)$

Length $OA = \sqrt{5^2 + 2^2} = \sqrt{29}$

Length $AB = \sqrt{(7-5)^2 + (3+2)^2} = \sqrt{29}$

Length $BC = \sqrt{(7-2)^2 + (5-3)^2} = \sqrt{29}$

Length $OC = \sqrt{29}$

Gradient $OA = -\dfrac{2}{5}$

Gradient $AB = \dfrac{5}{2}$ So OA and AB are perpendicular

Gradient $BC = -\dfrac{2}{5}$ so AB and BC are perpendicular

Gradient $OC = \dfrac{5}{2}$ so OC and OA are perpendicular.

Therefore it is a square.

b 29 square units

9 a Q represents $2+4i$, R represents $6-4i$,

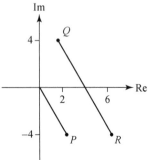

b OP has been enlarged by scale factor 2 centre the origin then translated by the vector $\begin{pmatrix} 2 \\ 4 \end{pmatrix}$.

Or alternatively: OP has been enlarged by scale factor 2 centre $(-2, -4)$.

10 a $-2+9i$

b CB is an enlargement of OA centre the origin, scale factor 2, then translated by the vector $\begin{pmatrix} -2 \\ 9 \end{pmatrix}$

11 $|a-c|(b+d)$

12

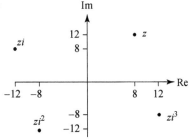

z is rotated around the origin by $\dfrac{\pi}{2}$ radians (anticlockwise) for zi, by π radians for zi^2 and by $\dfrac{3\pi}{2}$ radians (anticlockwise) for zi^3

13 Let $z = a + bi$,

Then $iz = ai + bi^2 = -b + ai$

Gradient of $OA = \dfrac{b}{a}$

Gradient of $OB = \dfrac{a}{-b}$

$\dfrac{b}{a} \times \dfrac{a}{-b} = -1$ so OA is perpendicular to OB

14 17 square units

15 3 square units

16 10

17 a $\dfrac{z}{w} = \dfrac{12-5i}{3+2i} = \dfrac{(12-5i)(3-2i)}{(3+2i)(3-2i)}$

$= \dfrac{36-15i-24i-10}{9+4}$

$= \dfrac{26-39i}{13} = 2-3i$

17 b

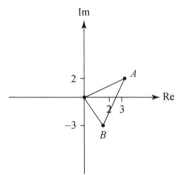

c Gradient of $OB = -\dfrac{3}{2}$

Gradient of $OA = \dfrac{2}{3}$

$-\dfrac{3}{2} \times \dfrac{2}{3} = -1$ therefore OB and OC are perpendicular so OAB is a right-angled triangle.

d $\dfrac{13}{2}$ square units

18

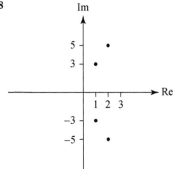

19 a i A kite

ii An isosceles triangle

iii A trapezium

b i $a=1, b=0, c=0, d=0, e=-16$

ii $a=1, b=-4, c=6, d=-4, e=-80$

1 a Modulus $=13$

Argument $= 0.395^c$

b Modulus $= 5$

Argument $= -0.644^c$

c Modulus $= 3$

Argument $= -\dfrac{\pi}{2}$

d Modulus $= 10$

Argument $= -2.21^c$.

e Modulus $= 5\sqrt{2}$

Argument $= 1.71$

f Modulus $= \sqrt{2^2+1^2} = \sqrt{5}$

Argument $= -2.68^c$

g Modulus $= 2$

Argument $= -\dfrac{3\pi}{4}$

h Modulus $= 3$

Argument $= 0.955^c$

2 a $zw = (1+3i)(-5+2i) = -11-13i$

$|zw| = \sqrt{11^2+13^2} = \sqrt{290}$

$|z| = \sqrt{1^2+3^2} = \sqrt{10}$

$|w| = \sqrt{5^2+2^2} = \sqrt{29}$

$|z||w| = \sqrt{10}\sqrt{29} = \sqrt{290} = |zw|$ as required.

$\dfrac{z}{w} = \dfrac{1+3i}{-5+2i}$

$= \dfrac{(1+3i)(5+2i)}{(-5+2i)(5+2i)} = \dfrac{1}{29} - \dfrac{17}{29}i$

$$\left|\frac{z}{w}\right| = \sqrt{\left(\frac{1}{29}\right)^2 + \left(\frac{17}{29}\right)^2} = \sqrt{\frac{10}{29}}$$

$$\frac{|z|}{|w|} = \frac{\sqrt{10}}{\sqrt{29}} = \sqrt{\frac{10}{29}} = \left|\frac{z}{w}\right| \text{ as required}$$

b $\ zw = (-2-i)(\sqrt{5}i) = \sqrt{5} - 2\sqrt{5}i$

$$|zw| = \sqrt{(\sqrt{5})^2 + (2\sqrt{5})^2} = 5$$

$$|z| = \sqrt{2^2 + 1^2} = \sqrt{5}$$

$$|w| = \sqrt{5}$$

$$|z||w| = \sqrt{5}\sqrt{5} = 5 = |zw| \text{ as required.}$$

$$\frac{z}{w} = \frac{-2-i}{\sqrt{5}i}$$

$$= \frac{(-2-i)i}{\sqrt{5}i \cdot i} = -\frac{1}{\sqrt{5}} + \frac{2}{\sqrt{5}}i$$

$$\left|\frac{z}{w}\right| = \sqrt{\left(\frac{1}{\sqrt{5}}\right)^2 + \left(\frac{2}{\sqrt{5}}\right)^2} = 1$$

$$\frac{|z|}{|w|} = \frac{\sqrt{5}}{\sqrt{5}} = 1 = \left|\frac{z}{w}\right| \text{ as required}$$

c $\ zw = (-\sqrt{3}+6i)(1-\sqrt{3}i) = 5\sqrt{3}+9i$

$$|zw| = \sqrt{(5\sqrt{3})^2 + (9)^2} = 2\sqrt{39}$$

$$|z| = \sqrt{(\sqrt{3})^2 + 6^2} = \sqrt{39}$$

$$|w| = \sqrt{1^2 + (\sqrt{3})^2} = 2$$

$$|z||w| = 2\sqrt{39} = |zw| \text{ as required.}$$

$$\frac{z}{w} = \frac{-\sqrt{3}+6i}{1-\sqrt{3}i}$$

$$= \frac{(-\sqrt{3}+6i)(1+\sqrt{3}i)}{(1-\sqrt{3}i)(1+\sqrt{3}i)} = -\frac{7\sqrt{3}}{4} + \frac{3}{4}i$$

$$\left|\frac{z}{w}\right| = \sqrt{\left(\frac{7\sqrt{3}}{4}\right)^2 + \left(\frac{3}{4}\right)^2} = \frac{\sqrt{39}}{2}$$

$$\frac{|z|}{|w|} = \frac{\sqrt{39}}{2} = \left|\frac{z}{w}\right| \text{ as required}$$

3 a $\ zw = (1+i)(3+\sqrt{3}i) = (3-\sqrt{3})+(3+\sqrt{3})i$

$$\arg(zw) = \tan^{-1}\left(\frac{3+\sqrt{3}}{3-\sqrt{3}}\right) = \frac{5\pi}{12}$$

$$\arg z = \tan^{-1}1 = \frac{\pi}{4}$$

$$\arg w = \tan^{-1}\frac{\sqrt{3}}{3} = \frac{\pi}{6}$$

$$\arg z + \arg w = \frac{\pi}{4} + \frac{\pi}{6} = \frac{5\pi}{12} \text{ as required}$$

$$\frac{z}{w} = \frac{1+i}{3+\sqrt{3}i}$$

$$= \frac{(1+i)(3-\sqrt{3}i)}{(3+\sqrt{3}i)(3-\sqrt{3}i)} = \frac{3+\sqrt{3}}{12} + \frac{3-\sqrt{3}}{12}i$$

$$\arg\left(\frac{z}{w}\right) = \tan^{-1}\frac{3-\sqrt{3}}{3+\sqrt{3}} = \frac{\pi}{12}$$

$$\arg z - \arg w = \frac{\pi}{4} - \frac{\pi}{6} = \frac{\pi}{12} \text{ as required}$$

b $\ zw = i(2-2i) = 2+2i$

$$\arg(zw) = \tan^{-1}\left(\frac{2}{2}\right) = \frac{\pi}{4}$$

$$\arg z = \frac{\pi}{2}$$

$$\arg w = -\tan^{-1}\frac{2}{2} = -\frac{\pi}{4}$$

$$\arg z + \arg w = \frac{\pi}{2} + \frac{-\pi}{4} = \frac{\pi}{4} \text{ as required}$$

$$\frac{z}{w} = \frac{i}{2-2i}$$

$$= \frac{i(2+2i)}{(2-2i)(2+2i)} = -\frac{1}{4} + \frac{1}{4}i$$

$$\arg\left(\frac{z}{w}\right) = \pi - \tan^{-1}\frac{0.25}{0.25} = \frac{3\pi}{4}$$

$$\arg z - \arg w = \frac{\pi}{2} - \frac{-\pi}{4} = \frac{3\pi}{4} \text{ as required}$$

c $\ zw = -2i(\sqrt{3}-3i) = -6-2\sqrt{3}i$

$$\arg(zw) = -\pi + \tan^{-1}\left(\frac{2\sqrt{3}}{6}\right) = -\frac{5\pi}{6}$$

$$\arg z = -\tan^{-1}\frac{3}{\sqrt{3}} = -\frac{\pi}{3}$$

$$\arg w = -\frac{\pi}{2}$$

$$\arg z + \arg w = -\frac{\pi}{3} + \frac{-\pi}{2} = -\frac{5\pi}{6} \text{ as required}$$

$$\frac{z}{w} = \frac{\sqrt{3}-3i}{-2i}$$

$$= \frac{i(\sqrt{3}-3i)}{-2i \cdot i} = \frac{3}{2} + \frac{\sqrt{3}}{2}i$$

$$\arg\left(\frac{z}{w}\right) = \tan^{-1}\frac{\sqrt{3}}{3} = \frac{\pi}{6}$$

$$\arg z - \arg w = -\frac{\pi}{3} - \frac{-\pi}{2} = \frac{\pi}{6} \text{ as required}$$

4 $\ |w| = \sqrt{3}$

$$\arg(w) = \frac{2\pi}{3}$$

5 $\ |z_2| = \frac{\sqrt{3}}{6}$

$$\arg(z_2) = -\frac{11}{12}\pi$$

6 $\ |w| = 8\sqrt{3}$

$$\arg(w) = \frac{\pi}{2}$$

7 a $\ 3i$ **b** $\ -5$ **c** $\ -5\sqrt{3}+5i$ **d** $\ -\dfrac{\sqrt{3}}{2} - \dfrac{3}{2}i$

8 a $\ z = 3\sqrt{2}\left(\cos\frac{\pi}{4} + i\sin\frac{\pi}{4}\right)$

b $\ z = 2\left(\cos\left(-\frac{\pi}{3}\right) + i\sin\left(-\frac{\pi}{3}\right)\right)$

c $\ z = 4\left(\cos\left(-\frac{5\pi}{6}\right) + i\sin\left(-\frac{5\pi}{6}\right)\right)$

d $\ z = \sqrt{97}\left(\cos(1.99) + i\sin(1.99)\right)$

9 a

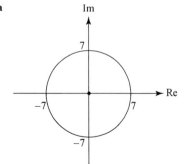

b

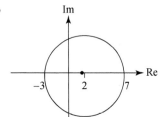

c

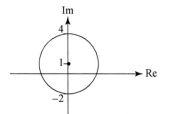

d

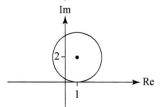

e

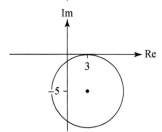

f

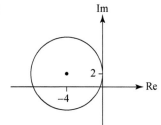

10 a

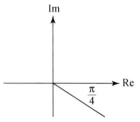

b

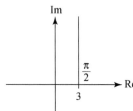

c

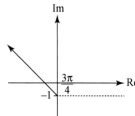

d

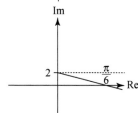

e

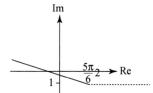

f

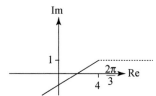

g

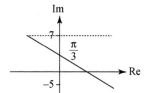

11 a

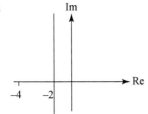

b

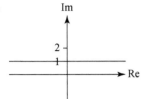

c

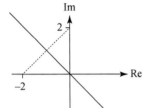

d

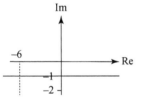

e

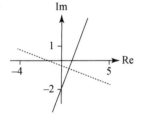

Exercise 1.4B

1 Let $z = |z|(\cos A + i\sin A)$ and $w = |w|(\cos B + i\sin B)$

Then $zw = |z|(\cos A + i\sin A)|w|(\cos B + i\sin B)$

$= |z||w|(\cos A\cos B + i\sin A\cos B + i\sin B\cos A$
$\quad + i^2\sin A\sin B)$

$= |z||w|(\cos A\cos B - \sin A\sin B$
$\quad + i(\sin A\cos B + \sin B\cos A))$

$= |z||w|(\cos(A+B) + i\sin(A+B))$

So $|zw| = |z||w|$

And $\arg(zw) = A + B = \arg z + \arg w$ as required

2 Let $z = x + iy$

Then $|x + iy + 3 - 2i| = 4$

$(x+3)^2 + (y-2)^2 = 16$

Circle centre $(-3, 2)$ and radius 4

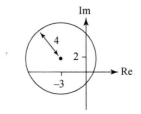

3 a Let $z = x + iy$

$|x + iy - 2 - i| = 2$

$(x-2)^2 + (y-1)^2 = 4$

Circle centre $(2, 1)$ and radius 2

b $\sqrt{3}$ square units

4 a $x = 2.5$　　　　　　　**b** $y = -x$

c $x + 2y = 3$　　　　　　**d** $8x - 2y = 7$

e $2x = 7$　　　　　　　　**f** $13x - 7y = 10$

5 a $y = x + 3$　　　　　　**b** $x = -5$

c $y = \sqrt{3}x + 1 + 2\sqrt{3}$　**d** $y = -\sqrt{3}x - 1 + 4\sqrt{3}$

6 $y = 5x + 3$

7 $y = \dfrac{\sqrt{2} - 1}{1 + \sqrt{2}}x$

8 $x - 2y + 3 = 0$

9 a

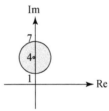

b

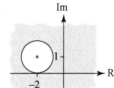

c

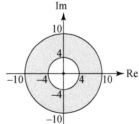

d

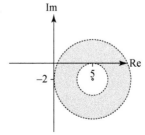

10 42π square units

11 a

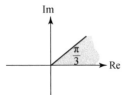

b

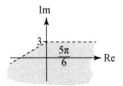

c

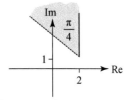

12 16π square units

13 a

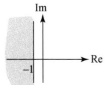

b

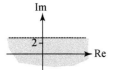

c

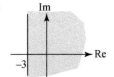

d

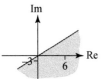

e

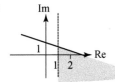

14

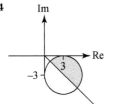

15

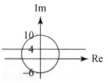

16

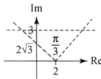

17

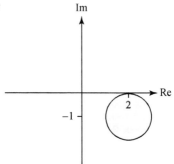

18

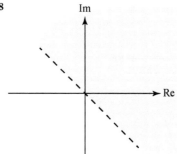

19

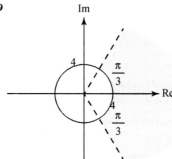

20

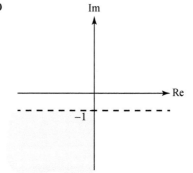

21 a Let $z = x + iy$

$$|x + iy + 3| = 2|x + iy - 6i|$$
$$|x + iy + 3|^2 = 4|x + iy - 6i|^2$$
$$(x + 3)^2 + y^2 = 4x^2 + 4(y - 6)^2$$
$$x^2 + 6x + 9 + y^2 = 4x^2 + 4y^2 - 48y + 144$$
$$3x^2 - 6x + 3y^2 - 48y + 135 = 0$$
$$x^2 - 2x + y^2 - 16y + 45 = 0$$
$$(x - 1)^2 - 1 + (y - 8)^2 - 64 + 45 = 0$$
$$(x - 1)^2 + (y - 8)^2 = 20$$

Circle centre (1, 8) radius $2\sqrt{5}$

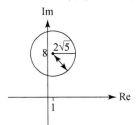

b

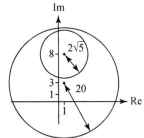

22 a $(x - 3)^2 + (y + 1)^2 = 9$

b $x = y$

c $z = \dfrac{2 + \sqrt{2}}{2} + \dfrac{2 + \sqrt{2}}{2}i$ or $z = \dfrac{2 - \sqrt{2}}{2} + \dfrac{2 - \sqrt{2}}{2}i$

23 $z = 2\sqrt{2} + (3 - 2\sqrt{2})i$

Review exercise 1

1 a $3 - 7i$ **b** $6 - 9i$

   c $12 - 5i$ **d** $-22 - 7i$

   e $4 - 28i$ **f** $-40 - 42i$

   g $5 + 4i$ **h** $-2 + 3i$

   i $\dfrac{2}{13} + \dfrac{23}{13}i$ **j** $\dfrac{3}{41} - \dfrac{69}{82}i$

   k $\dfrac{10}{41} + \dfrac{8}{41}i$ **l** $1 + \dfrac{2}{3}i$

2 a $x = 2 \pm 4i$ **b** $x = -3 \pm i$ **c** $x = 5 \pm \sqrt{2}i$

3 a $x = 2, 1 \pm 8i$ **b** $x = -5, 3 \pm i$

   c $x = 9 \pm i, 4$ **d** $x = 1 \pm i, -1 \pm i$

4 a $x = -\dfrac{5}{2} \pm \dfrac{\sqrt{3}}{2}i, 2 \pm i$

   b $x = \dfrac{1}{2} + \dfrac{\sqrt{3}}{2}i, 4, -1$

5 a Modulus $= \sqrt{85}$ **b** Modulus $= 3\sqrt{2}$

     Argument $= 1.35^c$ Argument $= -\dfrac{\pi}{4}^c$

   c Modulus $= 7$ **d** Modulus $= 2$

     Argument $= \dfrac{\pi}{2}^c$ Argument $= -\dfrac{\pi}{2}$

   e Modulus $= \sqrt{17}$ **f** Modulus $= 5$

     Argument $= 1.82^c$ Argument $= -2.21^c$

6 a $10(\cos(0.644) + i\sin(0.644))$

   b $13(\cos(2.75) + i\sin(2.75))$

   c $2\sqrt{2}\left(\cos\left(-\dfrac{3\pi}{4}\right) + i\sin\left(-\dfrac{3\pi}{4}\right)\right)$

   d $2\left(\cos\left(-\dfrac{\pi}{6}\right) + i\sin\left(-\dfrac{\pi}{6}\right)\right)$

   e $\sqrt{5}\left(\cos\left(-\dfrac{\pi}{3}\right) + i\sin\left(-\dfrac{\pi}{3}\right)\right)$

7 a $\sqrt{3} + i$ **b** $\dfrac{\sqrt{6}}{2} - \dfrac{\sqrt{6}}{2}i$

8 a $3\sqrt{2}\left(\cos\left(-\dfrac{\pi}{6}\right) + i\sin\left(-\dfrac{\pi}{6}\right)\right)$

   b $\sqrt{2}\left(\cos\left(-\dfrac{\pi}{2}\right) + i\sin\left(-\dfrac{\pi}{2}\right)\right)$

   c $\dfrac{\sqrt{2}}{2}\left(\cos\left(\dfrac{\pi}{2}\right) + i\sin\left(\dfrac{\pi}{2}\right)\right)$

9

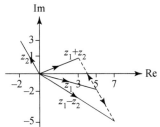

10 a

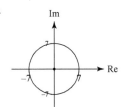

$x^2 + y^2 = 49$

b

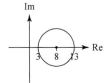

$(x - 8)^2 + y^2 = 25$

c

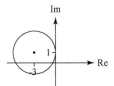

$(x + 3)^2 + (y - 1)^2 = 9$

d

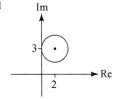

$(x - 2)^2 + (y - 3)^2 = 4$

11 a

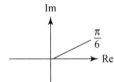

b

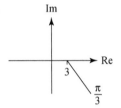

c

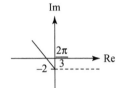

d

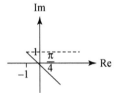

12 a

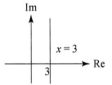

b

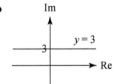

c

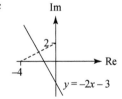

d

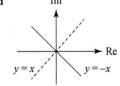

13 a

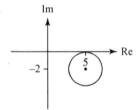

b

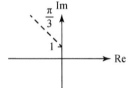

c

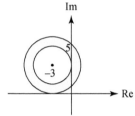

d

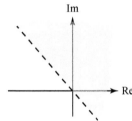

14

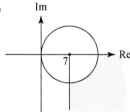

Assessment 1

1 **a** **i** $-2+14i$
       ii $96-50i$

   iii $\dfrac{15}{101}+\dfrac{52}{101}i$

   b $|z|=14.2$, $\arg z = -0.885^{c}$

2 **a** $-1+2i$, $-1-2i$

   b

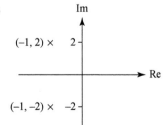

c Reflection in real axis

3 a $z = \dfrac{(5+2i)(3-i)}{(3+i)(3-i)}$

$= \dfrac{15 - 5i + 6i - 2i^2}{9 - i^2}$

$= \dfrac{17 + i}{10}$

$= \dfrac{17}{10} + \dfrac{1}{10}i$

$\left(a = \dfrac{17}{10}, b = \dfrac{1}{10} \right)$

b i $\dfrac{17}{5}$ **ii** $\dfrac{1}{5}i$ **iii** $\dfrac{29}{10}$

c i

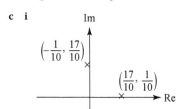

ii iz is a rotation of z 90° anticlockwise around origin.

4 $b = \pm 4, a = \pm 1$

5 $b = \pm 1, a = \pm\sqrt{2}$

6 a $2 \pm 3i, -2$

  b 12 square units

7 a $5 + i$, 7

  b $a = -17, b = 96$

8 $z = 6 + 9i$

9 $w = -3 - 7i, z = 2 - 9i$

10 a $|z| = 6, \arg z = \dfrac{\pi}{3}$

  b $|w| = 2, \arg z = \dfrac{3\pi}{4}$

  c i $|zw| = 12$

   ii $\left| \dfrac{z}{w} \right| = 3$

   iii $\arg(zw) = -\dfrac{11\pi}{12}$

   iv $\arg\left(\dfrac{z}{w} \right) = -\dfrac{5\pi}{12}$

11 a $w = \sqrt{3}\left(\cos(-0.615) + i\sin(-0.615) \right)$

  b $\sqrt{33}$

12 a $z = 2\left(\cos\left(-\dfrac{\pi}{6} \right) + i\sin\left(-\dfrac{\pi}{6} \right) \right)$

  b i $\arg(zw) = -\dfrac{\pi}{4}$

   ii $\arg\left(\dfrac{z}{w} \right) = -\dfrac{\pi}{12}$

  c $|w| = 5$

13 a i

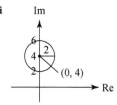

ii

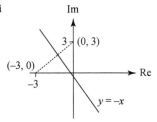

b i $x^2 + (y-4)^2 = 4$

  ii $y = -x$

14 a $|x + iy + 3 - 4i| = 4$

   $(x+3)^2 + (y-4)^2 = 16$

   Therefore a circle (with centre $(-3, 4)$ and radius 4)

b, c

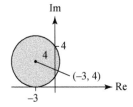

15 i

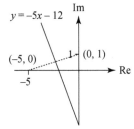

  ii $y = -5x - 12$

16 $-8 + 5i$ and $8 - 5i$

17 $z = \sqrt{5} - i$ or $-\sqrt{5} + i$

18 28 square units

19 a $-i - 4$

  b $x^3 + 11x^2 + 41x + 51 = 0$

20 a $z = -2 - i$ and $z = 3 + 5i$.

  b $z^4 - 2z^3 + 15x^2 + 106z + 170 = 0$

21 So $a = 1, b = -4, c = 56, d = -104, e = 676$

22 $w = 2 \pm i, z = -1 \mp 2i$

23 $w = 6 \pm i, z = 2 \mp 3i$

24 a $2(3i)^5 + (3i)^4 + 36(3i)^3 + 18(3i)^2 + 162(3i) + 81$

   $= 486i + 81 - 972i - 162 + 486i + 81$

   $= 0$ so $3i$ is a solution

  b $(x^2 + 9)^2(2x + 1)$

25 $x = -1 \pm i, 3$

26 a $\beta = 3\left(\cos\left(-\dfrac{5\pi}{6} \right) + i\sin\left(-\dfrac{5\pi}{6} \right) \right)$

  b i $|\alpha\beta| = 9$

   ii $\left| \dfrac{\alpha}{\beta} \right| = 1$

   iii $\arg(\alpha\beta) = 0$

  c $x^2 + 3\sqrt{3}x + 9 = 0$

27 a $\dfrac{z_1}{z_2} = \dfrac{(6-a)}{5} + \dfrac{(3+2a)}{5}i$

  b $a = \pm 9$

28 a $w = \dfrac{1}{2}\left(\cos\left(-\dfrac{\pi}{3}\right) + i\sin\left(-\dfrac{\pi}{3}\right)\right)$

b $z = \dfrac{\sqrt{3}}{24} + \dfrac{1}{24}i$

29 Circle, centre $(3, -1)$, radius 1

30 a

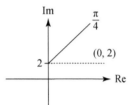

b $y = x + 2$

31

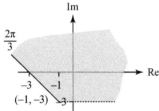

32

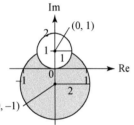

33 5 square units

34 $\arg(z - (\sqrt{3} + i)) = -\dfrac{5\pi}{6}$

35 $|x + yi + 2| = 3|x + yi - 2i|$

$(x + 2)^2 + y^2 = 9(x^2 + (y - 2)^2)$

$x^2 + 4x + 4 + y^2 = 9x^2 + 9y^2 - 36y + 36$

$8x^2 - 4x + 8y^2 - 36y + 32 = 0$

$x^2 - \dfrac{x}{2} + y^2 - \dfrac{9}{2}y + 4 = 0$

$\left(x - \dfrac{1}{4}\right)^2 + \left(y - \dfrac{9}{4}\right)^2 = \dfrac{9}{8}$

So a circle, centre $\left(\dfrac{1}{4}, \dfrac{9}{4}\right)$ and radius $\dfrac{3}{4}\sqrt{2}$

36 a

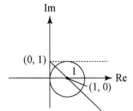

b $z = \left(\dfrac{2 - \sqrt{2}}{2}\right) + \dfrac{\sqrt{2}}{2}i$ or $z = \left(\dfrac{2 + \sqrt{2}}{2}\right) - \dfrac{\sqrt{2}}{2}i$

37 a $z = 6 - i$ or $z = -2 - i$

b

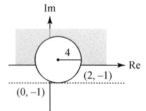

38 a

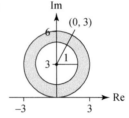

b 8π

39 a

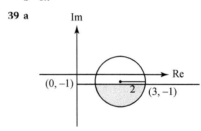

b 2π

1 a 5 and 9   **b** -6 and 7

   c 8 and -12 **d** -10 and -5

   e 4 and $\dfrac{8}{3}$ **f** $-\dfrac{1}{4}$ and $\dfrac{3}{2}$

2 a i $x^2 - 10x + 9 = 0$ **ii** $3x^2 - 4x + 1 = 0$

   iii $x^2 - 8x + 15 = 0$ **iv** $x^2 - 28x + 27 = 0$

   v $3x^2 - 10x + 3 = 0$

  b i $x^2 - 26x + 25 = 0$ **ii** $5x^2 - 4x - 1 = 0$

   iii $x^2 - 9 = 0$ **iv** $x^2 + 124x - 125 = 0$

   v $5x^2 + 26x + 5 = 0$

  c i $x^2 + x + 1 = 0$ **ii** $x^2 + x + 1 = 0$

   iii $x^2 - 3x + 3 = 0$ **iv** $x^2 - 2x + 1 = 0$

   v $x^2 + x + 1 = 0$

  d i $4x^2 - 81x + 64 = 0$ **ii** $8x^2 + 7x - 2 = 0$

   iii $2x^2 - 15x + 14 = 0$ **iv** $8x^2 - 679x - 512 = 0$

   v $16x^2 + 81x + 16 = 0$

  e i $25x^2 + 111x + 144 = 0$ **ii** $12x^2 + 3x + 5 = 0$

   iii $5x^2 - 17x + 26 = 0$ **iv** $125x^2 - 513x + 1728 = 0$

   v $20x^2 + 37x + 20 = 0$

  f i $256x^2 - 32x + 1 = 0$ **ii** $x^2 - 8x + 16 = 0$

   iii $16x^2 - 72x + 81 = 0$ **iv** $4096x^2 - 128x + 1 = 0$

   v $x^2 - 2x + 1 = 0$

3 a $-4, -9$ and 14   **b** $7, -11$ and -12

   c 13, 22 and 26   **d** $\dfrac{-5}{2}, \dfrac{17}{2}$ and $\dfrac{21}{2}$

   e $\dfrac{1}{4}, \dfrac{3}{4}$ and -2 **f** $-\dfrac{3}{4}, \dfrac{-3}{2}$ and $\dfrac{-5}{8}$

4 a i $x^3 - 12x^2 + 4x - 9 = 0$

   ii $3x^3 - 4x^2 + 2x + 1 = 0$

   iii $x^3 - 4x^2 + 11 = 0$

  b i $x^3 - 31x^2 + 235x - 121 = 0$

ii $11x^3 + 9x^2 - 7x + 1 = 0$
iii $x^3 - 13x^2 + 49x - 43 = 0$
c i $9x^3 - 121x^2 + 32x - 64 = 0$
ii $8x^3 + 12x^2 + 7x - 3 = 0$
iii $3x^3 - 25x^2 + 52x - 36 = 0$
d i $25x^3 - 284x^2 + 316x - 25 = 0$
ii $5x^3 + 14x^2 - 12x - 5 = 0$
iii $5x^3 - 18x^2 - 2x + 31 = 0$
5 a i 26 **ii** 0
b i 30 **ii** $\dfrac{-11}{12}$
6 $x = \pm 5$ or ± 8
7 $x = \pm 3$ whilst $x^2 = -1$ has no roots therefore only 2 real roots.
8 $x^4 + 5x^2 + 4 = 0 \equiv (x^2 + 4)(x^2 + 1) = 0 \rightarrow x^2 = -4$ or -1
Hence neither $x^2 = -4$ nor $x^2 = -1$ have any roots therefore no real roots.
9 $x = 2, 4$ or $\dfrac{-4}{3}$

Exercise 2.1B Reasoning and problem-solving

1 a $y^3 + 5y^2 + 3y + 1 = 0$
b $y^3 + 9y^2 + 45y + 27 = 0$
2 a $y^3 + 4y^2 + 3y = 0$
b $y = 0, -1$ or -3
c $x = 2, 1$ or -1
3 $8y^3 - 4y^2 - 14y + 9 = 0$
4 $y^4 + 16y^3 + 42y^2 + 96y + 197 = 0$
5 $x = -2, 4$ or 7
6 $x = -3, -5$ or 2
7 $x = -3, \dfrac{1}{2}$ or 4
8 $x = -5, -2, 0$ or 3
9 $b = 5$
10 $\alpha^3 = 4\alpha + 3$; $\beta^3 = 4\beta + 3$ and $\gamma^3 = 4\gamma + 3$
Hence $\alpha^3 + \beta^3 + \gamma^3 = 4(\alpha + \beta + \gamma) + 9$
But from $x^3 = 4x + 3$ or $x^3 - 4x - 3 = 0$, $\alpha + \beta + \gamma = 0$
Hence $\alpha^3 + \beta^3 + \gamma^3 = 4(0) + 9 = 9$
11 $2y^3 - 4y^2 - 4y + 9 = 0$
12 a $\alpha + \beta + \gamma = 4$
b $y^3 - 8y^2 + 15y - 3 = 0$
c 3

Exercise 2.2A Fluency and skills

1 a $1^3 + 2^3 + 3^3 + 4^3 + 5^3$
b $4^2 + 5^2 + ... + n^2$
c $-1 + 0 + 3 + 8 + ... + n^2 - 2n$
d $\dfrac{1}{3} + \dfrac{1}{4} + \dfrac{1}{5} + \dfrac{1}{6} + ... + \dfrac{1}{n+2}$
e $-1 + 8 - 27 + 64 - 125 + 216$
f $(n-3)(n-2) + (n-2)(n-1) + (n-1)(n) + (n)(n+1)$
2 a $\displaystyle\sum_{1}^{16}(2r-1)$ **b** $\displaystyle\sum_{1}^{n}r^5$
c $\displaystyle\sum_{1}^{n+1}\dfrac{1}{r}$ **d** $\displaystyle\sum_{1}^{14}(-1)^{n+1}3r$
e $\displaystyle\sum_{1}^{n}r(r+2)$ **f** $\displaystyle\sum_{1}^{n}(-1)^{r+1}\dfrac{(2r-1)(r+3)}{(r+2)}$
3 a $3n$ **b** $4n^2 + 2n + 3$
4 a $\dfrac{3n(3n+1)}{2}$ **b** $\dfrac{n(2n-1)(4n-1)}{3}$
c $n^2(2n-1)^2$ **d** n^2
e $\dfrac{n(n+1)(n+2)}{3}$ **f** $\dfrac{n(n+1)(n-1)}{3}$

5 a $\dfrac{n}{3}(n^2 - 1)$ **b** 330

6 a rth term $= r(r+1)(r+2)$
$$\sum_{1}^{n} r(r+1)(r+2) = \frac{n(n+1)(n+2)(n+3)}{4}$$
b rth term $= (2r-1)(n+3)$
$$\sum_{1}^{n}(2r-1)(2r+1) = \frac{n(4n^2 + 21n - 1)}{6}$$
c rth term $= r(r+2)(r+4)$
$$\sum_{1}^{n} r(r+2)(r+4) = \frac{n(n+1)(n+4)(n+5)}{4}$$
d rth term $= 4r - 2$
$$\sum_{1}^{n} 4r - 2 = 2n^2$$
e rth term $= r^2 + 1$
$$\sum_{1}^{n} r^2 + 1 = \frac{n(2n^2 + 3n + 7)}{6}$$

Exercise 2.2B Reasoning and problem-solving

1 a 250 720
b $n^2 + 4n - 32$
c $\dfrac{(n-2)(n-1)(n-3)}{3} - 2$
d $(2n+1)(n+1)(2n^2 + 3n + 4) - \dfrac{n(n-1)(n^2 - n + 6)}{4}$
e $\dfrac{4(2n-5)^2(n-2)^2 - (n-5)^2(n-4)^2}{4}$
2 a 6, 10, 14 **b** $4n + 2$
3 a $2n$ **b** $2n(2n-3)$
4 $\displaystyle\sum_{1}^{n}(2r+1)^2 \equiv \sum_{1}^{n}(4r^2 + 4r + 1)$
$$\equiv 4\sum_{1}^{n}r^2 + 4\sum_{1}^{n}r + \sum_{1}^{n}1$$
$$\equiv \frac{4n(n+1)(2n+1)}{6} + \frac{4n(n+1)}{2} + n$$
$$\equiv \frac{n[2(n+1)(2n+1) + 6(n+1) + 3)]}{3}$$
$$\equiv \frac{n[4n^2 + 12n + 11]}{3}$$
Hence $a = 12$ and $b = 11$.
5 a $\dfrac{n(n+1)(n-1)(n+2)}{2}$ **b** 5940
c The general term $2r(r-1)(r+1)$ is a multiple of 2. It must also be a multiple of 4 since at least one of the consecutive terms must be even. At least one of the three consecutive terms must be a multiple of 3. So every term must be a multiple of $4 \times 3 = 12$.
6 a $\dfrac{n(n+1)}{2}$
b $\dfrac{n(n+1)(n+2)}{6}$
c 599 kg (3sf)
7 $\dfrac{n(2n+1)(7n+1)}{6}$

8 The general term is $r(m-r+1)$

$S = \dfrac{m(m+1)(m+2)}{6}$

9 a $\dfrac{n(2n-1)(2n+1)}{3}$

b $1^2 + 3^2 + 5^2 + ... + (2r-1)^2$

$\equiv \displaystyle\sum_{1}^{2n} r^2 - 4\sum_{1}^{n} r^2$

$= \dfrac{2n(2n+1)(4n+1)}{6} - \dfrac{4n(n+1)(2n+1)}{6}$

$= \dfrac{n(2n+1)[(4n+1)-2(n+1)]}{3}$

$= \dfrac{n(2n+1)(2n-1)}{3}$

10 a $n(n+1)(2n+1) + \dfrac{(3^{n+1}-3)}{2}$

b 5

Exercise 2.3A

1 a When $n = 1$, $\displaystyle\sum_{r=1}^{n} 1 = 1$

and $n = 1$ so true for $n = 1$
Assume true for $n = k$ and consider $n = k + 1$:

$\displaystyle\sum_{r=1}^{k+1} 1 = \sum_{r=1}^{k} 1 + 1$

$= k + 1$

So true for $n = k + 1$
The statement is true for $n = 1$ and by assuming it is true for $n = k$ it is shown to be true for $n = k + 1$, therefore, by mathematical induction, it is true for all $n \in \mathbb{N}$

b When $n = 1$, $\displaystyle\sum_{r=1}^{n} r = 1$

and $\dfrac{1}{2}n(n+1) = \dfrac{1}{2} \times 1(1+1)$

$= \dfrac{1}{2}(2)$

$= 1$

so true for $n = 1$
Assume true for $n = k$ and consider $n = k + 1$:

$\displaystyle\sum_{r=1}^{k+1} r = \sum_{r=1}^{k} r + (k+1)$

$= \dfrac{1}{2}k(k+1) + k + 1$

$= \dfrac{1}{2}(k+1)(k+2)$

$= \dfrac{1}{2}(k+1)(k+1+1)$

So true for $n = k + 1$
The statement is true for $n = 1$ and by assuming it is true for $n = k$ it is shown to be true for $n = k + 1$, therefore, by mathematical induction, it is true for all $n \in \mathbb{N}$

c When $n = 1$, $\displaystyle\sum_{r=1}^{n} (2r+3) = 2 \times 1 + 3$

$= 5$

and $n(n+4) = 1(1+4)$

$= 5$

so true for $n = 1$

Assume true for $n = k$ and consider $n = k + 1$:

$\displaystyle\sum_{r=1}^{k+1} (2r+3) = \sum_{r=1}^{k} (2r+3) + 2(k+1) + 3$

$= k(k+4) + 2k + 5$

$= k^2 + 6k + 5$

$= (k+1)(k+5)$

$= (k+1)(k+1+4)$

So true for $n = k + 1$
The statement is true for $n = 1$ and by assuming it is true for $n = k$ it is shown to be true for $n = k + 1$, therefore, by mathematical induction, it is true for all $n \in \mathbb{N}$

d When $n = 1$, $\displaystyle\sum_{r=1}^{n} r(r+1) = 1(1+1)$

$= 2$

and $\dfrac{1}{3}n(n+1)(n+2) = \dfrac{1}{3} \times 1(1+1)(1+2)$

$= \dfrac{1}{3}(2)(3)$

$= 2$

so true for $n = 1$
Assume true for $n = k$ and consider $n = k + 1$:

$\displaystyle\sum_{r=1}^{k+1} r(r+1) = \sum_{r=1}^{k} r(r+1) + (k+1)(k+2)$

$= \dfrac{1}{3}k(k+1)(k+2) + (k+1)(k+2)$

$= \dfrac{1}{3}(k+1)(k+2)(k+3)$

$= \dfrac{1}{3}(k+1)(k+1+1)(k+1+2)$

So true for $n = k + 1$
The statement is true for $n = 1$ and by assuming it is true for $n = k$ it is shown to be true for $n = k + 1$, therefore, by mathematical induction, it is true for all $n \in \mathbb{N}$

e When $n = 1$, $\displaystyle\sum_{r=1}^{n} (r-1)^2 = (1-1)^2$

$= 0$

and $\dfrac{1}{6}n(n-1)(2n-1) = \dfrac{1}{6} \times 1(1-1)(2 \times 1 - 1)$

$= \dfrac{1}{6}(0)(1)$

$= 0$

so true for $n = 1$
Assume true for $n = k$ and consider $n = k + 1$:

$\displaystyle\sum_{r=1}^{k+1} (r-1)^2 = \sum_{r=1}^{k} (r-1)^2 + (k+1-1)^2$

$= \dfrac{1}{6}k(k-1)(2k-1) + k^2$

$= \dfrac{1}{6}k[(k-1)(2k-1) + 6k]$

$= \dfrac{1}{6}k[2k^2 - 2k - k + 1 + 6k]$

$= \dfrac{1}{6}k(2k^2 + 3k + 1)$

$= \dfrac{1}{6}k(2k+1)(k+1)$

$= \dfrac{1}{6}(k+1)(k+1-1)(2(k+1)-1)$

So true for $n = k+1$

The statement is true for $n = 1$ and by assuming it is true for $n = k$ it is shown to be true for $n = k + 1$, therefore, by mathematical induction, it is true for all $n \in \mathbb{N}$

f When $n = 1$, $\displaystyle\sum_{r=1}^{n}(r+1)(r-1) = (1+1)(1-1)$

$$= 0$$

and $\dfrac{1}{6}n(2n+5)(n-1) = \dfrac{1}{6}\times 1(2\times 1+5)(1-1)$

$$= \dfrac{1}{6}(7)(0)$$

$$= 0$$

so true for $n = 1$

Assume true for $n = k$ and consider $n = k + 1$:

$$\sum_{r=1}^{k+1}(r+1)(r-1) = \sum_{r=1}^{k}(r+1)(r-1)+(k+1+1)(k+1-1)$$

$$= \dfrac{1}{6}k(2k+5)(k-1)+k(k+2)$$

$$= \dfrac{1}{6}k[(2k+5)(k-1)+6(k+2)]$$

$$= \dfrac{1}{6}k(2k^2-2k+5k-5+6k+12)$$

$$= \dfrac{1}{6}k(2k^2+9k+7)$$

$$= \dfrac{1}{6}k(2k+7)(k+1)$$

$$= \dfrac{1}{6}(k+1)(2(k+1)+5)(k+1-1)$$

So true for $n = k+1$

The statement is true for $n = 1$ and by assuming it is true for $n = k$ it is shown to be true for $n = k + 1$, therefore, by mathematical induction, it is true for all $n \in \mathbb{N}$

2 a When $n = 1$, $\displaystyle\sum_{r=1}^{n}(r+1)^2 = 2^2 = 4$

and $\dfrac{1}{6}n(2n^2+9n+13) = \dfrac{1}{6}(1)(24) = 4$

So the statement is true when $n = 1$

Assume statement is true for $n = k$ and substitute $n = k+1$

into the formula:

$$\sum_{r=1}^{k+1}(r+1)^2 = \sum_{r=1}^{k}(r+1)^2+(k+1+1)^2$$

$$= \dfrac{1}{6}k(2k^2+9k+13)+(k+2)^2$$

$$= \dfrac{1}{6}(2k^3+9k^2+13k+6k^2+24k+24)$$

$$= \dfrac{1}{6}(2k^3+15k^2+37k+24)$$

$$= \dfrac{1}{6}(k+1)(2k^2+13k+24)$$

$$= \dfrac{1}{6}(k+1)(2(k+1)^2+9(k+1)+13)$$

So the statement is true when $n = k+1$

The statement is true for $n = 1$ and by assuming it is true for $n = k$ it is shown to be true for $n = k+1$, therefore, by mathematical induction, it is true for all $n \in \mathbb{N}$

b When $n = 1$, $\displaystyle\sum_{r=1}^{n}5^{n-1} = 5^0 = 1$

and $\dfrac{1}{4}(5^n-1) = \dfrac{1}{4}(4) = 1$

So the statement is true when $n = 1$

Assume statement is true for $n = k$ and substitute $n = k+1$

into the formula:

$$\sum_{r=1}^{k+1}5^{r-1} = \sum_{r=1}^{k}5^{r-1}+5^{k+1-1}$$

$$= \dfrac{1}{4}(5^k-1)+5^k$$

$$= \dfrac{1}{4}(5^k-1+4(5^k))$$

$$= \dfrac{1}{4}(5(5^k)-1)$$

$$= \dfrac{1}{4}(5^{k+1}-1)$$

So the statement is true when $n = k+1$

The statement is true for $n = 1$ and by assuming it is true for $n = k$ it is shown to be true for $n = k+1$, therefore, by mathematical induction, it is true for all $n \in \mathbb{N}$

3 When $n = 1$, $\displaystyle\sum_{r=1}^{n}r^3 = 1^3$

$$= 1$$

and $\dfrac{1}{4}n^2(n+1)^2 = \dfrac{1}{4}\times 1^2(1+1)^2$

$$= \dfrac{1}{4}(1)(2)^2$$

$$= 1$$

so true for $n = 1$

Assume true for $n = k$ and consider $n = k+1$:

$$\sum_{r=1}^{k+1}r^3 = \sum_{r=1}^{k}r^3+(k+1)^3$$

$$= \dfrac{1}{4}k^2(k+1)^2+(k+1)^3$$

$$= \dfrac{1}{4}(k+1)^2[k^2+4(k+1)]$$

$$= \dfrac{1}{4}(k+1)^2[k^2+4k+4]$$

$$= \dfrac{1}{4}(k+1)^2(k+2)^2$$

$$= \dfrac{1}{4}(k+1)^2(k+1+1)^2$$

So true for $n = k+1$

The statement is true for $n = 1$ and by assuming it is true for $n = k$ it is shown to be true for $n = k+1$, therefore, by mathematical induction, it is true for all $n \in \mathbb{N}$

4 When $n=1$, $\displaystyle\sum_{r=1}^{2n} r = 1+2$

$$= 3$$

and $n(2n+1) = 1(2\times1+1)$

$$= 3$$

so true for $n=1$

Assume true for $n=k$ and consider $n=k+1$:

$$\sum_{r=1}^{2(k+1)} r = \sum_{r=1}^{2k} r + (2k+1)+(2k+2)$$

$$= k(2k+1)+4k+3$$

$$= 2k^2+k+4k+3$$

$$= 2k^2+5k+3$$

$$= (2k+3)(k+1)$$

$$= (k+1)(2(k+1)+1)$$

So true for $n=k+1$

The statement is true for $n=1$ and by assuming it is true for $n=k$ it is shown to be true for $n=k+1$, therefore, by mathematical induction, it is true for all $n\in\mathbb{N}$

5 When $n=1$, $\displaystyle\sum_{r=1}^{2n} r^2 = 1^2+2^2 = 5$

and $\dfrac{1}{3}n(2n+1)(4n+1) = \dfrac{1}{3}(1)(3)(5) = 5$

So the statement is true when $n=1$

Assume statement is true for $n=k$ and substitute $n=k+1$

into the formula.

$$\sum_{r=1}^{2(k+1)} r^2 = \sum_{r=1}^{2k} r^2 + (2k+1)^2 + (2k+2)^2$$

$$= \dfrac{1}{3}k(2k+1)(4k+1)+(4k^2+4k+1)+(4k^2+8k+4)$$

$$= \dfrac{1}{3}(8k^3+6k^2+k+12k^2+12k+3+12k^2+24k+12)$$

$$= \dfrac{1}{3}(8k^3+30k^2+37k+15)$$

$$= \dfrac{1}{3}(k+1)(8k^2+22k+15)$$

$$= \dfrac{1}{3}(k+1)(2k+3)(4k+5)$$

$$= \dfrac{1}{3}(k+1)(2(k+1)+1)(4(k+1)+1)$$

So statement is true when $n=k+1$

The statement is true for $n=1$ and by assuming it is true for $n=k$ it is shown to be true for $n=k+1$, therefore, by

mathematical induction, it is true for all $n\in\mathbb{N}$

6 a When $n=1$, $\displaystyle\sum_{r=1}^{n} 2^r = 2^1$

$$= 2$$

and $2(2^n-1) = 2(2^1-1)$

$$= 2$$

so true for $n=1$

Assume true for $n=k$ and consider $n=k+1$:

$$\sum_{r=1}^{k+1} 2^r = \sum_{r=1}^{k} 2^r + 2^{k+1}$$

$$= 2(2^k-1)+2^{k+1}$$

$$= 2(2^k)-2+2(2^k)$$

$$= 4(2^k)-2$$

$$= 2(2^{k+1}-1)$$

So true for $n=k+1$

The statement is true for $n=1$ and by assuming it is true for $n=k$ it is shown to be true for $n=k+1$, therefore, by mathematical induction, it is true for all $n\in\mathbb{N}$

b When $n=1$, $\displaystyle\sum_{r=1}^{n} 3^r = 3^1$

$$= 3$$

and $\dfrac{3}{2}(3^n-1) = \dfrac{3}{2}(3^1-1)$

$$= \dfrac{3}{2}(2)$$

$$= 3$$

so true for $n=1$

Assume true for $n=k$ and consider $n=k+1$:

$$\sum_{r=1}^{k+1} 3^r = \sum_{r=1}^{k} 3^r + 3^{k+1}$$

$$= \dfrac{3}{2}(3^k-1)+3^{k+1}$$

$$= \dfrac{3}{2}\left(3^k-1+\dfrac{2}{3}(3^{k+1})\right)$$

$$= \dfrac{3}{2}\left(3^k-1+\dfrac{2}{3}(3)3^k\right)$$

$$= \dfrac{3}{2}(3^k-1+2(3^k))$$

$$= \dfrac{3}{2}(3(3^k)-1)$$

$$= \dfrac{3}{2}(3^{k+1}-1)$$

So true for $n=k+1$

The statement is true for $n=1$ and by assuming it is true for $n=k$ it is shown to be true for $n=k+1$, therefore, by mathematical induction, it is true for all $n\in\mathbb{N}$

c When $n=1$, $\displaystyle\sum_{r=1}^{n} 4^r = 4^1$

$$= 4$$

and $\dfrac{4}{3}(4^n-1) = \dfrac{4}{3}(4^1-1)$

$$= \dfrac{4}{3}(3)$$

$$= 4$$

so true for $n=1$

Assume true for $n=k$ and consider $n=k+1$:

$$\sum_{r=1}^{k+1} 4^r = \sum_{r=1}^{k} 4^r + 4^{k+1}$$

$$= \dfrac{4}{3}(4^k-1)+4^{k+1}$$

$$= \dfrac{4}{3}\left(4^k-1+\dfrac{3}{4}(4^{k+1})\right)$$

$$= \dfrac{4}{3}\left(4^k-1+\dfrac{3}{4}(4)4^k\right)$$

$$= \frac{4}{3}(4^k - 1 + 3(4^k))$$

$$= \frac{4}{3}(4(4^k) - 1)$$

$$= \frac{4}{3}(4^{k+1} - 1)$$

So true for $n = k + 1$

The statement is true for $n = 1$ and by assuming it is true for $n = k$ it is shown to be true for $n = k + 1$, therefore, by mathematical induction, it is true for all $n \in \mathbb{N}$

d When $n = 1$, $\displaystyle\sum_{r=1}^{n} 2^{r-1} = 2^{1-1}$

$$= 1$$

and $2^n - 1 = 2^1 - 1$

$$= 1$$

so true for $n = 1$

Assume true for $n = k$ and consider $n = k + 1$:

$$\sum_{r=1}^{k+1} 2^{r-1} = \sum_{r=1}^{k} 2^{r-1} + 2^{k+1-1}$$

$$= 2^k - 1 + 2^k$$

$$= 2(2^k) - 1$$

$$= 2^{k+1} - 1$$

So true for $n = k + 1$

The statement is true for $n = 1$ and by assuming it is true for $n = k$ it is shown to be true for $n = k + 1$, therefore, by mathematical induction, it is true for all $n \in \mathbb{N}$

e When $n = 1$, $\displaystyle\sum_{r=1}^{n} 3^{r-1} = 3^{1-1}$

$$= 1$$

and $\frac{1}{2}(3^n - 1) = \frac{1}{2}(3^1 - 1)$

$$= 1$$

so true for $n = 1$

Assume true for $n = k$ and consider $n = k + 1$:

$$\sum_{r=1}^{k+1} 3^{r-1} = \sum_{r=1}^{k} 3^{r-1} + 3^{k+1-1}$$

$$= \frac{1}{2}(3^k - 1) + 3^k$$

$$= \frac{1}{2}(3^k - 1 + 2(3^k))$$

$$= \frac{1}{2}(3(3^k) - 1)$$

$$= \frac{1}{2}(3^{k+1} - 1)$$

So true for $n = k + 1$

The statement is true for $n = 1$ and by assuming it is true for $n = k$ it is shown to be true for $n = k + 1$, therefore, by mathematical induction, it is true for all $n \in \mathbb{N}$

f When $n = 1$, $\displaystyle\sum_{r=1}^{n} \left(\frac{1}{2}\right)^r = \left(\frac{1}{2}\right)^1$

$$= \frac{1}{2}$$

and $1 - \left(\frac{1}{2}\right)^n = 1 - \left(\frac{1}{2}\right)^1$

$$= \frac{1}{2}$$

so true for $n = 1$

Assume true for $n = k$ and consider $n = k + 1$:

$$\sum_{r=1}^{k+1} \left(\frac{1}{2}\right)^r = \sum_{r=1}^{k} \left(\frac{1}{2}\right)^r + \left(\frac{1}{2}\right)^{k+1}$$

$$= 1 - \left(\frac{1}{2}\right)^k + \left(\frac{1}{2}\right)^{k+1}$$

$$= 1 - \left(\frac{1}{2}\right)^k \left(1 - \frac{1}{2}\right)$$

$$= 1 - \left(\frac{1}{2}\right)^k \left(\frac{1}{2}\right)$$

$$= 1 - \left(\frac{1}{2}\right)^{k+1}$$

So true for $n = k + 1$

The statement is true for $n = 1$ and by assuming it is true for $n = k$ it is shown to be true for $n = k + 1$, therefore, by mathematical induction, it is true for all $n \in \mathbb{N}$

7 When $n = 1$, $\displaystyle\sum_{r=1}^{n} \frac{1}{r(r+1)} = \frac{1}{1(1+1)}$

$$= \frac{1}{2}$$

and $\dfrac{n}{n+1} = \dfrac{1}{1+1}$

$$= \frac{1}{2}$$

so true for $n = 1$

Assume true for $n = k$ and consider $n = k + 1$:

$$\sum_{r=1}^{k+1} \frac{1}{r(r+1)} = \sum_{r=1}^{k} \frac{1}{r(r+1)} + \frac{1}{(k+1)(k+1+1)}$$

$$= \frac{k}{k+1} + \frac{1}{(k+1)(k+2)}$$

$$= \frac{k(k+2)+1}{(k+1)(k+2)}$$

$$= \frac{k^2 + 2k + 1}{(k+1)(k+2)}$$

$$= \frac{(k+1)^2}{(k+1)(k+2)}$$

$$= \frac{k+1}{k+2}$$

$$= \frac{k+1}{k+1+1}$$

So true for $n = k + 1$

The statement is true for $n = 1$ and by assuming it is true for $n = k$ it is shown to be true for $n = k + 1$, therefore, by mathematical induction, it is true for all $n \in \mathbb{N}$

8 When $n = 2$, $\displaystyle\sum_{r=2}^{n} \frac{1}{r(r-1)} = \frac{1}{2(2-1)}$

$$= \frac{1}{2}$$

and $\dfrac{n-1}{n} = \dfrac{2-1}{2}$

$$= \frac{1}{2}$$

so true for $n = 2$

Assume true for $n = k$ and consider $n = k + 1$:

$$\sum_{r=2}^{k+1} \frac{1}{r(r-1)} = \sum_{r=2}^{k} \frac{1}{r(r-1)} + \frac{1}{(k+1)(k+1-1)}$$

$$= \frac{k-1}{k} + \frac{1}{k(k+1)}$$

$$= \frac{(k-1)(k+1)+1}{k(k+1)}$$

$$= \frac{k^2-1+1}{k(k+1)}$$

$$= \frac{k^2}{k(k+1)}$$

$$= \frac{k}{k+1}$$

$$= \frac{(k+1)-1}{k+1}$$

So true for $n = k + 1$

The statement is true for $n = 2$ and by assuming it is true for $n = k$ it is shown to be true for $n = k + 1$, therefore, by mathematical induction, it is true for all $n \in \mathbb{N}$

9   When $n = 1$, $\displaystyle\sum_{r=1}^{n} \frac{1}{r^2+2r} = \frac{1}{1^2+2}$

$$= \frac{1}{3}$$

and $\dfrac{n(3n+5)}{4(n+1)(n+2)} = \dfrac{1(3+5)}{4(1+1)(1+2)}$

$$= \frac{8}{24}$$

$$= \frac{1}{3}$$

so true for $n = 1$

Assume true for $n = k$ and consider $n = k + 1$:

$$\sum_{r=1}^{k+1} \frac{1}{r^2+2r} = \sum_{r=1}^{k} \frac{1}{r^2+2r} + \frac{1}{(k+1)^2+2(k+1)}$$

$$= \frac{k(3k+5)}{4(k+1)(k+2)} + \frac{1}{k^2+2k+1+2k+2}$$

$$= \frac{k(3k+5)}{4(k+1)(k+2)} + \frac{1}{k^2+4k+3}$$

$$= \frac{k(3k+5)}{4(k+1)(k+2)} + \frac{1}{(k+3)(k+1)}$$

$$= \frac{k(3k+5)(k+3)+4(k+2)}{4(k+1)(k+2)(k+3)}$$

$$= \frac{k(3k^2+14k+15)+4(k+2)}{4(k+1)(k+2)(k+3)}$$

$$= \frac{3k^3+14k^2+15k+4k+8}{4(k+1)(k+2)(k+3)}$$

$$= \frac{(3k+8)(k+1)^2}{4(k+1)(k+2)(k+3)}$$

$$= \frac{(k+1)(3k+8)}{4(k+2)(k+3)}$$

$$= \frac{(k+1)(3(k+1)+5)}{4(k+1+1)(k+1+2)}$$

So true for $n = k + 1$

The statement is true for $n = 1$ and by assuming it is true for $n = k$ it is shown to be true for $n = k + 1$, therefore, by mathematical induction, it is true for all $n \in \mathbb{N}$

1   a   When $n = 1$, $n^2 + 3n = 1 + 3$

$$= 4$$

$$= 2 \times 2$$

so true for $n = 1$

Assume true for $n = k$ and consider $n = k + 1$:

$(k+1)^2 + 3(k+1) = k^2 + 2k + 1 + 3k + 3$

$$= k^2 + 5k + 4$$

$$= (k^2 + 3k) + 2k + 4$$

$$= 2A + 2k + 4 \text{ since } k^2 + 3k \text{ divisible by } 2$$

$$= 2(A + k + 2)$$

So true for $n = k + 1$

The statement is true for $n = 1$ and by assuming it is true for $n = k$ it is shown to be true for $n = k + 1$, therefore, by mathematical induction, it is true for all $n \in \mathbb{N}$

b   When $n = 1$, $5n^2 - n = 5 - 1$

$$= 4$$

$$= 2 \times 2$$

so true for $n = 1$

Assume true for $n = k$ and consider $n = k + 1$:

$5(k+1)^2 - (k+1) = 5k^2 + 10k + 5 - k - 1$

$$= 5k^2 + 9k + 4$$

$$= (5k^2 - k) + 10k + 4$$

$$= 2A + 10k + 4 \text{ since } 5k^2 - k \text{ divisible by } 2$$

$$= 2(A + 5k + 2)$$

So true for $n = k + 1$

The statement is true for $n = 1$ and by assuming it is true for $n = k$ it is shown to be true for $n = k + 1$, therefore, by mathematical induction, it is true for all $n \in \mathbb{N}$

c   When $n = 1$, $8n^3 + 4n = 8 + 4$

$$= 12$$

$$= 12 \times 1$$

so true for $n = 1$

Assume true for $n = k$ and consider $n = k + 1$:

$8(k+1)^3 + 4(k+1) = 8(k^3 + 3k^2 + 3k + 1) + 4k + 4$

$$= 8k^3 + 24k^2 + 24k + 8 + 4k + 4$$

$$= 8k^3 + 24k^2 + 28k + 12$$

$$= (8k^3 + 4k) + 24k^2 + 24k + 12$$

$$= 12A + 12(2k^2 + 2k + 1) \text{ since } 8k^3 + 4k$$

$$\text{divisible by } 12$$

$$= 12(A + 2k^2 + 2k + 1)$$

So true for $n = k + 1$

The statement is true for $n = 1$ and by assuming it is true for $n = k$ it is shown to be true for $n = k + 1$, therefore, by mathematical induction, it is true for all $n \in \mathbb{N}$

d   When $n = 1$, $11n^3 + 4n = 11 + 4$

$$= 15$$

$$= 3 \times 5$$

so true for $n = 1$

Assume true for $n = k$ and consider $n = k + 1$:

$11(k+1)^3 + 4(k+1) = 11(k^3 + 3k^2 + 3k + 1) + 4k + 4$

$$= 11k^3 + 33k^2 + 33k + 11 + 4k + 4$$

$$= 11k^3 + 33k^2 + 37k + 15$$

$$= (11k^3 + 4k) + 33k^2 + 33k + 15$$

$$= 3A + 3(11k^2 + 11k + 5) \text{ since } 11k^3 + 4k$$

$$\text{divisible by } 3$$

$$= 3(A + 11k^2 + 11k + 5)$$

So true for $n = k + 1$

The statement is true for $n = 1$ and by assuming it is true for $n = k$ it is shown to be true for $n = k + 1$, therefore, by mathematical induction, it is true for all $n \in \mathbb{N}$

2 When $n = 1$, $7n^2 + 25n - 4$
$$= 7 + 25 - 4$$
$$= 28$$
$$= 2 \times 14$$
so true for $n = 1$
Assume true for $n = k$ and consider $n = k + 1$:
$7(k+1)^2 + 25(k+1) - 4 = 7(k^2 + 2k + 1) + 25k + 25 - 4$
$$= 7k^2 + 14k + 7 + 25k + 25 - 4$$
$$= 7k^2 + 39k + 28$$
$$= (7k^2 + 25k - 4) + 14k + 32$$
$$= 2A + 2(7k + 16) \text{ since } 7k^2 + 25k - 4$$
$$\text{divisible by } 2$$
$$= 2(A + 7k + 16)$$
So true for $n = k + 1$
The statement is true for $n = 1$ and by assuming it is true for $n = k$ it is shown to be true for $n = k + 1$, therefore, by mathematical induction, it is true for all $n \in \mathbb{N}$

3 When $n = 2$, $n^3 - n = 2^3 - 2$
$$= 6$$
$$= 3 \times 2$$
so true for $n = 2$
Assume true for $n = k$ and consider $n = k + 1$:
$(k+1)^3 - (k+1) = (k^3 + 3k^2 + 3k + 1) - k - 1$
$$= k^3 + 3k^2 + 2k$$
$$= (k^3 - k) + 3k^2 + 3k$$
$$= 3A + 3(k^2 + k) \text{ since } k^3 - k \text{ divisible by } 3$$
$$= 3(A + k^2 + k)$$
So true for $n = k + 1$
The statement is true for $n = 2$ and by assuming it is true for $n = k$ it is shown to be true for $n = k + 1$, therefore, by mathematical induction, it is true for all $n \in \mathbb{N}$, $n \geq 2$

4 When $n = 1$, $10n^3 + 3n^2 + 5n - 6$
$$= 10 + 3 + 5 - 6$$
$$= 12$$
$$= 6 \times 2$$
so true for $n = 1$
Assume true for $n = k$ and consider $n = k + 1$:
$10(k+1)^3 + 3(k+1)^2 + 5(k+1) - 6$
$$= 10(k^3 + 3k^2 + 3k + 1) + 3(k^2 + 2k + 1) + 5(k+1) - 6$$
$$= 10k^3 + 30k^2 + 30k + 10 + 3k^2 + 6k + 3 + 5k + 5 - 6$$
$$= 10k^3 + 33k^2 + 41k + 12$$
$$= (10k^3 + 3k^2 + 5k - 6) + 30k^2 + 36k + 18$$
$$= 6A + 30k^2 + 36k + 18 \text{ since } 10k^3 + 3k^2 + 5k - 6$$
$$\text{divisible by } 6$$
$$= 6(A + 5k^2 + 6k + 3)$$
So true for $n = k + 1$
The statement is true for $n = 1$ and by assuming it is true for $n = k$ it is shown to be true for $n = k + 1$, therefore, by mathematical induction, it is true for all $n \in \mathbb{N}$

5 a When $n = 1$, $6^n + 9 = 6^1 + 9$
$$= 15$$
$$= 5 \times 3$$
so true for $n = 1$
Assume true for $n = k$ and consider $n = k + 1$:
$6^{k+1} + 9 = 6(6^k) + 9$
$$= (6^k + 9) + 5(6^k)$$
$$= 5A + 5(6^k) \text{ since } 6^k + 9 \text{ divisible by } 5$$
$$= 5(A + 6^k)$$
So true for $n = k + 1$
The statement is true for $n = 1$ and by assuming it is true for $n = k$ it is shown to be true for $n = k + 1$, therefore, by mathematical induction, it is true for all $n \in \mathbb{N}$

b When $n = 1$, $3^{2n} - 1 = 3^2 - 1$
$$= 8$$
$$= 8 \times 1$$
so true for $n = 1$

Assume true for $n = k$ and consider $n = k + 1$:
$3^{2(k+1)} - 1 = 3^{2k} 3^2 - 1$
$$= 9(3^{2k}) - 1$$
$$= (3^{2k} - 1) + 8(3^{2k})$$
$$= 8A + 8(3^{2k}) \text{ since } 3^{2k} - 1 \text{ divisible by } 8$$
$$= 8(A + 3^{2k})$$
So true for $n = k + 1$
The statement is true for $n = 1$ and by assuming it is true for $n = k$ it is shown to be true for $n = k + 1$, therefore, by mathematical induction, it is true for all $n \in \mathbb{N}$

c When $n = 1$, $2^{3n+1} - 2 = 2^4 - 2$
$$= 14$$
$$= 7 \times 2$$
so true for $n = 1$
Assume true for $n = k$ and consider $n = k + 1$:
$2^{3(k+1)+1} - 2 = 2^{3k+4} - 2$
$$= 2^3(2^{3k+1}) - 2$$
$$= 8(2^{3k+1}) - 2$$
$$= (2^{3k+1} - 2) + 7(2^{3k+1})$$
$$= 7A + 7(2^{3k+1}) \text{ since } 3^{2k} - 1 \text{ divisible by } 7$$
$$= 7(A + 2^{3k+1})$$
So true for $n = k + 1$
The statement is true for $n = 1$ and by assuming it is true for $n = k$ it is shown to be true for $n = k + 1$, therefore, by mathematical induction, it is true for all $n \in \mathbb{N}$

6 When $n = 1$, $5^n - 4n + 3 = 5^1 - 4 + 3$
$$= 4$$
$$= 4 \times 1$$
so true for $n = 1$
Assume true for $n = k$ and consider $n = k + 1$:
$5^{k+1} - 4(k+1) + 3 = 5(5^k) - 4k - 4 + 3$
$$= (5^k - 4k + 3) + 4(5^k) - 4$$
$$= 4A + 4(5^k - 1) \text{ since } 5^k - 4k + 3 \text{ divisible by } 4$$
$$= 4(A + 5^k - 1)$$
So true for $n = k + 1$
The statement is true for $n = 1$ and by assuming it is true for $n = k$ it is shown to be true for $n = k + 1$, therefore, by mathematical induction, it is true for all $n \in \mathbb{N}$

7 When $n = 1$, $3^n + 2n + 7 = 3^1 + 2 + 7$
$$= 12$$
$$= 4 \times 3$$
so true for $n = 1$
Assume true for $n = k$ and consider $n = k + 1$:
$3^{k+1} + 2(k+1) + 7 = 3(3^k) + 2k + 9$
$$= 3(3^k + 2k + 7) - 4k - 12$$
$$= 3(4A) - 4(k+3) \text{ since } 3^{k+1} + 2(k+1) + 7$$
$$\text{divisible by } 4$$
$$= 4(3A - k - 3)$$
So true for $n = k + 1$
The statement is true for $n = 1$ and by assuming it is true for $n = k$ it is shown to be true for $n = k + 1$, therefore, by mathematical induction, it is true for all $n \in \mathbb{N}$

8 When $n = 1$, $7^n - 3n + 5 = 7 - 3 + 5$
$$= 9$$
$$= 3 \times 3$$
so true for $n = 1$
Assume true for $n = k$ and consider $n = k + 1$:
$7^{k+1} - 3(k+1) + 5 = 7(7^k) - 3k + 2$
$$= (7^k - 3k + 5) + 6(7^k) - 3$$
$$= 3A + 3(2(7^k) - 1) \text{ since } 7^k - 3k + 5$$
$$\text{divisible by } 3$$
$$= 3(A + 2(7^k) - 1)$$
So true for $n = k + 1$
The statement is true for $n = 1$ and by assuming it is true for $n = k$ it is shown to be true for $n = k + 1$, therefore, by mathematical induction, it is true for all $n \in \mathbb{N}$

9 a When $n=1$, $\displaystyle\sum_{r=n+1}^{2n} r^2 = \sum_{r=2}^{2} r^2$

$$= 2^2$$
$$= 4$$

and $\dfrac{1}{6}n(2n+1)(7n+1) = \dfrac{1}{6}(3)(8)$

$$= 4$$

so true for $n=1$

Assume true for $n=k$ and consider $n=k+1$:

$$\sum_{r=(k+1)+1}^{2(k+1)} r^2 = \sum_{r=k+1}^{2k} r^2 + (2k+1)^2 + (2k+2)^2 - (k+1)^2$$

$$= \frac{1}{6}k(2k+1)(7k+1) + 4k^2 + 4k + 1 + 4k^2$$
$$+ 8k + 4 - k^2 - 2k - 1$$

$$= \frac{1}{6}(14k^3 + 9k^2 + k + 42k^2 + 60k + 24)$$

$$= \frac{1}{6}(14k^3 + 51k^2 + 61k + 24)$$

$$= \frac{1}{6}(k+1)(2k+3)(7k+8)$$

$$= \frac{1}{6}(k+1)(2(k+1)+1)(7(k+1)+1)$$

So true for $n=k+1$

The statement is true for $n=1$ and by assuming it is true for $n=k$ it is shown to be true for $n=k+1$, therefore, by mathematical induction, it is true for all $n \in \mathbb{N}$

b When $n=1$, $\displaystyle\sum_{r=n}^{2n} r^3 = \sum_{r=1}^{2} r^3$

$$= 1 + 8$$
$$= 9$$

and $\dfrac{3}{4}n^2(5n+1)(n+1) = \dfrac{3}{4}(1)(6)(2)$

$$= 9$$

so true for $n=1$

Assume true for $n=k$ and consider $n=k+1$:

$$\sum_{r=(k+1)}^{2(k+1)} r^3 = \sum_{r=k}^{2k} r^3 + (2k+1)^3 + (2k+2)^3 - k^3$$

$$= \frac{3}{4}k^2(5k+1)(k+1) + 8k^3 + 12k^2 + 6k + 1$$
$$+ 8k^3 + 24k^2 + 24k + 8 - k^3$$

$$= \frac{3}{4}(5k^4 + 6k^3 + k^2 + 20k^3 + 48k^2 + 40k + 12)$$

$$= \frac{3}{4}(5k^4 + 26k^3 + 49k^2 + 40k + 12)$$

$$= \frac{3}{4}(k+1)(5k^3 + 21k^2 + 28k + 12)$$

$$= \frac{3}{4}(k+1)(k+1)(k+2)(5k+6)$$

$$= \frac{3}{4}(k+1)^2(k+1+1)(5(k+1)+1)$$

So true for $n=k+1$

The statement is true for $n=1$ and by assuming it is true for $n=k$ it is shown to be true for $n=k+1$, therefore, by mathematical induction, it is true for all $n \in \mathbb{N}$

c When $n=1$, $8^n - 5^n = 8^1 - 5^1$

$$= 3$$
$$= 3 \times 1$$

so true for $n=1$

Assume true for $n=k$ and consider $n=k+1$:

$$8^{k+1} - 5^{k+1} = 8(8^k) - 5(5^k)$$
$$= 5(8^k - 5^k) + 3(8^k)$$
$$= 5(3A) + 3(8^k) \text{ since } 8^k - 5^k \text{ divisible by 3}$$
$$= 3(5A + 8^k)$$

So true for $n=k+1$

The statement is true for $n=1$ and by assuming it is true for $n=k$ it is shown to be true for $n=k+1$, therefore, by mathematical induction, it is true for all $n \in \mathbb{N}$

Review exercise 1

1 a $-5, 2, -1$ **b** $-9, -11, 8$

2 $y^3 - 8y^2 + 24y - 16 = 0$

3 $x = -1, 1, 4$

4 $5, -3, -5$

5 a -1

 b i $y^3 + 2y^2 - 2y - 2 = 0$ **ii** 2

6 a 3 **b** $-\dfrac{3}{2}$ **c** 6

7 a $3 + 4 + 5 + 6 + 7 + 8 + 9 + 10$

 b $n^2 + (n+1)^2 + (n+2)^2 + (n+3)^2 + (n+4)^2$

 c $6 + 14 + 24 + \ldots + ((n-1)^2 + 5(n-1)) + (n^2 + 5n)$

 d $-\dfrac{1}{2} - 1 + \infty + \ldots + \dfrac{1}{n-4} + \dfrac{1}{n-3}$

 e $4 - 9 + 16 - 25 + 36 - 49 + 64 - 81 + 100$

 f $2(n-1)(3n-5) + 2n(3n-2) + 2(n+1)(3n+1)$
 $+ 2(n+2)(3n+4) + 2(n+3)(3n+7)$

8 a $n(2-n)$

 b $\dfrac{n(n+1)(2n-5)}{3}$

 c $\dfrac{n(n+1)(n^2+n+10)}{4}$

9 $166\,650$

10 a $(2n+1)^2(n+1)^2 - 3(2n+1)(n+1) - \dfrac{n^2(n-1)^2}{4} + \dfrac{3n(n-1)}{2}$
 $+ 2n + 4$

 b $\displaystyle\sum_{n}^{2n+1}(r^3 - 3r + 2) = 7844$

11 a $\dfrac{n(n+1)(3n+2)(n-1)}{12}$ **b** $\dfrac{n(n+1)(2n+1)(6n+5)}{3}$

12 a When $n=1$, $\displaystyle\sum_{r=1}^{n} 3r^2 = 3(1^2)$

$$= 3$$

and $\dfrac{1}{2}n(n+1)(2n+1) = \dfrac{1}{2}(1)(1+1)(2+1)$

$$= 3$$

so true for $n=1$

Assume true for $n=k$ and consider $n=k+1$:

$$\sum_{r=1}^{k+1} 3r^2 = \sum_{r=1}^{k} 3r^2 + 3(k+1)^2$$

$$= \frac{1}{2}k(k+1)(2k+1) + 3(k+1)^2$$

$$= \frac{1}{2}(k+1)(2k^2 + k + 6k + 6)$$

$$= \frac{1}{2}(k+1)(k+2)(2k+3)$$

$$= \frac{1}{2}(k+1)(k+1+1)(2(k+1)+1)$$

So true for $n=k+1$

The statement is true for $n=1$ and by assuming it is true for $n=k$ it is shown to be true for $n=k+1$, therefore, by mathematical induction, it is true for all $n \in \mathbb{N}$

b When $n = 1$, $\displaystyle\sum_{r=1}^{n} r^2(r-1) = 1^2(1-1)$

$$= 0$$

and $\dfrac{1}{12}n(n+1)(3n+2)(n-1) = \dfrac{1}{12}(1)(1+1)(3+2)(1-1)$

$$= 0$$

so true for $n = 1$

Assume true for $n = k$ and consider $n = k + 1$:

$$\sum_{r=1}^{k+1} r^2(r-1) = \sum_{r=1}^{k} r^2(r-1) + (k+1)^2(k+1-1)$$

$$= \frac{1}{12}k(k+1)(3k+2)(k-1) + k(k+1)^2$$

$$= \frac{1}{12}k(k+1)\left[(3k+2)(k-1) + 12(k+1)\right]$$

$$= \frac{1}{12}k(k+1)\left[3k^2 - k - 2 + 12k + 12\right]$$

$$= \frac{1}{12}k(k+1)(3k^2 + 11k + 10)$$

$$= \frac{1}{12}k(k+1)(3k+5)(k+2)$$

$$= \frac{1}{12}(k+1)(k+1+1)(3(k+1)+2)(k+1-1))$$

So true for $n = k + 1$

The statement is true for $n = 1$ and by assuming it is true for $n = k$ it is shown to be true for $n = k + 1$, therefore, by mathematical induction, it is true for all $n \in \mathbb{N}$

c Let $n = 1$, then $\displaystyle\sum_{r=1}^{n}(r+3)(2r-1) = 4 \times 1 = 4$

and $\dfrac{1}{6}n(4n^2 + 21n - 1) = \dfrac{1}{6}(4 + 21 - 1) = 4$

So true for $n = 1$

Assume true for $n = k$ and let $n = k + 1$

$$\sum_{r=1}^{k+1}(r+3)(2r-1)$$

$$= \sum_{r=1}^{k}(r+3)(2r-1) + (k+1+3)(2(k+1)-1)$$

$$= \frac{1}{6}k(4k^2 + 21k - 1) + (k+4)(2k+1)$$

$$= \frac{1}{6}(4k^3 + 21k^2 - k) + (2k^2 + 9k + 4)$$

$$= \frac{1}{6}(4k^3 + 21k^2 - k + 12k^2 + 54k + 24)$$

$$= \frac{1}{6}(4k^3 + 33k^2 + 53k + 24)$$

$$= \frac{1}{6}(k+1)(4k^2 + 29k + 24)$$

$$= \frac{1}{6}(k+1)(4(k+1)^2 + 21(k+1) - 1)$$

True for $n = 1$ and true for $n = k$ implies true for $n = k + 2$ therefore true for all $n \in \mathbb{N}$

13 Let $n = 1$, then $\displaystyle\sum_{r=1}^{2n} r^2 = 1^2 + 2^2 = 5$

and $\dfrac{1}{3}n(2x+1)(4n+1) = \dfrac{1}{3}(1)(3)(5) = 5$ So true for $n = 1$.

Assume true for $n = k$ and let $n = k + 1$

$$\sum_{r=1}^{2(k+1)} r^2 = \sum_{r=1}^{2k} r^2 + (2k+1)^2 + (2k+2)^2$$

$$= \frac{1}{3}k(2k+1)(4k+1) + 4k^2 + 4k + 1 + 4k^2 + 8k + 4$$

$$= \frac{1}{3}(8k^3 + 6k^2 + k) + 8k^2 + 12k + 5$$

$$= \frac{1}{3}(8k^3 + 6k^2 + k + 24k^2 + 36k + 15)$$

$$= \frac{1}{3}(8k^3 + 30k^2 + 37k + 15)$$

$$= \frac{1}{3}(k+1)(8k^2 + 22k + 15)$$

$$= \frac{1}{3}(k+1)(2k+3)(4k+5)$$

$$= \frac{1}{3}(k+1)(2(k+1)+1)(4(k+1)+1)$$

True for $n = 1$ and true for $n = k$ implies true for $n = k + 2$ therefore true for all $n \in \mathbb{N}$

14 a When $n = 1$, $\displaystyle\sum_{r=1}^{n} 5^r = 5^1$

$$= 5$$

and $\dfrac{5}{4}(5^n - 1) = \dfrac{5}{4}(5^1 - 1)$

$$= 5$$

so true for $n = 1$

Assume true for $n = k$ and consider $n = k + 1$:

$$\sum_{r=1}^{k+1} 5^r = \sum_{r=1}^{k} 5^r + 5^{k+1}$$

$$= \frac{5}{4}(5^k - 1) + 5^{k+1}$$

$$= \frac{5}{4}(5^k - 1 + 4(5^k))$$

$$= \frac{5}{4}(5(5^k) - 1)$$

$$= \frac{5}{4}(5^{k+1} - 1)$$

So true for $n = k + 1$

The statement is true for $n = 1$ and by assuming it is true for $n = k$ it is shown to be true for $n = k + 1$, therefore, by mathematical induction, it is true for all $n \in \mathbb{N}$

b When $n = 1$, $\displaystyle\sum_{r=1}^{n} 2^{r+1} = 2^{1+1}$

$$= 4$$

and $4(2^n - 1) = 4(2^1 - 1)$

$$= 4$$

so true for $n = 1$

Assume true for $n = k$ and consider $n = k + 1$:

$$\sum_{r=1}^{k+1} 2^{r+1} = \sum_{r=1}^{k} 2^{r+1} + 2^{k+1+1}$$

$$= 4(2^k - 1) + 2^{k+2}$$

$$= 4(2^k - 1) + 2^2 2^k$$

$$= 4(2^k - 1 + 2^k)$$

$$= 4(2(2^k) - 1)$$

$$= 4(2^{k+1} - 1)$$

So true for $n = k + 1$

The statement is true for $n = 1$ and by assuming it is true for $n = k$ it is shown to be true for $n = k + 1$, therefore, by mathematical induction, it is true for all $n \in \mathbb{N}$

c Let $n = 1$, then $\displaystyle\sum_{r=1}^{2n} 5^{r-1} = 5^0 + 5^1 = 6$

and $\dfrac{1}{4}(25^n - 1) = \dfrac{1}{4} \times 24 = 6$ So true for $n = 1$

Assume true for $n = k$ and let $n = k + 1$

$$\sum_{r=1}^{2(k+1)} 5^{r-1} = \sum_{r=1}^{2k} 5^{r-1} + 5^{2k+1-1} + 5^{2k+2-1}$$

$$= \dfrac{1}{4}(25^k - 1) + 5^{2k} + 5^{2k+1}$$

$$= \dfrac{1}{4}(25^k - 1) + 25^k + 5(25^k)$$

$$= \dfrac{1}{4}(25^k - 1 + 4(25^k) + 20(25^k))$$

$$= \dfrac{1}{4}(25(25^k) - 1)$$

$$= \dfrac{1}{4}(25^{(k+1)} - 1)$$

True for $n = 1$ and true for $n = k$ implies true for $n = k + 2$ therefore true for all $n \in \mathbb{N}$

15 When $n = 1$, $3^{2n+1} + 1 = 3^{2+1} + 1$

$$= 28$$
$$= 4 \times 7$$

so true for $n = 1$
Assume true for $n = k$ and consider $n = k + 1$:
$3^{2(k+1)+1} + 1 = 3^{2k+3} + 1$

$$= 3^2 3^{2k+1} + 1$$
$$= 9(3^{2k+1}) + 1$$
$$= (3^{2k+1} + 1) + 8(3^{2k+1})$$
$$= 4A + 8(3^{2k+1}) \text{ since true for } n = k$$
$$= 4(A + 2(3^{2k+1}))$$

So true for $n = k + 1$
The statement is true for $n = 1$ and by assuming it is true for $n = k$ it is shown to be true for $n = k + 1$, therefore, by mathematical induction, it is true for all $n \in \mathbb{N}$

16 When $n = 1$, $2^{2n} - 3n + 2 = 2^2 - 3 + 2$

$$= 3$$

so true for $n = 1$
Assume true for $n = k$ and consider $n = k + 1$:
$2^{2(k+1)} - 3(k+1) + 2 = 2^{2k+2} - 3k - 3 + 2$

$$= 2^2 2^{2k} - 3k + 2 - 3$$
$$= (2^{2k} - 3k + 2) + 3(2^{2k}) - 3$$
$$= 3A + 3(2^{2k} - 1) \text{ since true for } n = k$$
$$= 3(A + 2^{2k} - 1)$$

So true for $n = k + 1$
The statement is true for $n = 1$ and by assuming it is true for $n = k$ it is shown to be true for $n = k + 1$, therefore, by mathematical induction, it is true for all $n \in \mathbb{N}$

17 Let $n = 1$, then $6^n - 1 = 5$ which is divisible by 5
So true for $n = 1$
Assume true for $n = k$ and let $n = k + 1$
$6^{k+1} - 1 = 6(6^k) - 1$

$$= 5(6^k) + 6^k - 1$$
$$= 5(6^k) + 5A \text{ for some integer } A \text{ since } 6^k - 1 \text{ is divisibly by 5}$$
$$= 5(6^k + A)$$

So divisible by 5 and therefore true for $n = k + 1$
The statement is true for $n = 1$ and by assuming it is true for $n = k$ it is shown to be true for $n = k + 1$, therefore, by mathematical induction, it is true for all $n \in \mathbb{N}$

1 $-\dfrac{4}{3}$

2 -1

3 $2n(n+1)^2$

4 a $u^3 = 1$ **b** $u = 1$ **c** $x = 1$

5 a $\dfrac{n}{2}(2n+3)(n-1)$ **b** 4200

6 a $\dfrac{n}{6}(4n^2 + 21n - 1)$

   b 3035

   c 3695

7 a Let $f(n) = 3^n + 2n + 3$

$f(0) = 3^0 + 2(0) + 3 = 4$ which is divisible by 4

So the statement is true when $n = 0$
Assume $f(n)$ is divisible by 4 for $n = k$ and substitute

$n = k + 1$ into the expression.

$f(k+1) = 3^{k+1} + 2(k+1) + 3$

$$= 3(3^k) + 2k + 5$$

$$= 3(3^k + 2k + 3) - (4k + 4)$$

$$= 3(3^k + 2k + 3) - 4(k + 1)$$

$$= 3f(k) - 4(k + 1)$$

$f(k)$ divisible by 4, therefore $f(k+1)$ is divisible by 4

So the statement is true when $n = k + 1$

The statement is true for $n = 0$ and by assuming it is true for $n = k$ it is shown to be true for $n = k + 1$, therefore, by

mathematical induction, it is true for all integers $n \geq 0$

   b Let $f(n) = 3^n + 2n + 3$

When $n = 2$, $f(2) = 3^3 + 2(3) + 3 = 36$
36 is not divisible by 8, so $3^n + 2n + 3$ is not divisible by

8 for all integers $n \geq 0$

8 When $n = 1$, $\displaystyle\sum_{r=1}^{n} 1 + 3r = 4$

$\dfrac{1}{2}n(3n+5) = \dfrac{1}{2}(3+5) = 4$

So true for $n = 1$
Assume true for $n = k$

$$\sum_{r=1}^{k+1} 1 + 3r = 1 + 3(k+1) + \sum_{r=1}^{k} 1 + 3r$$

$$= 1 + 3(k+1) + \dfrac{1}{2}k(3k+5)$$

$$= 1 + 3k + 3 + \dfrac{3}{2}k^2 + \dfrac{5}{2}k$$

$$= \dfrac{3}{2}k^2 + \dfrac{11}{2}k + 4$$

$$= \dfrac{1}{2}(3k^2 + 11k + 8)$$

$$= \frac{1}{2}(3k+8)(k+1)$$

$$= \frac{1}{2}(k+1)(3(k+1)+5)$$

True for $n = 1$ and assuming true for $n = k$ implies true for $n = k + 1$, hence true for all $n \in \mathbb{N}$

9 When $n = 1$, $\displaystyle\sum_{r=1}^{n} (r-1)^3 = 0$

$$\frac{1}{4}(n-1)^2 n^2 = \frac{1}{4}(1-1)^2(1^2) = 0$$

So true for $n = 1$

Assume true for $n = k$

$$\sum_{r=1}^{k+1} (r-1)^3 = (k+1-1)^3 + \sum_{r=1}^{k} (r-1)^3$$

$$= k^3 + \frac{1}{4}(k-1)^2 k^2$$

$$= \frac{1}{4}k^2 (4k + (k-1)^2)$$

$$= \frac{1}{4}(k)^2 (4k + k^2 - 2k + 1)$$

$$= \frac{1}{4}k^2 (k^2 + 2k + 1)$$

$$= \frac{1}{4}k^2 (k+1)^2$$

True for $n = 1$ and assuming true for $n = k$ implies true for $n = k + 1$, hence true for all $n \in \mathbb{N}$

10 a $f(x) = \frac{1}{2}(x-2)(x-3)(x+4)$

b $g(x) = 4x^3 - 14x^2 + 2x + 20$

11 a $\alpha\beta\gamma = -1$, $\displaystyle\sum \alpha\beta = -\frac{3}{2}$, $\displaystyle\sum \alpha = \frac{1}{2}$

b $u^3 + 2u^2 - 5u + 2 = 0$

12 a 5   **b** $25 + 2k$

13 $\displaystyle\sum_{r=1}^{n} r^2 = \frac{n}{6}(n+1)(2n+1)$

14 a $\displaystyle\sum_{r=1}^{n} 3r^2 - 2r$

$$= \frac{3}{6}n(n+1)(2n+1) - \frac{2}{2}n(n+1)$$

$$= \frac{1}{2}n(n+1)(2n+1) - n(n+1)$$

$$= \frac{1}{2}n(n+1)[(2n+1)-2]$$

$$= \frac{1}{2}n(n+1)(2n-1) \text{ as required}$$

b $\displaystyle\sum_{r=1}^{n} (r+1)^2 = \sum_{r=1}^{n} r^2 + 2r + 1$

$$= \frac{1}{6}n(n+1)(2n+1) + \frac{2}{2}n(n+1) + n$$

$$= \frac{1}{6}n[(n+1)(2n+1) + 6(n+1) + 6]$$

$$= \frac{1}{6}n[2n^2 + 3n + 1 + 6n + 6 + 6]$$

$$= \frac{1}{6}n(2n^2 + 9n + 13) \text{ as required}$$

15 a $\displaystyle\sum_{r=1}^{2n} r^2(r+2) = \sum_{r=1}^{2n} r^3 + 2r^2$

$$= \frac{1}{4}(2n)^2(2n+1)^2 + \frac{2}{6}(2n)(2n+1)(2(2n)+1)$$

$$= \frac{1}{4}(4n^2)(4n^2+4n+1) + \frac{1}{3}(2n)(8n^2+6n+1)$$

$$= \frac{1}{3}(3n^2(4n^2+4n+1) + 2n(8n^2+6n+1))$$

$$= \frac{1}{3}n(3n(4n^2+4n+1) + 2(8n^2+6n+1))$$

$$= \frac{1}{3}n(12n^3+12n^2+3n+16n^2+12n+2)$$

$$= \frac{1}{3}n(12n^3+28n^2+15n+2)$$

$$= \frac{1}{3}n(2n+1)(6n^2+11n+2) \text{ as required}$$

b 1704

16 When $n = 1$, $\displaystyle\sum_{r=1}^{n} 4^{r-1} = 1$

$$\frac{1}{3}(4^n - 1) = \frac{1}{3}(4-1) = 1$$

So true for $n = 1$

Assume true for $n = k$

$$\sum_{r=1}^{k+1} 4^{r-1} = 4^{(k+1)-1} + \sum_{r=1}^{k} 4^{r-1}$$

$$= 4^k + \frac{1}{3}(4^k - 1)$$

$$= \frac{4}{3}(4^k) - \frac{1}{3}$$

$$= \frac{1}{3}(4^{k+1} - 1)$$

True for $n = 1$ and assuming true for $n = k$ implies true for $n = k + 1$, hence true for all $n \in \mathbb{N}$

17 When $n = 1$, $4n^3 + 8n = 12$ which is divisible by 12 so true for $n = 1$

Assume true for $n = k$

$4(k+1)^3 + 8(k+1)$

$= 4(k^3 + 3k^2 + 3k + 1) + 8(k+1)$

$= 4k^3 + 8k + 12k^2 + 12k + 4 + 8$

$= 4k^3 + 8k + 12k^2 + 12k + 12$

$= 4k^3 + 8k + 12(k^2 + k + 1)$

So divisible by 12 since $4k^3 + 8k$ and $12(k^2 + k + 1)$ are divisible by 12

True for $n = 1$ and assuming true for $n = k$ implies true for $n = k + 1$, hence true for all $n \in \mathbb{N}$

18 When $n = 1$, $4^{2n+1} - 1 = 4^3 - 1 = 63$

$63 = 21 \times 3$ hence a multiple of 3

Assume true for $n = k$

$4^{2(k+1)+1} - 1 = 4^{2k+3} - 1$

$$= 4^2 4^{2k+1} - 1$$

$$= 16(4^{2k+1} - 1) + 15$$

So a multiple of 3 since $4^{2k+1} - 1$ assumed to be a multiple of 3 and $15 = 5 \times 3$ so a multiple of 3.

True for $n = 1$ and assuming true for $n = k$ implies true for $n = k + 1$, hence true for all $n \in \mathbb{N}$

19 a $\alpha + \beta = -\dfrac{k}{4}, \alpha\beta = \dfrac{3(k+1)}{4}$

$(\alpha + \beta)^2 = \alpha^2 + 2\alpha\beta + \beta^2$

$\alpha^2 + \beta^2 = (\alpha + \beta)^2 - 2\alpha\beta$

$\qquad = \left(-\dfrac{k}{4}\right)^2 - 2\left(\dfrac{3(k+1)}{4}\right)$

$\qquad = \dfrac{k^2}{16} - \dfrac{3k+3}{2}$

$\qquad = \dfrac{1}{16}(k^2 - 24k - 24)$

b i $4y^2 + (k-24)y + 39 = 0$

ii $\alpha + 3, \beta + 3$

20 a $u^4 + 27u + 81 = 0$

b $0 = u^4 + 2u^2 - u + 1$

21 a $\displaystyle\sum_{r=n}^{2n} r(r-1) = \sum_{r=n}^{2n} r^2 - r$

$= \left[\dfrac{2n}{6}(2n+1)(2(2n)+1) - \dfrac{2n}{2}(2n+1)\right]$

$\quad - \left[\dfrac{(n-1)}{6}(n-1+1)(2(n-1)+1) - \dfrac{(n-1)}{2}(n-1+1)\right]$

$= \left[\dfrac{n}{3}(2n+1)(4n+1) - n(2n+1)\right]$

$\quad - \left[\dfrac{n(n-1)}{6}(2n-1) - \dfrac{n(n-1)}{2}\right]$

$= \dfrac{n}{6}\left[2(8n^2 + 6n + 1) - 6(2n+1)\right]$

$\quad - \dfrac{n}{6}\left[(2n^2 - 3n + 1) - 3(n-1)\right]$

$= \dfrac{n}{6}\left[16n^2 + 12n + 2 - 12n - 6 - 2n^2 + 3n - 1 + 3n - 3\right]$

$= \dfrac{n}{6}\left[14n^2 + 6n - 8\right]$

$= \dfrac{n}{3}\left[7n^2 + 3n - 4\right]$

$= \dfrac{n}{3}(n+1)(7n-4)$ as required $\left(A = \dfrac{1}{3}\right)$

b 9792

22 a $U_n = (n+1)\ln 3^n$

b $S_n = \displaystyle\sum_{r=1}^{n} r(r+1)\ln 3$

$= \ln 3 \displaystyle\sum_{r=1}^{n} n^2 + n$

$= \ln 3\left[\dfrac{n}{6}(n+1)(2n+1) + \dfrac{n}{2}(n+1)\right]$

$= \dfrac{n}{6}(n+1)[2n+1+3]\ln 3$

$= \dfrac{n}{6}(n+1)[2n+4]\ln 3$

$= \dfrac{n}{3}(n+1)[n+2]\ln 3$

23 When $n = 1$, $\dfrac{1 + 2^{n-1}}{2^{n-1}} = \dfrac{1 + 2^0}{2^0} = 2$ so true for $n = 1$

Assume true for $n = k$.

$u_{k+1} = \dfrac{1 + u_k}{2}$

$\qquad = \dfrac{1 + \left(\dfrac{1 + 2^{k-1}}{2^{k-1}}\right)}{2}$

$\qquad = \dfrac{2^{k-1} + (1 + 2^{k-1})}{2(2^{k-1})}$

$\qquad = \dfrac{2(2^{k-1}) + 1}{2(2^{k-1})}$

$\qquad = \dfrac{2^k + 1}{2^k}$ so true for $n = k+1$ since $2^{k+1-1} = 2^k$

True for $n = 1$ and assuming true for $n = k$ implies true for $n = k+1$, hence true for all $n \in \mathbb{N}$

24 When $n = 1, u_1 = 5$

And $u_1 = \dfrac{1}{2}(7 + 3) = 5$

So the statement is true when $n = 1$

Assume true for $n = k$, i.e. assume $u_k = \dfrac{1}{2}(7 + 3^k)$

$u_{k+1} = 3u_k - 7$

$\qquad = 3\left(\dfrac{1}{2}(7 + 3^k)\right) - 7$

$\qquad = \dfrac{1}{2}(3(7 + 3^k) - 14)$

$\qquad = \dfrac{1}{2}(21 + 3(3^k) - 14)$

$\qquad = \dfrac{1}{2}(7 + 3^{k+1})$

So the statement is true when $n = k+1$

The statement is true for $n = 1$ and by assuming it is true for $n = k$ it is shown to be true for $n = k+1$, therefore, by mathematical induction, it is true for all integers $n \geq 1$

25 a When $n = 2$, $\displaystyle\sum_{r=2}^{n} \dfrac{1}{r^2 - 1} = \dfrac{1}{3}$

$\dfrac{3n^2 - n - 2}{4n(n+1)} = \dfrac{12 - 2 - 2}{4(2)(3)} = \dfrac{8}{24} = \dfrac{1}{3}$

So true for $n = 2$

Assume true for $n = k$

$\displaystyle\sum_{r=2}^{k+1} \dfrac{1}{r^2 - 1} = \dfrac{1}{(k+1)^2 - 1} + \sum_{r=2}^{k} \dfrac{1}{r^2 - 1}$

$\qquad = \dfrac{1}{(k+1)^2 - 1} + \dfrac{3k^2 - k - 2}{4k(k+1)}$

$\qquad = \dfrac{1}{k^2 + 2k} + \dfrac{3k^2 - k - 2}{4k(k+1)}$

$\qquad = \dfrac{4(k+1)}{4k(k+2)(k+1)} + \dfrac{(3k^2 - k - 2)(k+2)}{4k(k+1)(k+2)}$

$\qquad = \dfrac{4(k+1) + (3k^2 - k - 2)(k+2)}{4k(k+1)(k+2)}$

$$= \frac{4k+4+3k^3+6k^2-k^2-2k-2k-4}{4k(k+1)(k+2)}$$

$$= \frac{3k^3+5k^2}{4k(k+1)(k+2)}$$

$$= \frac{k^2(3k+5)}{4k(k+1)(k+2)}$$

$$= \frac{k(3k+5)}{4(k+1)(k+2)}$$

$$\left[\frac{3(k+1)^2-(k+1)-2}{4(k+1)(k+1+1)}=\frac{3k^2+6k+3-k-1-2}{4(k+1)(k+2)}\right.$$

$$\left.= \frac{3k^2+5k}{4(k+1)(k+2)} \text{ as required}\right]$$

True for $n=2$ and assuming true for $n=k$ implies true for $n=k+1$, hence true for all $n \geq 2$

b Since $\dfrac{1}{r^2-1}$ not defined at $r=1$

Chapter 3
Exercise 3.1A

1 $\dfrac{21}{2}\pi$

2 $\dfrac{9207}{5}\pi$

3 $\dfrac{477}{7}\pi$

4 139π

5 a $\displaystyle\int_0^2 8-x^3 \ dx = \left[8x-\frac{x^4}{4}\right]_0^2$

$$= (16-4)-0$$
$$= 12 \text{ as required}$$

b $\dfrac{576}{7}\pi$

6 a $\dfrac{1}{20}$ square units **b** $\dfrac{\pi}{252}$

7 $\dfrac{59}{30}\pi$

8 $V = \pi \displaystyle\int_a^2 \left(\frac{2}{\sqrt{x^3}}\right)^2 dx$

$$= \pi \int_a^2 4x^{-3} dx$$

$$= \pi\left[-2x^{-2}\right]_a^2$$

$$= -2\pi\left(\frac{1}{4}-\frac{1}{a^2}\right)$$

$$= \frac{-2(a^2-4)\pi}{4a^2}$$

$$= \frac{(4-a^2)\pi}{2a^2} \text{ as required}$$

9 $\pi\displaystyle\int_{0.5}^1 \left(\frac{1}{x^2}\right)^2 dx = \pi\int_{0.5}^1 x^{-4} dx$

$$= \pi\left[-\frac{1}{3}x^{-3}\right]_{0.5}^1$$

$$= \pi\left(-\frac{1}{3}-\frac{8}{3}\right)$$

$$= \frac{7}{3}\pi$$

10 a $\displaystyle\int_1^3 \frac{2}{5}x^{-2}+\frac{1}{5}x^{-\left(\frac{3}{2}\right)} \ dx = \left[-\frac{2}{5}x^{-1}-\frac{2}{5}x^{-\frac{1}{2}}\right]_1^3$

$$= -\frac{2}{15}-\frac{2}{5\sqrt{3}}-\left(-\frac{2}{5}-\frac{2}{5}\right)$$

$$= -\frac{2}{15}-\frac{2\sqrt{3}}{15}+\frac{4}{5}$$

$$= \frac{1}{15}(-2-2\sqrt{3}+12)$$

$$= \frac{10-2\sqrt{3}}{15}$$

b i 0.405 (3sf)
ii 0.101 (3sf)

11 a 0, 16 **b** $\dfrac{512}{3}$ **c** 6620 cubic units

12 $\dfrac{256}{15}\pi$

13 $\dfrac{36}{35}\sqrt{3}\,\pi$

Exercise 3.1B

1 a $\dfrac{19}{12}$ **b** $\dfrac{109}{30}\pi$

2 a 12 **b** $\dfrac{912}{7}\pi$

3 a $\dfrac{10}{3}$ square units **b** $\dfrac{8}{3}\pi$

4 $16-x^4=7 \Rightarrow x=\pm\sqrt{3}$

a $\displaystyle\int_0^{\sqrt{3}} 16-x^4 dx = \left[16x-\frac{x^5}{5}\right]_0^{\sqrt{3}}$

$$= 16\sqrt{3}-\frac{9}{5}\sqrt{3}-0$$

$$= \frac{71}{5}\sqrt{3}$$

Area of R $= \dfrac{71}{5}\sqrt{3}-7\times\sqrt{3}$

$$= \frac{36}{5}\sqrt{3} \text{ as required}$$

b $\dfrac{792}{5}\sqrt{3}\pi$

5

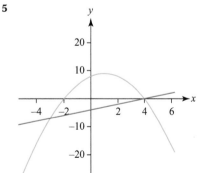

a $\displaystyle\int_0^4 8+2x-x^2\,dx = \left[8x+x^2-\dfrac{x^3}{3}\right]_0^4$

$\qquad\qquad\qquad = 32+16-\dfrac{64}{3}$

$\qquad\qquad\qquad = \dfrac{80}{3}$

Area required $= \dfrac{80}{3}+\dfrac{1}{2}\times 4\times 4$

$\qquad\qquad = \dfrac{104}{3}$ as required.

b $\dfrac{1696}{15}\pi$

6 $V=\pi\displaystyle\int_0^b (ax)^2\,dx$

$\qquad = \pi\displaystyle\int_0^b a^2x^2\,dx$

$\qquad = \pi\left[\dfrac{a^2}{3}x^3\right]_0^b$

$\qquad = \dfrac{\pi a^2}{3}(b^3-0)$

$\qquad = \dfrac{\pi(ab)^2}{3}b$

$\qquad = \dfrac{1}{3}\pi r^2 h$ where $r=ab$ and $h=b$

7 $V=\pi\displaystyle\int_0^h r^2\,dx$

$\qquad = \pi\left[r^2x\right]_0^h$

$\qquad = \pi(r^2h-0)$

$\qquad = \pi r^2 h$ as required

8 $V=\pi\displaystyle\int_0^r \sqrt{r^2-x^2}^{\,2}\,dx$

$\qquad = \pi\displaystyle\int_0^r r^2-x^2\,dx$

$\qquad = \pi\left[r^2x-\dfrac{x^3}{3}\right]_0^r$

$\qquad = \pi\left(r^3-\dfrac{r^3}{3}-0\right)$

$\qquad = \pi\left(\dfrac{2}{3}r^3\right)$

$\qquad = \dfrac{2}{3}\pi r^3$

9 a A has coordinates $(7, 14)$, B has coordinates $(7, -14)$

  b 686π

10 a $\dfrac{38}{3}$ square units   **b** $\dfrac{4848}{35}\pi$

Exercise 3.2A

1 $\dfrac{625}{4}\pi$

2 i $\dfrac{\pi}{15}$ **ii** $\dfrac{\pi}{75}$

3 $\dfrac{5261}{105}\pi$

4 $\dfrac{9}{10}\pi$

5 $\dfrac{7}{3}\pi$

6 a $x=-\sqrt{y}+2$ **b** $\dfrac{4}{3}$ square units   **c** $\dfrac{5}{6}\pi$

7 a $4\sqrt{3}$ square units **b** $\dfrac{9}{2}\pi$ cubic units

8 i $\dfrac{24}{5}\pi\sqrt{3}$ **ii** $\dfrac{9}{2}\pi$

9 8π

10 $A=\dfrac{14}{3}$

11 a $A=\dfrac{\sqrt{6}}{3}$ **b** $\dfrac{\sqrt{2}}{15}\pi(27-\sqrt{3})$

Exercise 3.2B

1 $\dfrac{48}{5}\pi$

2 a $\dfrac{1376}{15}\pi$ cubic units

  b 24π cubic units

3 a $\dfrac{32}{3}\pi$ **b** $\dfrac{128}{15}\pi$

4 a $\dfrac{288}{7}\pi$ **b** $\dfrac{48}{5}\pi$

5 $\dfrac{1024}{21}\pi$ cubic units

6 $\left(\dfrac{23}{6}+\dfrac{8}{3}\sqrt{2}\right)\pi$ cubic units

7 $k=\dfrac{10}{9}\pi$

8 a 2.57 square units

  b $\dfrac{5504}{3645}\pi$ or 4.74 cubic units

Review exercise 3

1 a $3-\dfrac{1}{x^2}=0 \Rightarrow x=\pm\dfrac{1}{\sqrt{3}}$

$\displaystyle\int_{\frac{1}{\sqrt{3}}}^{2} 3-x^{-2}\,dx = \left[3x+\dfrac{1}{x}\right]_{\frac{1}{\sqrt{3}}}^{2}$

$\qquad\qquad = \left(6+\dfrac{1}{2}\right)-\left(\dfrac{3}{\sqrt{3}}+\sqrt{3}\right)$

$\qquad\qquad = \dfrac{13}{2}-2\sqrt{3}$

$\qquad\qquad = \dfrac{13-4\sqrt{3}}{2}$

  b $\dfrac{503-192\sqrt{3}}{24}\pi$

2 291π

3 50π

4 $\pi\displaystyle\int_0^{\frac{1}{2}} 16y^4\,dy = \pi\left[\dfrac{16}{5}y^5\right]_0^{\frac{1}{2}}$

$\qquad\qquad = \pi\left(\dfrac{1}{10}-0\right)$

$\qquad\qquad = \dfrac{\pi}{10}$

5 a $\displaystyle\int_{0}^{2} y^{4}+1 \, dy = \left[\dfrac{y^{5}}{5}+y\right]_{0}^{2}$

$$= \dfrac{32}{5}+2-0$$

$$= \dfrac{42}{5} \text{ as required}$$

b $\dfrac{3226}{45}\pi$

6 i $\dfrac{128}{5}\pi$ **ii** $\dfrac{24}{11}\pi$

7 a

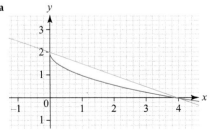

b $\dfrac{8}{3}\pi$ **c** $\dfrac{64}{15}\pi$

8 a $\dfrac{49}{48}$ **b** $\dfrac{901}{1920}\pi$

9 $\dfrac{9}{4}\pi$ cubic units

10 $\dfrac{13}{15}\pi$

11 $\left(\dfrac{24}{5}\sqrt{3}-\dfrac{88}{15}\right)\pi$

Assessment 3

1 a $\alpha=2$ **b** $\dfrac{20}{3}$ **c** $\dfrac{496}{15}\pi$

2 18π

3 $\dfrac{619}{21}\pi$

4 1170 cubic units

5 $\dfrac{512}{105}\pi$ cubic units

6 2 or $-\dfrac{2}{5}$

7 $\dfrac{38}{5}\pi$

8 $\dfrac{7}{6}\pi$

9 $\dfrac{8}{9}\sqrt{2}\pi$

10 a $\dfrac{8}{3}$ **b** $\pi\left(\dfrac{3}{2}+\ln 4\right)$

11 a $a=3$ **b** 4π **c** 6π
$\quad\quad b=2$

12 1.32 litres

13 $\left(\dfrac{5+4\sqrt{2}}{3}\right)\pi$ cubic units

14 a $\dfrac{104}{3}$ square units

b $\dfrac{160}{3}\pi$ cubic units

1 a i 3×2 **ii** 3×1 **iii** 2×2 **iv** 2×4
b the matrix in part **iii** is a square matrix

2 a $\begin{pmatrix} 4 & 3 & 2 \\ 5 & -1 & 6 \end{pmatrix}$ **b** $\begin{pmatrix} -4 & -6 \\ 1 & -3 \end{pmatrix}$

c $\begin{pmatrix} -4 & 0 & 16 \\ 8 & 20 & -12 \\ 32 & -24 & 0 \end{pmatrix}$ **d** $\begin{pmatrix} -2 & 12 \\ 6 & 10.5 \\ 1 & -4 \end{pmatrix}$

3 a i $\begin{pmatrix} 9 & -3 \\ -5 & 2 \end{pmatrix}$ **ii** Not possible as different order matrices

iii $\begin{pmatrix} 13 & 6 \\ -5 & -6 \\ -5 & -1 \end{pmatrix}$ **iv** $\begin{pmatrix} 24 & 12 \\ 6 & -18 \\ -6 & 0 \end{pmatrix}$

b i $\begin{pmatrix} 45 & -20 \\ 0 & -10 \end{pmatrix}-\begin{pmatrix} 0 & 1 \\ -5 & 4 \end{pmatrix}=\begin{pmatrix} 45 & -21 \\ 5 & -14 \end{pmatrix}$

ii $\begin{pmatrix} -10 & -4 \\ 14 & 0 \\ 6 & 2 \end{pmatrix}+\begin{pmatrix} 56 & 28 \\ 14 & -42 \\ -14 & 0 \end{pmatrix}=\begin{pmatrix} 46 & 24 \\ 28 & -42 \\ -8 & 2 \end{pmatrix}$

$$=2\begin{pmatrix} 23 & 12 \\ 14 & -21 \\ -4 & 1 \end{pmatrix}$$

4 a $\begin{pmatrix} -6 & -15 & 3 & -18 \\ -10 & 9 & -1 & -10 \end{pmatrix}$ **b** $\begin{pmatrix} 16 & 2 \\ 36 & 72 \\ 30 & -15 \end{pmatrix}$

c $\begin{pmatrix} 0 \\ 13 \end{pmatrix}$ **d** $(-8 \;\; -5 \;\; -10)$

e $\begin{pmatrix} -4 & 8 & 20 \\ -3 & 6 & 15 \end{pmatrix}$ **f** $\begin{pmatrix} -5 & 6 & 4 \\ -2 & -12 & 4 \\ 6 & -3 & -5 \end{pmatrix}$

5 $\begin{pmatrix} 3 & 2 & 0 \\ -1 & 0 & -2 \end{pmatrix}\begin{pmatrix} 4 & 1 \\ 0 & 3 \\ -3 & 0 \end{pmatrix}$

$$=\begin{pmatrix} (3\times 4)+(2\times 0)+(0\times -3) & (3\times 1)+(2\times 3)+(0\times 0) \\ (-1\times 4)+(0\times 0)+(-2\times -3) & (-1\times 1)+(0\times 3)+(-2\times 0) \end{pmatrix}$$

$$=\begin{pmatrix} 12+0+0 & 3+6+0 \\ -4+0+6 & -1+0+0 \end{pmatrix}$$

$$=\begin{pmatrix} 12 & 9 \\ 2 & -1 \end{pmatrix}$$

6 a Not possible since 1 column in **A** but 2 rows in **B**
b $\begin{pmatrix} 51 \\ -7 \end{pmatrix}$
c $\begin{pmatrix} -14 \\ -72 \end{pmatrix}$
d Not possible since 3 columns in **C** but 2 rows in **B**
e $\begin{pmatrix} 36 \\ 40 \end{pmatrix}$
f Not possible as **C** not a square matrix.

7 a $\begin{pmatrix} 7a & 8a+6 & 3a \\ 2-a & 2 & 3+a \end{pmatrix}$ **b** $\begin{pmatrix} 3a & -a^{2} \\ 6 & -2a \\ -3a & a^{2} \end{pmatrix}$

c $(a \quad a)$

d $\begin{pmatrix} a^2-6 & 6a \\ -4a & a^2-6 \end{pmatrix}$

8 $\mathbf{A}^3 = \begin{pmatrix} 2 & 5 \\ -1 & 2 \end{pmatrix}\begin{pmatrix} 2 & 5 \\ -1 & 2 \end{pmatrix}\begin{pmatrix} 2 & 5 \\ -1 & 2 \end{pmatrix}$

$= \begin{pmatrix} 2 & 5 \\ -1 & 2 \end{pmatrix}\begin{pmatrix} 4-5 & 10+10 \\ -2-2 & -5+4 \end{pmatrix}$

$= \begin{pmatrix} 2 & 5 \\ -1 & 2 \end{pmatrix}\begin{pmatrix} -1 & 20 \\ -4 & -1 \end{pmatrix}$

$= \begin{pmatrix} -2-20 & 40-5 \\ 1-8 & -20-2 \end{pmatrix}$

$= \begin{pmatrix} -22 & 35 \\ -7 & -22 \end{pmatrix} \qquad k=35$

Exercise 4.1B

1 $a=-3$
   $b=-5$
   $c=-1$
   $d=2$

2 $a=-3$
   $b=4$
   $c=-3$

3 a $x=-1,\ y=4$
   b $x=-2,\ y=4$

4 a $x=1,\ -\dfrac{3}{5}$
   b $x=-2,\ -5$
   c $x=\pm3$
   d $x=-2$

5 $a=7,\ b=-3,\ c=2$

6 $x=3,\ y=4,\ z=-1$

7 $a=-1,\ b=-2,\ c=6$

8 a $\begin{pmatrix} 2 & 5 \\ 6 & -1 \end{pmatrix}\begin{pmatrix} x \\ y \end{pmatrix} = \begin{pmatrix} 6 \\ 3 \end{pmatrix}$

b $\begin{pmatrix} 1 & 1 & -1 \\ 2 & 1 & 1 \\ 3 & 2 & 2 \end{pmatrix}\begin{pmatrix} x \\ y \\ z \end{pmatrix} = \begin{pmatrix} -4 \\ 4 \\ 10 \end{pmatrix}$

9 $\begin{pmatrix} 1 & 5 \\ 0 & 1 \end{pmatrix}^1 = \begin{pmatrix} 1 & 5 \\ 0 & 1 \end{pmatrix}$ and $\begin{pmatrix} 1 & 5\times1 \\ 0 & 1 \end{pmatrix} = \begin{pmatrix} 1 & 5 \\ 0 & 1 \end{pmatrix}$ so true for $n=1$

Assume true for $n=k$

$\begin{pmatrix} 1 & 5 \\ 0 & 1 \end{pmatrix}^{k+1} = \begin{pmatrix} 1 & 5 \\ 0 & 1 \end{pmatrix}^k\begin{pmatrix} 1 & 5 \\ 0 & 1 \end{pmatrix}$

$= \begin{pmatrix} 1 & 5k \\ 0 & 1 \end{pmatrix}\begin{pmatrix} 1 & 5 \\ 0 & 1 \end{pmatrix}$

$= \begin{pmatrix} 1 & 5+5k \\ 0 & 1 \end{pmatrix}$

$= \begin{pmatrix} 1 & 5(k+1) \\ 0 & 1 \end{pmatrix}$ so true for $n=k+1$

Since true for $n=1$ and assuming true for $n=k$ implies true for $n=k+1$, therefore true for all positive integers n

10 $\begin{pmatrix} 5 & 4 \\ 0 & 1 \end{pmatrix}^1 = \begin{pmatrix} 5 & 4 \\ 0 & 1 \end{pmatrix}$ and $\begin{pmatrix} 5^1 & 5^1-1 \\ 0 & 1 \end{pmatrix} = \begin{pmatrix} 5 & 4 \\ 0 & 1 \end{pmatrix}$ so true for $n=1$

Assume true for $n=k$

$\begin{pmatrix} 5 & 4 \\ 0 & 1 \end{pmatrix}^{k+1} = \begin{pmatrix} 5 & 4 \\ 0 & 1 \end{pmatrix}^k\begin{pmatrix} 5 & 4 \\ 0 & 1 \end{pmatrix}$

$= \begin{pmatrix} 5^k & 5^k-1 \\ 0 & 1 \end{pmatrix}\begin{pmatrix} 5 & 4 \\ 0 & 1 \end{pmatrix}$

$= \begin{pmatrix} 5(5^k) & 4(5^k)+5^k-1 \\ 0 & 1 \end{pmatrix}$

$= \begin{pmatrix} 5^{k+1} & 5(5^k)-1 \\ 0 & 1 \end{pmatrix}$

$= \begin{pmatrix} 5^{k+1} & 5^{k+1}-1 \\ 0 & 1 \end{pmatrix}$ so true for $n=k+1$

Since true for $n=1$ and assuming true for $n=k$ implies true for $n=k+1$, therefore true for all positive integers n

11 $\begin{pmatrix} -2 & -1 \\ 9 & 4 \end{pmatrix}^1 = \begin{pmatrix} -2 & -1 \\ 9 & 4 \end{pmatrix}$ and $\begin{pmatrix} 1-3\times1 & -1 \\ 9\times1 & 3\times1+1 \end{pmatrix} = \begin{pmatrix} -2 & -1 \\ 9 & 4 \end{pmatrix}$

so true for $n=1$

Assume true for $n=k$

$\begin{pmatrix} -2 & -1 \\ 9 & 4 \end{pmatrix}^{k+1} = \begin{pmatrix} -2 & -1 \\ 9 & 4 \end{pmatrix}^k\begin{pmatrix} -2 & -1 \\ 9 & 4 \end{pmatrix}$

$= \begin{pmatrix} 1-3k & -k \\ 9k & 3k+1 \end{pmatrix}\begin{pmatrix} -2 & -1 \\ 9 & 4 \end{pmatrix}$

$= \begin{pmatrix} -2+6k-9k & -1+3k-4k \\ -18k+27k+9 & -9k+12k+4 \end{pmatrix}$

$= \begin{pmatrix} -2-3k & -1-k \\ 9k+9 & 3k+4 \end{pmatrix}$

$= \begin{pmatrix} 1-3(k+1) & -(k+1) \\ 9(k+1) & 3(k+1)+1 \end{pmatrix}$ so true for $n=k+1$

Since true for $n=1$ and assuming true for $n=k$ implies true for $n=k+1$, therefore true for all positive integers n

12 $\begin{pmatrix} 1 & 4 \\ 0 & 2 \end{pmatrix}^1 = \begin{pmatrix} 1 & 4 \\ 0 & 2 \end{pmatrix}$ and $\begin{pmatrix} 1 & 4(2^1-1) \\ 0 & 2^1 \end{pmatrix} = \begin{pmatrix} 1 & 4 \\ 0 & 2 \end{pmatrix}$ so true for $n=1$

Assume true for $n=k$

$\begin{pmatrix} 1 & 4 \\ 0 & 2 \end{pmatrix}^{k+1} = \begin{pmatrix} 1 & 4 \\ 0 & 2 \end{pmatrix}^k\begin{pmatrix} 1 & 4 \\ 0 & 2 \end{pmatrix}$

$= \begin{pmatrix} 1 & 4(2^k-1) \\ 0 & 2^k \end{pmatrix}\begin{pmatrix} 1 & 4 \\ 0 & 2 \end{pmatrix}$

$= \begin{pmatrix} 1 & 4+8.2^k-8 \\ 0 & 2.2^k \end{pmatrix}$

$= \begin{pmatrix} 1 & 4(2.2^k-1) \\ 0 & 2.2^k \end{pmatrix}$

$= \begin{pmatrix} 1 & 4(2^{k+1}-1) \\ 0 & 2^{k+1} \end{pmatrix}$ so true for $n=k+1$

Since true for $n=1$ and assuming true for $n=k$ implies true for $n=k+1$, therefore true for all positive integers n

13 Let $\mathbf{A} = \begin{pmatrix} a_1 & a_2 \\ a_3 & a_4 \\ a_5 & a_6 \end{pmatrix}$, $\mathbf{B} = \begin{pmatrix} b_1 & b_2 \\ b_3 & b_4 \\ b_5 & b_6 \end{pmatrix}$, $\mathbf{C} = \begin{pmatrix} c_1 & c_2 \\ c_3 & c_4 \\ c_5 & c_6 \end{pmatrix}$

$$\mathbf{A}+(\mathbf{B}+\mathbf{C})=\begin{pmatrix} a_1 & a_2 \\ a_3 & a_4 \\ a_5 & a_6 \end{pmatrix}+\begin{pmatrix} b_1+c_1 & b_2+c_2 \\ b_3+c_3 & b_4+c_4 \\ b_5+c_5 & b_6+c_6 \end{pmatrix}$$

$$=\begin{pmatrix} a_1+b_1+c_1 & a_2+b_2+c_2 \\ a_3+b_3+c_3 & a_4+b_4+c_4 \\ a_5+b_5+c_5 & a_6+b_6+c_6 \end{pmatrix}$$

$$=\begin{pmatrix} (a_1+b_1)+c_1 & (a_2+b_2)+c_2 \\ (a_3+b_3)+c_3 & (a_4+b_4)+c_4 \\ (a_5+b_5)+c_5 & (a_6+b_6)+c_6 \end{pmatrix}$$

$$=\begin{pmatrix} a_1+b_1 & a_2+b_2 \\ a_3+b_3 & a_4+b_4 \\ a_5+b_5 & a_6+b_6 \end{pmatrix}+\begin{pmatrix} c_1 & c_2 \\ c_3 & c_4 \\ c_5 & c_6 \end{pmatrix}$$

$$=(\mathbf{A}+\mathbf{B})+\mathbf{C} \text{ as required}$$

14 Let $\mathbf{A}=\begin{pmatrix} a_1 & a_2 \\ a_3 & a_4 \end{pmatrix}$, $\mathbf{B}=\begin{pmatrix} b_1 & b_2 \\ b_3 & b_4 \end{pmatrix}$

$$\mathbf{A}+\mathbf{B}=\begin{pmatrix} a_1 & a_2 \\ a_3 & a_4 \end{pmatrix}+\begin{pmatrix} b_1 & b_2 \\ b_3 & b_4 \end{pmatrix}$$

$$=\begin{pmatrix} a_1+b_1 & a_2+b_2 \\ a_3+b_3 & a_4+b_4 \end{pmatrix}$$

$$=\begin{pmatrix} b_1+a_1 & b_2+a_2 \\ b_3+a_3 & b_4+a_4 \end{pmatrix}$$

$$=\begin{pmatrix} b_1 & b_2 \\ b_3 & b_4 \end{pmatrix}+\begin{pmatrix} a_1 & a_2 \\ a_3 & a_4 \end{pmatrix}$$

$$=\mathbf{B}+\mathbf{A} \text{ as required}$$

15 e.g. let $\mathbf{A}=\begin{pmatrix} 2 & 0 \\ 1 & 1 \end{pmatrix}$, $\mathbf{B}=\begin{pmatrix} 0 & 1 \\ 2 & 2 \end{pmatrix}$

$$\mathbf{AB}=\begin{pmatrix} 2 & 0 \\ 1 & 1 \end{pmatrix}\begin{pmatrix} 0 & 1 \\ 2 & 2 \end{pmatrix}$$

$$=\begin{pmatrix} 0 & 2 \\ 2 & 3 \end{pmatrix}$$

$$\mathbf{BA}=\begin{pmatrix} 0 & 1 \\ 2 & 2 \end{pmatrix}\begin{pmatrix} 2 & 0 \\ 1 & 1 \end{pmatrix}$$

$$=\begin{pmatrix} 1 & 1 \\ 6 & 2 \end{pmatrix}$$

So $\mathbf{AB}\neq\mathbf{BA}$

16 Let $\mathbf{A}=\begin{pmatrix} a_1 & a_2 \\ a_3 & a_4 \end{pmatrix}$, $\mathbf{B}=\begin{pmatrix} b_1 & b_2 \\ b_3 & b_4 \end{pmatrix}$, $\mathbf{C}=\begin{pmatrix} c_1 & c_2 \\ c_3 & c_4 \end{pmatrix}$

$$\mathbf{A}(\mathbf{B}+\mathbf{C})$$

$$=\begin{pmatrix} a_1 & a_2 \\ a_3 & a_4 \end{pmatrix}\left[\begin{pmatrix} b_1 & b_2 \\ b_3 & b_4 \end{pmatrix}+\begin{pmatrix} c_1 & c_2 \\ c_3 & c_4 \end{pmatrix}\right]$$

$$=\begin{pmatrix} a_1 & a_2 \\ a_3 & a_4 \end{pmatrix}\begin{pmatrix} b_1+c_1 & b_2+c_2 \\ b_3+c_3 & b_4+c_4 \end{pmatrix}$$

$$=\begin{pmatrix} a_1(b_1+c_1)+a_2(b_3+c_3) & a_1(b_2+c_2)+a_2(b_4+c_4) \\ a_3(b_1+c_1)+a_4(b_3+c_3) & a_3(b_2+c_2)+a_4(b_4+c_4) \end{pmatrix}$$

$$=\begin{pmatrix} (a_1b_1+a_2b_3)+(a_1c_1+a_2c_3) & (a_1b_2+a_2b_4)+(a_1c_2+a_2c_4) \\ (a_3b_1+a_4b_3)+(a_3c_1+a_4c_3) & (a_3b_2+a_4b_4)+(a_3c_2+a_4c_4) \end{pmatrix}$$

$$=\begin{pmatrix} a_1b_1+a_2b_3 & a_1b_2+a_2b_4 \\ a_3b_1+a_4b_3 & a_3b_2+a_4b_4 \end{pmatrix}+\begin{pmatrix} a_1c_1+a_2c_3 & a_1c_2+a_2c_4 \\ a_3c_1+a_4c_3 & a_3c_2+a_4c_4 \end{pmatrix}$$

$$=\begin{pmatrix} a_1 & a_2 \\ a_3 & a_4 \end{pmatrix}\begin{pmatrix} b_1 & b_2 \\ b_3 & b_4 \end{pmatrix}+\begin{pmatrix} a_1 & a_2 \\ a_3 & a_4 \end{pmatrix}\begin{pmatrix} c_1 & c_2 \\ c_3 & c_4 \end{pmatrix}$$

$$=\mathbf{AB}+\mathbf{AC}$$

17 Let $\mathbf{A}=\begin{pmatrix} a_1 & a_2 \\ a_3 & a_4 \end{pmatrix}$

$$\mathbf{A}\begin{pmatrix} 1 & 0 \\ 0 & 1 \end{pmatrix}=\begin{pmatrix} a_1 & a_2 \\ a_3 & a_4 \end{pmatrix}\begin{pmatrix} 1 & 0 \\ 0 & 1 \end{pmatrix}$$

$$=\begin{pmatrix} a_1 & a_2 \\ a_3 & a_4 \end{pmatrix}$$

$$\begin{pmatrix} 1 & 0 \\ 0 & 1 \end{pmatrix}\mathbf{A}=\begin{pmatrix} 1 & 0 \\ 0 & 1 \end{pmatrix}\begin{pmatrix} a_1 & a_2 \\ a_3 & a_4 \end{pmatrix}$$

$$=\begin{pmatrix} a_1 & a_2 \\ a_3 & a_4 \end{pmatrix}$$

So $\mathbf{A}\begin{pmatrix} 1 & 0 \\ 0 & 1 \end{pmatrix}=\begin{pmatrix} 1 & 0 \\ 0 & 1 \end{pmatrix}\mathbf{A}$ as required

18 Let $\mathbf{B}=\begin{pmatrix} b_1 & b_2 & b_3 \\ b_4 & b_5 & b_6 \\ b_7 & b_8 & b_9 \end{pmatrix}$

$$\mathbf{B}\begin{pmatrix} 1 & 0 & 0 \\ 0 & 1 & 0 \\ 0 & 0 & 1 \end{pmatrix}=\begin{pmatrix} b_1 & b_2 & b_3 \\ b_4 & b_5 & b_6 \\ b_7 & b_8 & b_9 \end{pmatrix}\begin{pmatrix} 1 & 0 & 0 \\ 0 & 1 & 0 \\ 0 & 0 & 1 \end{pmatrix}$$

$$=\begin{pmatrix} b_1 & b_2 & b_3 \\ b_4 & b_5 & b_6 \\ b_7 & b_8 & b_9 \end{pmatrix}$$

$$\begin{pmatrix} 1 & 0 & 0 \\ 0 & 1 & 0 \\ 0 & 0 & 1 \end{pmatrix}\mathbf{B}=\begin{pmatrix} 1 & 0 & 0 \\ 0 & 1 & 0 \\ 0 & 0 & 1 \end{pmatrix}\begin{pmatrix} b_1 & b_2 & b_3 \\ b_4 & b_5 & b_6 \\ b_7 & b_8 & b_9 \end{pmatrix}$$

$$=\begin{pmatrix} b_1 & b_2 & b_3 \\ b_4 & b_5 & b_6 \\ b_7 & b_8 & b_9 \end{pmatrix}$$

So $\mathbf{B}\begin{pmatrix} 1 & 0 & 0 \\ 0 & 1 & 0 \\ 0 & 0 & 1 \end{pmatrix}=\begin{pmatrix} 1 & 0 & 0 \\ 0 & 1 & 0 \\ 0 & 0 & 1 \end{pmatrix}\mathbf{B}$ as required

19 a $\begin{pmatrix} 8 & 14 & 15 & 17 \\ 6 & 11 & 7 & 9 \\ 9 & 18 & 19 & 12 \end{pmatrix}$

$$\begin{pmatrix} 5 \\ 2 \\ 3 \\ 1 \end{pmatrix} \text{ or } \begin{pmatrix} 0.05 \\ 0.02 \\ 0.03 \\ 0.01 \end{pmatrix}$$

$$\textbf{b} \quad \begin{pmatrix} 8 & 14 & 15 & 17 \\ 6 & 11 & 7 & 9 \\ 9 & 18 & 19 & 12 \end{pmatrix} \begin{pmatrix} 0.05 \\ 0.02 \\ 0.03 \\ 0.01 \end{pmatrix}$$

$$= \begin{pmatrix} 1.3 \\ 0.82 \\ 1.5 \end{pmatrix}$$

c £3620

$$\textbf{20 a} \quad \begin{pmatrix} 5 & 20 & 7 & 6 & 50 \\ 4 & 15 & 15 & 8 & 100 \\ 12 & 4 & 2 & 30 & 50 \end{pmatrix} \begin{pmatrix} 12 \\ 8 \\ 18 \\ 30 \\ 1 \end{pmatrix}$$

$$\begin{pmatrix} 576 \\ 778 \\ 1162 \end{pmatrix}$$

b £25.16

21 a $\quad ae + bg = a\left(\dfrac{d}{ad - bc}\right) + b\left(-\dfrac{c}{ad - bc}\right)$

$$= \dfrac{ad}{ad - bc} - \dfrac{bc}{ad - bc}$$

$$= \dfrac{ad - bc}{ad - bc}$$

$$= 1 \text{ as required.}$$

b $\quad h = \dfrac{a}{ad - bc}, \; f = -\dfrac{b}{ad - bc}$

22 When $n = 1$, $\begin{pmatrix} 0 & 1 \\ 2 & 0 \end{pmatrix}^{2n} = \begin{pmatrix} 0 & 1 \\ 2 & 0 \end{pmatrix}^{2}$

$$= \begin{pmatrix} 2 & 0 \\ 0 & 2 \end{pmatrix}$$

$$2^n \textbf{I} = 2\begin{pmatrix} 1 & 0 \\ 0 & 1 \end{pmatrix}$$

$$= \begin{pmatrix} 2 & 0 \\ 0 & 2 \end{pmatrix}$$

So true for $n = 1$

Assume true for $n = k$ and consider $n = k + 1$:

$$\begin{pmatrix} 0 & 1 \\ 2 & 0 \end{pmatrix}^{2(k+1)} = \begin{pmatrix} 0 & 1 \\ 2 & 0 \end{pmatrix}^{2k}\begin{pmatrix} 0 & 1 \\ 2 & 0 \end{pmatrix}^{2}$$

$$= 2^k \begin{pmatrix} 1 & 0 \\ 0 & 1 \end{pmatrix}\begin{pmatrix} 2 & 0 \\ 0 & 2 \end{pmatrix}$$

$$= 2^k \begin{pmatrix} 2 & 0 \\ 0 & 2 \end{pmatrix}$$

$$= 2^k \cdot 2\begin{pmatrix} 1 & 0 \\ 0 & 1 \end{pmatrix}$$

$$= 2^{k+1}\textbf{I}$$

So true for $n = k + 1$

The statement is true for $n = 1$ and by assuming it is true for $n = k$ it is shown to be true for $n = k + 1$, therefore, by mathematical induction, it is true for all $n \in \mathbb{N}$

1 a

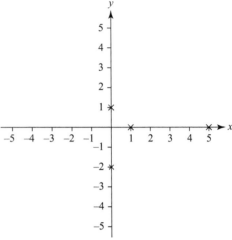

$$\textbf{b} \quad \begin{pmatrix} 5 & 0 \\ 0 & -2 \end{pmatrix}$$

2 a

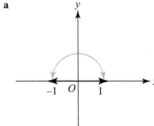

$$\textbf{b} \quad \begin{pmatrix} 0 \\ -1 \end{pmatrix}$$

$$\textbf{c} \quad \begin{pmatrix} -1 & 0 \\ 0 & -1 \end{pmatrix}$$

$$\textbf{3 a} \quad \begin{pmatrix} 0 & 1 \\ 1 & 0 \end{pmatrix} \qquad \textbf{b} \quad \begin{pmatrix} \dfrac{\sqrt{3}}{2} & -\dfrac{1}{2} \\ \dfrac{1}{2} & \dfrac{\sqrt{3}}{2} \end{pmatrix}$$

$$\textbf{c} \quad \begin{pmatrix} 5 & 0 \\ 0 & 5 \end{pmatrix} \qquad \textbf{d} \quad \begin{pmatrix} 1 & 0 \\ 0 & 4 \end{pmatrix}$$

4 a $(0, -2)$, $(4, 3)$ and $(-8, 3)$

b Stretch parallel to x-axis of scale factor 2

5 $\theta = 315°$

$$\textbf{6 a} \quad \begin{pmatrix} 6 \\ -1 \end{pmatrix}, \begin{pmatrix} -1 \\ 2 \end{pmatrix}$$

b Stretch scale factor 0.5 parallel to y-axis

7 a Enlargement scale factor 3 centre the origin

b Stretch scale factor 2 parallel to y-axis

c Rotation by 135° anticlockwise about origin

d Rotation by 270° anticlockwise about origin

e Reflection in y-axis

f Rotation by 20° clockwise about origin or 340° anticlockwise.

$$\textbf{8 a} \quad \textbf{i} \;\; \textbf{BA} = \begin{pmatrix} 0 & -1 \\ -1 & 0 \end{pmatrix} \qquad \textbf{ii} \;\; \textbf{AB} = \begin{pmatrix} -1 & 0 \\ -1 & 1 \end{pmatrix}$$

b $\textbf{BA} \neq \textbf{AB}$ since matrix multiplication is not commutative.

9 a $\textbf{A}$ is reflection in x-axis, $\textbf{B}$ is rotation about origin of 90° clockwise or 270° anticlockwise

b i $BA = \begin{pmatrix} 0 & -1 \\ -1 & 0 \end{pmatrix}$ **ii** $AB = \begin{pmatrix} 0 & 1 \\ 1 & 0 \end{pmatrix}$

c i Reflection in $y = -x$ **ii** Reflection in $y = x$

10 a $\begin{pmatrix} -\dfrac{\sqrt{2}}{2} & \dfrac{\sqrt{2}}{2} \\ \dfrac{\sqrt{2}}{2} & \dfrac{\sqrt{2}}{2} \end{pmatrix}$ **b** $\begin{pmatrix} 0 & -2 \\ -2 & 0 \end{pmatrix}$ **c** $\begin{pmatrix} 0 & 1 \\ 2 & 0 \end{pmatrix}$

11 $\left(\dfrac{3}{2}\sqrt{2}, 7\sqrt{2} \right), \left(-\dfrac{\sqrt{2}}{2}, 3\sqrt{2} \right), (2\sqrt{2}, -2\sqrt{2}), (4\sqrt{2}, 2\sqrt{2})$

12 a $\begin{pmatrix} -\dfrac{\sqrt{2}}{2} & -\dfrac{\sqrt{2}}{2} \\ \dfrac{\sqrt{2}}{2} & -\dfrac{\sqrt{2}}{2} \end{pmatrix}$

b $A^2 = \begin{pmatrix} -\dfrac{\sqrt{2}}{2} & -\dfrac{\sqrt{2}}{2} \\ \dfrac{\sqrt{2}}{2} & -\dfrac{\sqrt{2}}{2} \end{pmatrix}\begin{pmatrix} -\dfrac{\sqrt{2}}{2} & -\dfrac{\sqrt{2}}{2} \\ \dfrac{\sqrt{2}}{2} & -\dfrac{\sqrt{2}}{2} \end{pmatrix}$

$= \begin{pmatrix} \dfrac{1}{2} - \dfrac{1}{2} & \dfrac{1}{2} + \dfrac{1}{2} \\ -\dfrac{1}{2} - \dfrac{1}{2} & -\dfrac{1}{2} + \dfrac{1}{2} \end{pmatrix}$

$= \begin{pmatrix} 0 & 1 \\ -1 & 0 \end{pmatrix}$

c $\sin\theta = -1 \Rightarrow \theta = -90$
and $\cos(-90) = 0$
So a clockwise rotation of $90°$,
or show geometrically.

d An anticlockwise rotation by $45°$ about the origin.

13 a $\begin{pmatrix} -1 & 0 \\ 0 & -1 \end{pmatrix}$; rotation of $180°$ around origin

b $\begin{pmatrix} -\dfrac{1}{2} & -\dfrac{\sqrt{3}}{2} \\ \dfrac{\sqrt{3}}{2} & -\dfrac{1}{2} \end{pmatrix}$; rotation of $120°$ anticlockwise around origin

14 a $\begin{pmatrix} 1 & 0 \\ 0 & -3 \end{pmatrix}$

b A stretch of scale factor -3 parallel to the y-axis.

15 a $M = \begin{pmatrix} k\cos\theta & -k\sin\theta \\ k\sin\theta & k\cos\theta \end{pmatrix}$

b $\theta = (-45°), 135°, k = 2\sqrt{2}$

16 a $(4, 1, 0)$ and $(2, -5, 3)$

b Reflection in $y = 0$

17 a $\begin{pmatrix} 1 & 0 & 0 \\ 0 & \dfrac{\sqrt{3}}{2} & -\dfrac{1}{2} \\ 0 & \dfrac{1}{2} & \dfrac{\sqrt{3}}{2} \end{pmatrix}$ **b** $\begin{pmatrix} 1 & 0 & 0 \\ 0 & 1 & 0 \\ 0 & 0 & -1 \end{pmatrix}$

c $\begin{pmatrix} \dfrac{\sqrt{2}}{2} & 0 & \dfrac{\sqrt{2}}{2} \\ 0 & 1 & 0 \\ -\dfrac{\sqrt{2}}{2} & 0 & \dfrac{\sqrt{2}}{2} \end{pmatrix}$

18 a Rotation around x-axis, $\theta = 300°$

b $\begin{pmatrix} 3 \\ 2 - \dfrac{\sqrt{3}}{2} \\ -\dfrac{1}{2} - 2\sqrt{3} \end{pmatrix}$

19 Anticlockwise rotation of $300°$ around z-axis.

20 a Reflection in plane $z = 0$

b $M = \begin{pmatrix} 1 & 0 & 0 \\ 0 & 1 & 0 \\ 0 & 0 & -1 \end{pmatrix}$

21 a Rotation of $90°$ (anticlockwise) around the x-axis.
b Reflection in $x = 0$
c Rotation of $180°$ around the z-axis.
d Rotation of $120°$ anticlockwise around y-axis.

22 a $\begin{pmatrix} -\dfrac{\sqrt{2}}{2} & 0 & -\dfrac{\sqrt{2}}{2} \\ 0 & 1 & 0 \\ \dfrac{\sqrt{2}}{2} & 0 & -\dfrac{\sqrt{2}}{2} \end{pmatrix}$ **b** $\left(-1 - \dfrac{\sqrt{2}}{2}, 0, 1 - \dfrac{\sqrt{2}}{2} \right)$

23 a $\begin{pmatrix} \dfrac{\sqrt{3}}{2} & \dfrac{1}{2} & 0 \\ -\dfrac{1}{2} & \dfrac{\sqrt{3}}{2} & 0 \\ 0 & 0 & 1 \end{pmatrix}$ **b** $\left(\dfrac{3}{2}, -\dfrac{\sqrt{3}}{2}, 1 \right)$

Exercise 4.2B

1 a $y = 0$
b $y = -2x$

2 a $\begin{pmatrix} 3 & -2 \\ 2 & 3 \end{pmatrix}\begin{pmatrix} x \\ y \end{pmatrix} = \begin{pmatrix} x \\ y \end{pmatrix}$

$3x - 2y = x \Rightarrow x = y$
$2x + 3y = y \Rightarrow x = -y$
$(0, 0)$ is only point that satisfies both of these equations, therefore $(0, 0)$ is the only invariant point.

b $\begin{pmatrix} 2 & 0 & 1 \\ 0 & 3 & -2 \\ 1 & 0 & -4 \end{pmatrix}\begin{pmatrix} x \\ y \\ z \end{pmatrix} = \begin{pmatrix} x \\ y \\ z \end{pmatrix}$

$2x + z = x \Rightarrow z = -x$
$3y - 2z = y \Rightarrow y = z$
$x - 4z = z \Rightarrow x = 5z$
$(0, 0, 0)$ is only point that satisfies all of these equations, therefore $(0, 0, 0)$ is the only invariant point.

3 a No difference, e.g. reflection in y-axis effects only the x-coordinates and stretch parallel to y-axis effects only the y-coordinates. Or use a sketch. Or show that combined transformation is $\begin{pmatrix} -1 & 0 \\ 0 & k \end{pmatrix}$ for both orders.

b No difference, e.g. use a sketch or show that combined

transformation is $k\begin{pmatrix} \cos\theta & -\sin\theta \\ \sin\theta & \cos\theta \end{pmatrix}$ in both cases.

c Different. Reflection in $y = x$ followed by stretch along

x-axis will be $\begin{pmatrix} 0 & k \\ 1 & 0 \end{pmatrix}$, other way around is $\begin{pmatrix} 0 & 1 \\ k & 0 \end{pmatrix}$.

4 a Sketch parallel to x-axis of scale factor 3 and stretch parallel to y-axis of scale factor -2.

b Enlargement of scale factor 4 centre the origin and reflection in $y = x$.

5 a $y = \dfrac{1}{5}x$ **b** $y = 3x$ and $y = -x$

6 a A stretch parallel to the y-axis and a reflection in the y-axis (or rotation by 180° or 360° about origin)

b Reflection in y-axis

7 Reflection in line $y = x$ (or $y = x + k$)

8 Enlargement, centre the origin

9 a $y = \left(\dfrac{-1 \pm \sqrt{5}}{2}\right)x$ **b** $y = mx$ for any m

c $y = 2x$ and $y = -\dfrac{1}{2}x + c$ for any c.

10 a $y = 0$; line of invariant points

b $(0, 0)$

c $x = 0$; line of invariant points

d $y = x + c$ for any c; not lines of invariant points, only $(0, 0)$ is invariant.

11 a $y = x$.

b $\begin{pmatrix} -2 & 3 \\ -3 & 4 \end{pmatrix}\begin{pmatrix} x \\ x+c \end{pmatrix} = \begin{pmatrix} -2x+3(x+c) \\ -3x+4(x+c) \end{pmatrix}$

$= \begin{pmatrix} x+3c \\ x+4c \end{pmatrix}$

$= \begin{pmatrix} x+3c \\ (x+3c)+c \end{pmatrix}$

So invariant lines are $y = x + c$

12 $\begin{pmatrix} 1 & 0 \\ 0 & -1 \end{pmatrix}$

13 a Plane $x = 0$ **b** Plane $y = z$

c z-axis   **d** Plane $x = y$

Exercise 4.3A

1 $\det\begin{pmatrix} -7 & 3 \\ 5 & -4 \end{pmatrix} = -7 \times -4 - 3 \times 5$

$= 28 - 15$

$= 13$ as required

2 a -18 **b** -3

c $-5a$ **d** 2

3 $\det\begin{pmatrix} a+b & 2a \\ 2b & a+b \end{pmatrix} = (a+b)(a+b) - (2a)(2b)$

$= a^2 + 2ab + b^2 - 4ab$

$= a^2 - 2ab + b^2$

$= (a-b)^2$ as required

4 $k = 3, 4$

5 a Non-singular

b Singular

c Non-singular unless $a = 0$

d Singular

6 a $x = \dfrac{5}{2}$

b $x = \pm\sqrt{3}$

d $x = 0, \dfrac{-3}{2}$

e $x = 0, 4$

7 $y = -1, -2$

8 a i -30 **ii** 1

b i $AB = \begin{pmatrix} -1 & 2 \\ 0 & 5 \end{pmatrix}\begin{pmatrix} 0 & -3 \\ 2 & 4 \end{pmatrix} = \begin{pmatrix} 4 & 11 \\ 10 & 20 \end{pmatrix}$

$\det(AB) = 4 \times 20 - 11 \times 10 = -30$ as required

ii $A + B = \begin{pmatrix} -1 & 2 \\ 0 & 5 \end{pmatrix} + \begin{pmatrix} 0 & -3 \\ 2 & 4 \end{pmatrix} = \begin{pmatrix} -1 & -1 \\ 2 & 9 \end{pmatrix}$

$\det(A + B) = -1 \times 9 - -1 \times 2 = -7$

9 a 50 **b** $12b - 15ab$

10 $k = 0, -1$

11 a $x = \dfrac{4}{3}$ **b** $x = \pm 2\sqrt{\dfrac{6}{19}}$

12 a i 21 **ii** 7

b i Orientation is preserved as determinant is positive.

ii Determinant negative so orientation not preserved.

13 a 1.5

b Determinant is positive so orientation not changed.

14 a 80

b Determinant is negative so orientation changes.

15 a $\dfrac{1}{27}\begin{pmatrix} 5 & -2 \\ -4 & 7 \end{pmatrix}$ **b** $\dfrac{1}{20}\begin{pmatrix} 1 & 3 \\ 8 & 4 \end{pmatrix}$

c $-\dfrac{1}{x}\begin{pmatrix} -3 & -5 \\ x & 2x \end{pmatrix}$ **d** $\dfrac{1}{50}\begin{pmatrix} 2\sqrt{5} & \sqrt{5} \\ -2\sqrt{5} & 4\sqrt{5} \end{pmatrix}$

16 $\begin{pmatrix} 6 & 1 \\ -9 & -1 \end{pmatrix}^{-1} = \dfrac{1}{6 \times -1 - 1 \times -9}\begin{pmatrix} -1 & -1 \\ 9 & 6 \end{pmatrix}$

$= \dfrac{1}{3}\begin{pmatrix} -1 & -1 \\ 9 & 6 \end{pmatrix}$

$= \begin{pmatrix} -\dfrac{1}{3} & -\dfrac{1}{3} \\ 3 & 2 \end{pmatrix}$ as required.

Alternatively show that

$\begin{pmatrix} 6 & 1 \\ -9 & -1 \end{pmatrix}\begin{pmatrix} -\dfrac{1}{3} & -\dfrac{1}{3} \\ 3 & 2 \end{pmatrix} = I$

17 a $\begin{pmatrix} \dfrac{3}{b} & -\dfrac{2}{b} \\ -\dfrac{1}{a} & \dfrac{1}{a} \end{pmatrix}$

b $\det(A^{-1}) = \dfrac{3}{b} \cdot \dfrac{1}{a} - \left(-\dfrac{2}{b}\right)\left(-\dfrac{1}{a}\right)$

$= \dfrac{3}{ab} - \dfrac{2}{ab}$

$= \dfrac{1}{ab}$

$= \dfrac{1}{\det(A)}$ as required

18 $\begin{pmatrix} 3 & -2 \\ 4 & -3 \end{pmatrix}\begin{pmatrix} 3 & -2 \\ 4 & -3 \end{pmatrix} = \begin{pmatrix} 3 \times 3 + -2 \times 4 & 3 \times -2 + -2 \times -3 \\ 4 \times 3 + -3 \times 4 & 4 \times -2 + -3 \times -3 \end{pmatrix}$

$= \begin{pmatrix} 1 & 0 \\ 0 & 1 \end{pmatrix}$ therefore it is self-inverse.

19 $a = \pm\sqrt{7}$

20 $a = 3$

21 a $x = \dfrac{5}{3}, -2$ **b** $\dfrac{1}{10}\begin{pmatrix} 2x & x \\ 1-x & x+1 \end{pmatrix}$

22 a $\dfrac{1}{6}\begin{pmatrix} -2 & 3 & -2 \\ -2 & -6 & 4 \\ 4 & -3 & 4 \end{pmatrix}$

   b $\begin{pmatrix} \dfrac{2}{a} & -\dfrac{3}{a} & \dfrac{4}{a} \\ 5 & -7 & 9 \\ -1 & 2 & -2 \end{pmatrix}$

23 $\mathbf{B}^{-1} = \dfrac{1}{k^3}\begin{pmatrix} -k & 1+k^2 & -1 \\ -k & 1 & k^2-1 \\ k^2 & -k & k \end{pmatrix}$

24 a $\mathbf{T}^{-1} = \dfrac{1}{4}\begin{pmatrix} 2 & 0 \\ 0 & 2 \end{pmatrix}$

        $= \dfrac{1}{2}\begin{pmatrix} 1 & 0 \\ 0 & 1 \end{pmatrix}$

        $= \dfrac{1}{2}\mathbf{I}$ as required

   b Enlargements scale factor $2\left(\dfrac{1}{2}\right)$ centre the origin.

25 $(7, -2)$

26 a $\dfrac{\sqrt{2}}{2}\begin{pmatrix} -1 & -1 \\ 1 & -1 \end{pmatrix}$

   b $\mathbf{A}^{-1}\mathbf{A} = \dfrac{1}{\sqrt{2}}\begin{pmatrix} -1 & 1 \\ -1 & -1 \end{pmatrix}\dfrac{\sqrt{2}}{2}\begin{pmatrix} -1 & -1 \\ 1 & -1 \end{pmatrix}$

        $= \dfrac{1}{2}\begin{pmatrix} 1+1 & 1-1 \\ 1-1 & 1+1 \end{pmatrix}$

        $= \dfrac{1}{2}\begin{pmatrix} 2 & 0 \\ 0 & 2 \end{pmatrix}$

        $= \begin{pmatrix} 1 & 0 \\ 0 & 1 \end{pmatrix}$ as required, therefore the inverse is as given.

27 $(1, -2, 0)$

28 a Always self-inverse.

   b Only self-inverse if enlargement of scale factor ± 1

   c Only self-inverse when rotated by a multiple of $180°$

   d Always self-inverse.

Exercise 4.3B

1 a $x = 2, y = -3$ **b** $x = -5, y = -1$

  c $x = 3, y = \dfrac{1}{2}$ **d** $x = -\dfrac{1}{4}, y = \dfrac{1}{3}$

2 a $x = 2, y = -5, z = -3$

  b $x = \dfrac{1}{4}, y = -\dfrac{1}{2}, z = 1$

3 $x = -\dfrac{1}{a} - 6, y = 2 + 2a$

4 $x = \dfrac{3k-1}{3(k+1)}, y = \dfrac{4k}{3(k+1)}$

5 $\begin{pmatrix} -2 & 4 \\ -1 & -2 \end{pmatrix}$

6 $\begin{pmatrix} 4 & -2 \\ 0 & -1 \end{pmatrix}$

7 $\begin{pmatrix} -3 & -1 & -4 \\ 0 & 0 & -5 \\ 2 & 3 & -1 \end{pmatrix}$

8 $\begin{pmatrix} -6 & 3 & 3 \\ 2 & -1 & 0 \\ 4 & -5 & -3 \end{pmatrix}$

9 $\dfrac{b+2a}{b}$

10 $\begin{pmatrix} 17 & 16 \\ -9 & -8 \end{pmatrix}$

11 $\begin{pmatrix} 5 & 0 \\ 4 & -1 \end{pmatrix}$

12 $\begin{pmatrix} 17 & 9 & -1 \\ 22 & 0 & 4 \\ 15 & 6 & 2 \end{pmatrix}$

13 a $x = 1 - 2z$, substitute into second equation to give

   $y = 2 - 8z$,

   Check in third equation $9(1-2z)+(2-8z)+2z=5$

   So they form a sheaf (line has equations e.g. $x = 1 - 2z$,

   $y = 2 - 8z$)

  b Adding first two equations gives $z = 7x - 7$

   Substitute into second equation to give $y = 12x - 7$

   Check in third equation: $36x + 11(12x - 7) - 24(7x - 7) = 91$

   So they form a sheaf (line has equations e.g. $z = 7x - 7$,

   $y = 12x - 7$)

14 $a = -3, 2$

15 a Multiplying the first equation by 2 gives $6x - 4y + 2z = 14$,

   so this is parallel to the second as they are the same except

   the constant term. Therefore the three planes do not meet

   at a point.

  b If we multiply the first equation by -3 we get

   $6x - 9y + 3z = 12$ so this plane is parallel to the second as

   they are the same except the constant term. Therefore the

   three planes do not meet at a point.

16 a $\begin{pmatrix} 1 & 1 & 1 \\ 1 & -1 & 0 \\ 0 & 1 & -1 \end{pmatrix}\begin{pmatrix} r \\ h \\ f \end{pmatrix} = \begin{pmatrix} 17 \\ 1 \\ -10 \end{pmatrix}$

  b so 3 rabbits, 2 hamsters and 12 fish

17 a $\begin{pmatrix} 1 & 1 & 1 \\ 2 & 3 & 4 \\ 1 & 0 & -1 \end{pmatrix}\begin{pmatrix} x \\ y \\ z \end{pmatrix} = \begin{pmatrix} 12 \\ 36 \\ 0 \end{pmatrix}$

  b $\det\begin{pmatrix} 1 & 1 & 1 \\ 2 & 3 & 4 \\ 1 & 0 & -1 \end{pmatrix} = 1(-3-0) - 1(-2-4) + 1(0-3) = 0$,

   therefore not a unique solution.

  c 5 possible solutions

18 $\sqrt{2}\begin{pmatrix} 1 & 0 \\ 0 & 1 \end{pmatrix}$

19 a $\mathbf{B}^{-1}$ **b** $\mathbf{A}$

20 $\mathbf{A} = \mathbf{C}^{-1}\mathbf{B}\mathbf{C} \Rightarrow \mathbf{CA} = \mathbf{CC}^{-1}\mathbf{BC}$

                 $\Rightarrow \mathbf{CA} = \mathbf{BC}$

                 $\Rightarrow \mathbf{CAC}^{-1} = \mathbf{BCC}^{-1}$

                 $\Rightarrow \mathbf{CAC}^{-1} = \mathbf{B}$ as required

21 $\mathbf{ABA}^{-1} = \mathbf{I} \Rightarrow \mathbf{A}^{-1}\mathbf{ABA}^{-1} = \mathbf{A}^{-1}\mathbf{I}$

              $\Rightarrow \mathbf{BA}^{-1} = \mathbf{A}^{-1}$

              $\Rightarrow \mathbf{BA}^{-1}\mathbf{A} = \mathbf{A}^{-1}\mathbf{A}$

              $\Rightarrow \mathbf{B} = \mathbf{I}$

22 $(\mathbf{ABC})^{-1}(\mathbf{ABC})=\mathbf{I}$

$\Rightarrow(\mathbf{ABC})^{-1}\mathbf{ABCC}^{-1}=\mathbf{IC}^{-1}$

$\Rightarrow(\mathbf{ABC})^{-1}\mathbf{AB}=\mathbf{C}^{-1}$

$\Rightarrow(\mathbf{ABC})^{-1}\mathbf{ABB}^{-1}=\mathbf{C}^{-1}\mathbf{B}^{-1}$

$\Rightarrow(\mathbf{ABC})^{-1}\mathbf{A}=\mathbf{C}^{-1}\mathbf{B}^{-1}$

$\Rightarrow(\mathbf{ABC})^{-1}\mathbf{AA}^{-1}=\mathbf{C}^{-1}\mathbf{B}^{-1}\mathbf{A}^{-1}$

$\Rightarrow(\mathbf{ABC})^{-1}=\mathbf{C}^{-1}\mathbf{B}^{-1}\mathbf{A}^{-1}$ as required

23 P self-inverse $\Rightarrow \mathbf{P}=\mathbf{P}^{-1}$

$\Rightarrow \mathbf{PP}=\mathbf{P}^{-1}\mathbf{P}$

$\Rightarrow \mathbf{P}^2=\mathbf{I}$ as required

24 $\mathbf{PQP}=\mathbf{I}\Rightarrow \mathbf{P}^{-1}\mathbf{PQP}=\mathbf{P}^{-1}\mathbf{I}$

$\Rightarrow \mathbf{QP}=\mathbf{P}^{-1}$

$\Rightarrow \mathbf{QPP}^{-1}=\mathbf{P}^{-1}\mathbf{P}^{-1}$

$\Rightarrow \mathbf{Q}=(\mathbf{P}^{-1})^2$ as required

25 $(5,-1)$

26 a $\dfrac{16}{k}$ **b** $k<0$

c $A\left(-\dfrac{6}{k},4\right)$, $B(0,0)$, $C\left(\dfrac{2}{k},4\right)$

d $k=-2$

27 Let $\mathbf{A}=\begin{pmatrix} a_1 & a_2 \\ a_3 & a_4 \end{pmatrix}$ and $\mathbf{B}=\begin{pmatrix} b_1 & b_2 \\ b_3 & b_4 \end{pmatrix}$

$\mathbf{AB}=\begin{pmatrix} a_1 & a_2 \\ a_3 & a_4 \end{pmatrix}\begin{pmatrix} b_1 & b_2 \\ b_3 & b_4 \end{pmatrix}=\begin{pmatrix} a_1b_1+a_2b_3 & a_1b_2+a_2b_4 \\ a_3b_1+a_4b_3 & a_3b_2+a_4b_4 \end{pmatrix}$

Then

$\det(\mathbf{AB})=(a_1b_1+a_2b_3)(a_3b_2+a_4b_4)-(a_1b_2+a_2b_4)(a_3b_1+a_4b_3)$

$=(a_1b_1a_3b_2+a_1b_1a_4b_4+a_2b_3a_3b_2+a_2b_3a_4b_4)-$
$\quad(a_1b_2a_3b_1+a_1b_2a_4b_3+a_2b_4a_3b_1+a_2b_4a_4b_3)$

$=a_1b_1a_4b_4-a_1b_2a_3b_4-a_2b_1a_3b_4+a_2b_2a_3b_3$

$=(a_1a_4-a_2a_3)(b_1b_4-b_2b_3)$

$=\det(\mathbf{A})\det(\mathbf{B})$ as required

28 Let $\mathbf{A}=\begin{pmatrix} a_1 & a_2 \\ a_3 & a_4 \end{pmatrix}$ then $\mathbf{A}^{-1}=\dfrac{1}{a_1a_4-a_2a_3}\begin{pmatrix} a_4 & -a_2 \\ -a_3 & a_1 \end{pmatrix}$

$\det(\mathbf{A}^{-1})=\dfrac{a_4}{a_1a_4-a_2a_3}\cdot\dfrac{a_1}{a_1a_4-a_2a_3}-\dfrac{-a_2}{a_1a_4-a_2a_3}\cdot\dfrac{-a_3}{a_1a_4-a_2a_3}$

$=\dfrac{a_4a_1-a_2a_3}{(a_1a_4-a_2a_3)^2}$

$=\dfrac{1}{a_1a_4-a_2a_3}$

$=\dfrac{1}{\det(\mathbf{A})}$

$=[\det(\mathbf{A})]^{-1}$ as required

29 a i $\begin{pmatrix} \cos\theta & -\sin\theta \\ \sin\theta & \cos\theta \end{pmatrix}$ **ii** $\begin{pmatrix} \cos\theta & \sin\theta \\ -\sin\theta & \cos\theta \end{pmatrix}$

b $\begin{pmatrix} \cos\theta & -\sin\theta \\ \sin\theta & \cos\theta \end{pmatrix}\begin{pmatrix} \cos\theta & \sin\theta \\ -\sin\theta & \cos\theta \end{pmatrix}=\begin{pmatrix} 1 & 0 \\ 0 & 1 \end{pmatrix}$

$\Rightarrow \cos\theta\cos\theta-\sin\theta(-\sin\theta)\equiv 1$

$\Rightarrow \cos^2\theta+\sin^2\theta\equiv 1$ as required.

30 a Square with vertices at $(0,0)$, $(x,0)$, $(0,y)$ and (x,y)

$\mathbf{T}=\begin{pmatrix} a & 0 \\ 0 & a \end{pmatrix}$ is a linear enlargement

$\begin{pmatrix} a & 0 \\ 0 & a \end{pmatrix}\begin{pmatrix} 0 & x & 0 & x \\ 0 & 0 & y & y \end{pmatrix}=\begin{pmatrix} 0 & ax & 0 & ax \\ 0 & 0 & ay & ay \end{pmatrix}$

So a square with vertices at $(0,0)$, $(ax,0)$, $(0,ay)$ and (ax,ay)

b Area of original square is xy

Area of image is $(ax)(ay)=a^2xy$

$=\det(\mathbf{T})\times xy$ as required since
$\det(\mathbf{T})=a^2-0=a^2$

31 a Square with vertices at $(0,0)$, $(x,0)$, $(0,y)$ and (x,y)

$\mathbf{T}=\begin{pmatrix} a & 0 \\ 0 & b \end{pmatrix}$ is a stretch of a in x-direction and b in y-direction.

$\begin{pmatrix} a & 0 \\ 0 & b \end{pmatrix}\begin{pmatrix} 0 & x & 0 & x \\ 0 & 0 & y & y \end{pmatrix}=\begin{pmatrix} 0 & ax & 0 & ax \\ 0 & 0 & by & by \end{pmatrix}$

So a rectangle with vertices at $(0,0)$, $(ax,0)$, $(0,by)$ and (ax,by).

b Area of original square is xy

Area of image is $(ax)(by)=abxy$

$=\det(\mathbf{T})\times xy$ as required since
$\det(\mathbf{T})=ab-0=ab$

1 a A has order 2×2, B has order 2×3, C has order 3×2, D has order 2×3

b i $\begin{pmatrix} 8 & 0 & 2 \\ 5 & 3 & 0 \end{pmatrix}$ **ii** $\begin{pmatrix} -3 & -15 \\ 12 & 0 \\ 0 & 6 \end{pmatrix}$

iii $\begin{pmatrix} 8 & -6 & 5 \\ -10 & -9 & 0 \end{pmatrix}$ **iv** $\begin{pmatrix} 3 & 1 \\ -2 & 6 \end{pmatrix}$

c i $\begin{pmatrix} 10 & 20 & -6 \\ 60 & 40 & 4 \end{pmatrix}$

ii BA not possible as 3 columns in B but 2 rows in A

iii $\begin{pmatrix} 8 & -2 \\ 11 & -25 \end{pmatrix}$ **iv** $\begin{pmatrix} 14 & -62 \\ 24 & 8 \\ -8 & 24 \end{pmatrix}$

v AC not possible as 2 columns in A but 3 rows in C

vi $\begin{pmatrix} -8 & 7 & -3 \\ 32 & -8 & 12 \\ 0 & -2 & 0 \end{pmatrix}$ **vii** $\begin{pmatrix} 28 & 36 \\ -72 & 136 \end{pmatrix}$

viii $\mathbf{C}^2$ not possible as C not a square matrix

2 a $\begin{pmatrix} a^2-a & 3a-2 \\ 3a & 6 \\ 2a^2-2a & 4a-6 \end{pmatrix}$

b $\begin{pmatrix} -a-6 & 1+9a \\ -2a & 3+2a^2 \\ 2a^2 & -2a-9 \end{pmatrix}$

c $\begin{pmatrix} a^2+3a & 3a+6 \\ a^2+2a & 3a+4 \end{pmatrix}$

d $\begin{pmatrix} 1+6a & -9 & a^2-3 \\ 2a^2 & 1-3a & a \\ -2a & 2a^2-3 & 3a \end{pmatrix}$

3 a i $(2,-1)$ reflection in line $y=x$

ii $(-7,2)$ stretch scale factor 7 parallel to x-axis

iii $(-2,-1)$ rotation of 90° anticlockwise about origin

iv $\left(\dfrac{1}{2}(2+\sqrt{3}),\dfrac{1}{2}(1-2\sqrt{3})\right)$; rotation of 210° anticlockwise about origin

b i 5   **ii** 35   **iii** 5   **iv** 5

c **i** determinant negative so does not preserve orientation

ii, iii, iv have positive determinants so do preserve orientation

4 a $(3, -3\sqrt{2}, \sqrt{2})$; rotation of 45° anticlockwise around x-axis.

b $(3, -2, -4)$; reflection in plane $z = 0$

5 a $\begin{pmatrix} -1 & 0 \\ 0 & 1 \end{pmatrix}$

b $\begin{pmatrix} \dfrac{\sqrt{2}}{2} & \dfrac{\sqrt{2}}{2} \\ -\dfrac{\sqrt{2}}{2} & \dfrac{\sqrt{2}}{2} \end{pmatrix}$

c $\begin{pmatrix} -3 & 0 \\ 0 & -3 \end{pmatrix}$

6 a $\begin{pmatrix} \dfrac{\sqrt{2}}{2} & 0 & \dfrac{\sqrt{2}}{2} \\ 0 & 1 & 0 \\ -\dfrac{\sqrt{2}}{2} & 0 & \dfrac{\sqrt{2}}{2} \end{pmatrix}$

b $\begin{pmatrix} 1 & 0 & 0 \\ 0 & -1 & 0 \\ 0 & 0 & 1 \end{pmatrix}$

7 a $\begin{pmatrix} 0 & 1 \\ -2 & 0 \end{pmatrix}$

b $\begin{pmatrix} 0 & -1 \\ 1 & 0 \end{pmatrix}$

c $\begin{pmatrix} -\dfrac{\sqrt{3}}{2} & \dfrac{1}{2} \\ \dfrac{1}{2} & \dfrac{\sqrt{3}}{2} \end{pmatrix}$

8 a Invariant point is $(0, 0)$

b Line of invariant points is $x = 3y$

9 a $y = -x$ and $y = \dfrac{3}{2}x$

b $y = 3x$ and $y = -\dfrac{1}{3}x + c$

10 a $\det \begin{pmatrix} 2 & -1 \\ 4 & -3 \end{pmatrix} = 2 \times -3 - -1 \times 4 = -2$

b $\det \begin{pmatrix} 3a & b \\ -2a & b \end{pmatrix} = 3ab - -2ab = 5ab$

11 a 25

b $-44a - 21b$

12 a $x = \dfrac{4}{3}$

b $x = \pm\sqrt{2}$

13 a $x = -\dfrac{4}{3}$

b $2 \pm \sqrt{6}$

14 a $-\dfrac{1}{9}\begin{pmatrix} -2 & -3 \\ 1 & 6 \end{pmatrix}$

b $-\dfrac{1}{3}\begin{pmatrix} -2 & -1 \\ -3 & 0 \end{pmatrix}$

c Does not exist since $\det \begin{pmatrix} 3 & -6 \\ -4 & 8 \end{pmatrix} = 3 \times 8 - -6 \times -4 = 0$

d $\dfrac{1}{ab}\begin{pmatrix} 3b & -2b \\ -7a & 5a \end{pmatrix}$

15 a $\dfrac{1}{25}\begin{pmatrix} -5 & 5 & 10 \\ 22 & -7 & -19 \\ 4 & 1 & -8 \end{pmatrix}$

b det = 0, so inverse does not exist

16 $\begin{pmatrix} 2 & 3 \\ 0 & 1 \end{pmatrix}$

17 $\begin{pmatrix} 2 & -7 \\ -5 & -3 \end{pmatrix}$

18 $\begin{pmatrix} 5 & -1 & 1 \\ 2 & 0 & -4 \\ 1 & 3 & 0 \end{pmatrix}$

19 $\begin{pmatrix} 12 & -8 & 4 \\ -3 & -2 & 2 \\ 0 & 7 & 4 \end{pmatrix}$

20 a $x = -5, y = 2$ **b** $x = \dfrac{1}{5}, y = -1$

21 a $x = 2, y = -5, z = -10$

b $x = 4, y = -1, z = -3$

22 a Multiplying equation 3 by -2 gives equation 1 with a different constant so these planes are parallel so do not intersect.

b Inconsistent equations, they form a triangular prism.

23 a $\det \begin{pmatrix} 3 & -1 & 3 \\ 0 & 2 & -1 \\ 3 & 1 & 2 \end{pmatrix} = 15 + 3 - 18 = 0$

b Second equation gives $z = 2y + 9$

Substitute into first equation:

$3x - y + 3(2y + 9) = 14 \Rightarrow 3x = -13 - 5y$

Check in third equation: $(-13 - 5y) + y + 2(2y + 9) = 5 \neq 10$

Inconsistent (but no planes parallel) so they form a triangular prism.

24 a $\begin{pmatrix} 1 & 1 & 1 \\ 1 & -3 & 0 \\ 0.1 & 0.1 & -0.05 \end{pmatrix}\begin{pmatrix} a \\ b \\ c \end{pmatrix} = \begin{pmatrix} 12000 \\ 0 \\ 555 \end{pmatrix}$

b He bought £5775 of shares from company A, £1925 of company B and £4300 of company C

c The three planes represented by the equations meet at a single point.

Assessment 4

1 a $\begin{pmatrix} 15 & 6 \\ -9 & 3 \end{pmatrix}$

b $\begin{pmatrix} 8 & -6 \\ 0 & -4 \end{pmatrix}$

c $\begin{pmatrix} 9 & -1 \\ -3 & -1 \end{pmatrix}$

d $\begin{pmatrix} -20 & 19 \\ 12 & -7 \end{pmatrix}$

2 a $\begin{pmatrix} 12 & 9 \\ -7 & 25 \end{pmatrix}$

b $\begin{pmatrix} 30 & -17 \\ 9 & 7 \end{pmatrix}$

c **M** and **N** do not commute

3 a $\begin{pmatrix} 2 & 7 \\ 9 & 11 \end{pmatrix}$ **b** $\begin{pmatrix} 4 & 14 \\ 18 & 22 \end{pmatrix}$ **c** $\begin{pmatrix} 67 & 91 \\ 117 & 184 \end{pmatrix}$

4 $a = 4, b = 1, c = 16$

5 a $a = 10, b = -\dfrac{4}{35}, c = 1$

b They are inverses of one another

6 a Enlargement, 6

b the origin, 90 degrees

c reflection, the x-axis

7 a $\begin{pmatrix} \dfrac{\sqrt{3}}{2} & 0 & -\dfrac{1}{2} \\ 0 & 1 & 0 \\ \dfrac{1}{2} & 0 & \dfrac{\sqrt{3}}{2} \end{pmatrix}$

b $\left(\dfrac{5}{2}, 5, -\dfrac{\sqrt{3}}{2} \right)$

8 a $\mathbf{M} = \begin{pmatrix} -\dfrac{\sqrt{2}}{2} & -\dfrac{\sqrt{2}}{2} \\ \dfrac{\sqrt{2}}{2} & -\dfrac{\sqrt{2}}{2} \end{pmatrix}$

b $\mathbf{N} = \begin{pmatrix} \dfrac{\sqrt{2}}{2} & \dfrac{\sqrt{2}}{2} \\ -\dfrac{\sqrt{2}}{2} & \dfrac{\sqrt{2}}{2} \end{pmatrix}$

c $\begin{pmatrix} 0 & -1 \\ 1 & 0 \end{pmatrix}$

A rotation by 90 degrees anticlockwise about the origin. This the combined effect of the two rotations represented by **M** and **N**.

9 a The point $(0, 0)$

b Any point $(\lambda, -2\lambda)$

c Any point $(-3\lambda, \lambda)$

10 a The y-axis

b The z-axis

11 $\begin{pmatrix} x \\ y \end{pmatrix} = \begin{pmatrix} \dfrac{57}{25} \\ \dfrac{22}{25} \end{pmatrix}$

12 $\begin{pmatrix} x \\ y \end{pmatrix} = \begin{pmatrix} 4 \\ 3 \end{pmatrix}$

13 a 2

b $(0, 0)$

14 a 12

b 0

c 164

15 a $\begin{pmatrix} 9 & -16 \\ -5 & 9 \end{pmatrix}$

b $\dfrac{1}{22}\begin{pmatrix} 3 & 2 \\ 5 & 4 \end{pmatrix}$

c $\dfrac{1}{2}\begin{pmatrix} 2 & 1 & 1 \\ 16 & 8 & 10 \\ -7 & -3 & -4 \end{pmatrix}$

16 $a = 4, -7$

17 a $\dfrac{1}{28}\begin{pmatrix} 7 & 3 & 1 \\ -14 & 2 & 10 \\ 7 & 7 & -7 \end{pmatrix}$

b $x = 3, y = -5, z = 12$

18 a $\det \begin{pmatrix} 1 & 7 & -2 \\ 3 & 10 & -2 \\ 2 & -30 & 12 \end{pmatrix} = 0$

no inconsistencies, there are an infinite number of solutions described by the line $2x + 3y = 6, 11y - 4z = 12$, so the planes form a sheaf

b $\det \begin{pmatrix} 9 & -4 & 2 \\ 14 & -4 & -1 \\ 1 & 4 & -8 \end{pmatrix} = 0$

eliminating y from equations 1 and 3: $5x - 3z = 15$
eliminating y from equations 2 and 1: $5x - 3z = 18$
These are inconsistent and no planes are parallel, therefore there are no solutions and the planes form a triangular prism

19 a $\mathbf{A}^{-1} = \begin{pmatrix} 2 & -1 \\ -5 & 3 \end{pmatrix}$

b $\mathbf{B}^{-1} = \dfrac{1}{2}\begin{pmatrix} 1 & 4 \\ 1 & 2 \end{pmatrix}$

c $\begin{pmatrix} -9 & 5.5 \\ -4 & 2.5 \end{pmatrix}$

d $\begin{pmatrix} -5 & 11 \\ -8 & 18 \end{pmatrix}$

$(\mathbf{B}^{-1}\mathbf{A}^{-1})^{-1} = \mathbf{AB}$

Chapter 5

Exercise 5.1A

1 $\mathbf{r} = \begin{pmatrix} 0 \\ 4 \\ 3 \end{pmatrix} + \lambda \begin{pmatrix} 1 \\ -2 \\ 0 \end{pmatrix}$

2 $\mathbf{r} = 7\mathbf{i} - \mathbf{j} + 2\mathbf{k} + \lambda(5\mathbf{i} - 3\mathbf{j} + \mathbf{k})$

3 $\mathbf{r} = \begin{pmatrix} -7 \\ 2 \\ -5 \end{pmatrix} + \lambda \begin{pmatrix} 11 \\ -11 \\ 3 \end{pmatrix}$ or alternatively $\mathbf{r} = \begin{pmatrix} 4 \\ -9 \\ -2 \end{pmatrix} + \lambda \begin{pmatrix} 11 \\ -11 \\ 3 \end{pmatrix}$

4 $\mathbf{r} = 2\mathbf{i} + 3\mathbf{k} + \lambda(\mathbf{j} - 4\mathbf{k})$ or alternatively $\mathbf{r} = 2\mathbf{i} + \mathbf{j} - \mathbf{k} + \lambda(\mathbf{j} - 4\mathbf{k})$

5 $x = 5; \dfrac{y - 2}{2} = \dfrac{z + 7}{-1}$

6 $\dfrac{x - 4}{8} = \dfrac{y + 2}{-2} = \dfrac{z - 0}{3}$

7 $\dfrac{x - 9}{2} = \dfrac{y - 4}{1} = \dfrac{z + 3}{-2}$ or alternatively $\dfrac{x - 1}{2} = \dfrac{y - 0}{1} = \dfrac{z - 5}{-2}$

8 $\dfrac{x - 0}{-2} = \dfrac{y - 5}{3} = \dfrac{z - 1}{1}$ or alternatively $\dfrac{x - 4}{-2} = \dfrac{y + 1}{3} = \dfrac{z + 1}{1}$

9 a $y = 1, \dfrac{x - 0}{-4} = \dfrac{z + 2}{8}$

b $\dfrac{x + 3}{5} = \dfrac{y - 0}{-1} = \dfrac{z - 7}{3}$

10 a $\mathbf{r} = \begin{pmatrix} 5 \\ 1 \\ -3 \end{pmatrix} + \lambda \begin{pmatrix} 2 \\ -3 \\ 1 \end{pmatrix}$

b $\mathbf{r} = \begin{pmatrix} -2 \\ 0 \\ 5 \end{pmatrix} + \lambda \begin{pmatrix} 4 \\ -2 \\ -3 \end{pmatrix}$

11 a (3, 2, 1) lies on the line.

b (3, 2, 1) does not lie on the line

12 a (−2, 0, 4) does not line on the line

b (−2, 0, 4) lies on the line.

Exercise 5.1B

1 a Not the same line

b Same line

c Same line

2 a Not the same line.

b Not the same line.

c Same line.

d Same line.

e Not the same line.

3 a Intersect (−11, 16)

b Intersect at (−2, 10)

c Intersect at (1, −1)

4 a Intersect *at* (7, −6, −2)

b Parallel

c Skew

d Intersect at (−4, 2, 0)

5 $3\sqrt{10}$

6 $\mathbf{r} = \begin{pmatrix} 5 \\ -6 \end{pmatrix} + \lambda \begin{pmatrix} -3 \\ -2 \end{pmatrix}$

7 $\mathbf{r} = \begin{pmatrix} 1 \\ 2 \end{pmatrix} + \lambda \begin{pmatrix} -1 \\ 2 \end{pmatrix}$

8 a $\mathbf{r} = \begin{pmatrix} 3 \\ 1 \\ 0 \end{pmatrix} + \lambda \begin{pmatrix} 1 \\ 2 \\ -1 \end{pmatrix}$

b $\mathbf{r} = \begin{pmatrix} -3 \\ 0 \\ 1 \end{pmatrix} + \lambda \begin{pmatrix} -1 \\ -1 \\ 2 \end{pmatrix}$

c $\mathbf{r} = \begin{pmatrix} 3 \\ 0 \\ 1 \end{pmatrix} + \lambda \begin{pmatrix} 1 \\ -1 \\ 2 \end{pmatrix}$

9 $\mathbf{r} = \begin{pmatrix} 1 \\ 3 \\ 5 \end{pmatrix} + \lambda \begin{pmatrix} 2 \\ 1 \\ 4 \end{pmatrix}$

Exercise 5.2A

1 a 1

b 0

c 0

d 1

e 2

f 0

g 1

h 3

2 a 4

b −5

c $-\sqrt{2}$

d −3

3 $6a^2 - 4a$

4 k −6

5 a 46.4°

b 26.5°

c 63.9°

d 101.6°

6 a $\dfrac{a}{a^2 + 1}$

b $\dfrac{5 - 4a^2}{5a^2 + 5}$

7 $\begin{pmatrix} 6 \\ -4 \end{pmatrix} \cdot \begin{pmatrix} -8 \\ -12 \end{pmatrix} = -48 + 48 = 0$ so perpendicular

8 $\begin{pmatrix} 4 \\ -1 \\ 5 \end{pmatrix} \cdot \begin{pmatrix} -2 \\ -3 \\ 1 \end{pmatrix} = (4 \times -2) + (-1 \times -3) + (5 \times 1)$

$\qquad = 0$ therefore they are perpendicular

9 $3 \times 2 - 5 \times 4 - 2 \times -7 = 0$ so perpendicular

10 $2 \times 4 + 0 + 1 \times -3 = 5$ so not perpendicular

11 $a = 1$

12 $b = 2$

13 $c = 1, 2$

Exercise 5.2B

1 $\begin{pmatrix} -1 \\ 3 \end{pmatrix} \cdot \begin{pmatrix} 6 \\ 2 \end{pmatrix} = -6 + 6 = 0$ so perpendicular

2 $\begin{pmatrix} 6 \\ 7 \\ -5 \end{pmatrix} \cdot \begin{pmatrix} 2 \\ -1 \\ 1 \end{pmatrix} = 12 - 7 - 5 = 0$ so perpendicular

3 $\begin{pmatrix} 3 \\ -2 \\ 6 \end{pmatrix} \cdot \begin{pmatrix} 2 \\ 6 \\ 1 \end{pmatrix} = 6 - 12 + 6 = 0$ therefore they are perpendicular.

4 $6 + \lambda = -\mu \Rightarrow \lambda = -\mu - 6$

$6 + 2\lambda = 4 + 3\mu \Rightarrow 2(-\mu - 6) = -2 + 3\mu \Rightarrow \mu = -2, \lambda = -4$

So intersect at (−2, 2)

$\theta = 45°$

5 $\begin{pmatrix} 2 \\ 1 \\ -3 \end{pmatrix} + s \begin{pmatrix} 0 \\ 2 \\ 1 \end{pmatrix} = \begin{pmatrix} 5 \\ 11 \\ 3 \end{pmatrix} + t \begin{pmatrix} 3 \\ -4 \\ -1 \end{pmatrix}$

$2 = 5 + 3t \Rightarrow t = -1$

$1 + 2s = 11 + 4 \Rightarrow s = 7$

Check: $-3 + s = 4$

$3 - t = 4$ so they intersect

$\theta = 38°$

6 93°

7 $-1 + \lambda = -2 \Rightarrow \lambda = -1$

$2 + \lambda = 2 - \mu \Rightarrow \mu = 1$

Check: $3 - \lambda = 4$

$1 + 3\mu = 4$ hence they intersect

$\theta = 43.1°$

8 $a = \pm 2\sqrt{2}$

9 $a = \pm 2$

10 a $\left(2, 3, \dfrac{7}{2} \right)$

b $\cos \theta = \dfrac{67}{9\sqrt{161}}$

11 $k = 2 \pm \sqrt{3}$

12 $k = -\dfrac{1}{4}$

13 $\dfrac{\sqrt{89}}{2}$

14 $\dfrac{\sqrt{11}}{2}$

15 $\mathbf{a} \cdot \mathbf{b} = (a_1\mathbf{i} + a_2\mathbf{j} + a_3\mathbf{k}) \cdot (b_1\mathbf{i} + b_2\mathbf{j} + b_3\mathbf{k})$

$= a_1b_1\mathbf{i} \cdot \mathbf{i} + a_1b_2\mathbf{i} \cdot \mathbf{j} + a_1b_3\mathbf{i} \cdot \mathbf{k} + a_2b_1\mathbf{j} \cdot \mathbf{i} + a_2b_2\mathbf{j} \cdot \mathbf{j} + a_2b_3\mathbf{j} \cdot \mathbf{k}$
$\quad + a_3b_1\mathbf{k} \cdot \mathbf{i} + a_3b_2\mathbf{k} \cdot \mathbf{j} + a_3b_3\mathbf{k} \cdot \mathbf{k}$

$= a_1b_1\mathbf{i} \cdot \mathbf{i} + a_2b_2\mathbf{j} \cdot \mathbf{j} + a_3b_3 \ \mathbf{k} \cdot \mathbf{k}$ Using fact that $\mathbf{i}$, $\mathbf{j}$ and $\mathbf{k}$ are perpendicular

$= a_1b_1 + a_2b_2 + a_3b_3$ as required since $\mathbf{i} \cdot \mathbf{i} = \mathbf{j} \cdot \mathbf{j} = \mathbf{k} \cdot \mathbf{k} = 1$

16 Let $\mathbf{a} = a_1\mathbf{i} + a_2\mathbf{j} + a_3\mathbf{k}$, $\mathbf{b} = b_1\mathbf{i} + b_2\mathbf{j} + b_3\mathbf{k}$ and $\mathbf{c} = c_1\mathbf{i} + c_2\mathbf{j} + c_3\mathbf{k}$

$\mathbf{b} + \mathbf{c} = (b_1\mathbf{i} + b_2\mathbf{j} + b_3\mathbf{k}) + (c_1\mathbf{i} + c_2\mathbf{j} + c_3\mathbf{k})$

$= (b_1 + c_1)\mathbf{i} + (b_2 + c_2)\mathbf{j} + (b_3 + c_3)\mathbf{k}$

$\mathbf{a} \cdot (\mathbf{b} + \mathbf{c}) = (a_1 \mathbf{i} + a_2 \mathbf{j} + a_3 \mathbf{k})((b_1 + c_1)\mathbf{i} + (b_2 + c_2)\mathbf{j} + (b_3 + c_3)\mathbf{k})$

$= a_1(b_1 + c_1)\mathbf{i} \cdot \mathbf{i} + a_1(b_2 + c_2)\mathbf{i} \cdot \mathbf{j} + a_1(b_3 + c_3)\ \mathbf{i} \cdot \mathbf{k}$
$\quad + a_2(b_1 + c_1)\mathbf{j} \cdot \mathbf{i} + a_2(b_2 + c_2)\mathbf{j} \cdot \mathbf{j} + a_2(b_3 + c_3)\mathbf{j} \cdot \mathbf{k}$
$\quad + a_3(b_1 + c_1)\mathbf{k} \cdot \mathbf{i} + a_3(b_2 + c_2)\mathbf{k} \cdot \mathbf{j} + a_3(b_3 + c_3)\mathbf{k} \cdot \mathbf{k}$

$= a_1(b_1 + c_1)\mathbf{i} \cdot \mathbf{i} + a_2(b_2 + c_2)\mathbf{j} \cdot \mathbf{j} + a_3(b_3 + c_3)\mathbf{k} \cdot \mathbf{k}$

$= a_1(b_1 + c_1) + a_2(b_2 + c_2) + a_3(b_3 + c_3)$

$= a_1b_1 + a_1c_1 + a_2b_2 + a_2c_2 + a_3b_3 + a_3c_3$

$= (a_1b_1 + a_2b_2 + a_3b_3) + (a_1c_1 + a_2c_2 + a_3c_3)$

$= \mathbf{a} \cdot \mathbf{b} + \mathbf{a} \cdot \mathbf{c}$ (using qu 8 or prove as above) as required

17 a Let $\mathbf{a} = \begin{pmatrix} a_1 \\ a_2 \\ a_3 \end{pmatrix}$

Then $\mathbf{a} \cdot \mathbf{a} = \begin{pmatrix} a_1 \\ a_2 \\ a_3 \end{pmatrix} \cdot \begin{pmatrix} a_1 \\ a_2 \\ a_3 \end{pmatrix}$

$= a_1^2 + a_2^2 + a_3^2$

$= |\mathbf{a}|^2$

since $|\mathbf{a}| = \sqrt{a_1^2 + a_2^2 + a_3^2}$

b $a = b - c$ using the 'triangle rule'

$|\mathbf{a}|^2 = \mathbf{a} \cdot \mathbf{a}$

$= (\mathbf{b} - \mathbf{c}) \cdot (\mathbf{b} - \mathbf{c})$

$= \mathbf{b} \cdot \mathbf{b} + \mathbf{c} \cdot \mathbf{c} - 2\mathbf{b} \cdot \mathbf{c} \cos\theta$

$|\mathbf{b}|^2 + |\mathbf{c}|^2 - 2|\mathbf{b}||\mathbf{c}| \cos\theta$ as required

18 a When $y = 0$, $\dfrac{y - 5}{5} = -1$

So $\dfrac{z - 3}{3} = -1 \Rightarrow z = 0$

Therefore it cuts through x–axis

$\dfrac{x + 2}{-4} = -1 \Rightarrow x = 2$

So intercept is $(2, 0, 0)$

b $55.6°$

19 $\dfrac{27}{2} \sqrt{6}$

Exercise 5.3A

1 $\mathbf{r} = \begin{pmatrix} 1 \\ 5 \\ 2 \end{pmatrix} + s\begin{pmatrix} 1 \\ 1 \\ 0 \end{pmatrix} + t\begin{pmatrix} 0 \\ 1 \\ 2 \end{pmatrix}$

2 $\mathbf{r} = \begin{pmatrix} 0 \\ 6 \\ 2 \end{pmatrix} + s\begin{pmatrix} 2 \\ 0 \\ -3 \end{pmatrix} + t\begin{pmatrix} 5 \\ -2 \\ 4 \end{pmatrix}$

3 e.g. $\mathbf{r} = \begin{pmatrix} 3 \\ 1 \\ 0 \end{pmatrix} + s\begin{pmatrix} 1 \\ -3 \\ 2 \end{pmatrix} + t\begin{pmatrix} 8 \\ 1 \\ -4 \end{pmatrix}$

4 e.g. $\mathbf{r} = \begin{pmatrix} -2 \\ 0 \\ 0 \end{pmatrix} + s\begin{pmatrix} 7 \\ 1 \\ 1 \end{pmatrix} + t\begin{pmatrix} 8 \\ 2 \\ -5 \end{pmatrix}$

5 e.g. $\mathbf{r} = \begin{pmatrix} 0 \\ 3 \\ 2 \end{pmatrix} + s\begin{pmatrix} 1 \\ -5 \\ -2 \end{pmatrix} + t\begin{pmatrix} -1 \\ -1 \\ -3 \end{pmatrix}$

6 $\mathbf{r} \cdot \begin{pmatrix} 1 \\ 0 \\ 2 \end{pmatrix} = 11$

7 $\mathbf{r} \cdot \begin{pmatrix} 3 \\ -1 \\ 4 \end{pmatrix} = 10$

8 $\mathbf{r} \cdot \begin{pmatrix} 0 \\ 2 \\ -3 \end{pmatrix} = -1$

9 $5x + 4y - 2z = 18$

10 $2x - 5z = -8$

11 $x - y + 4z = 31$

12 a $\mathbf{r} \cdot \begin{pmatrix} 2 \\ 1 \\ -3 \end{pmatrix} = -2$

b $2x + y - 3z = -2$

13 a $\mathbf{r} \cdot \begin{pmatrix} 3 \\ -2 \\ 5 \end{pmatrix} = -4$

b $3x - 2y + 5z = -4$

14 a $3x + 5y - z = -2$

b $7x - y + 8z = 3$

15 a $\mathbf{r} \cdot \begin{pmatrix} 9 \\ 3 \\ -1 \end{pmatrix} = 5$

b $\mathbf{r} \cdot \begin{pmatrix} 2 \\ -7 \\ -15 \end{pmatrix} = -4$

16 $\begin{pmatrix} -11 \\ -4 \\ 10 \end{pmatrix} + t\begin{pmatrix} 4 \\ -2 \\ 0 \end{pmatrix} = \begin{pmatrix} 3 \\ -5 \\ -10 \end{pmatrix} + \mu\begin{pmatrix} 7 \\ -10 \\ 0 \end{pmatrix} + \lambda\begin{pmatrix} 1 \\ -3 \\ 4 \end{pmatrix}$

$\mathbf{k}: 10 = -10 + 4\lambda \Rightarrow \lambda = 5$

$\mathbf{i}: -11 + 4t = 3 + 7\mu + 5$

$\Rightarrow 4t = 19 + 7\mu$

$\mathbf{j}: -4 - 2t = -5 - 10\mu - 15$

$\Rightarrow 2t = 16 + 10\mu$

$\times 2$ to give $4t = 32 + 20\mu$

Substitute into $4t = 19 + 7\mu$ to give $32 + 20\mu = 19 + 7\mu$

$\Rightarrow \mu = -1$

$\Rightarrow t = 3$

$\begin{pmatrix} -11 \\ -4 \\ 10 \end{pmatrix} + 3\begin{pmatrix} 4 \\ -2 \\ 0 \end{pmatrix} = \begin{pmatrix} 1 \\ -10 \\ 10 \end{pmatrix}$

and $\begin{pmatrix} 3 \\ -5 \\ -10 \end{pmatrix} - 1\begin{pmatrix} 7 \\ -10 \\ 0 \end{pmatrix} + 5\begin{pmatrix} 1 \\ -3 \\ 4 \end{pmatrix} = \begin{pmatrix} 1 \\ -10 \\ 10 \end{pmatrix}$

So they intersect at the point $(1, -10, 10)$

17 $(6, -2, 20)$

18 $(21, -6, -3)$

19 $(-10, 0, 11)$

20 $(11, 12, -5)$

Exercise 5.3B

1 a $(-7, 10, 0)$ does lie on this plane.
 b $(-7, 10, 0)$ does not lie on this plane.
 c $(-7, 10, 0)$ does not lie on this plane.
 d $(-7, 10, 0)$ does lie on this plane.

2 $k = -9, 4$

3 $k = -15$

4 a Same plane **b** Not the same plane
 c Not the same plane **d** Not the same plane
 e Same plane **f** Not the same plane

5 $21.1°$

6 $173°$

7 $2.72°$

8 $164.6°$

9 $k = -\dfrac{2}{3}$

10 $\begin{pmatrix} 1 \\ -3+4\lambda \\ -2+6\lambda \end{pmatrix} \cdot \begin{pmatrix} 5 \\ 9 \\ -6 \end{pmatrix} = 5 + 9(-3+4\lambda) - 6(-2+6\lambda)$

$= 5 - 27 + 36\lambda + 12 - 36\lambda$

$= -10$ so the line lies on the plane.

11 $\begin{pmatrix} 1 \\ 2\lambda \\ -3+3\lambda \end{pmatrix} \cdot \begin{pmatrix} 1 \\ 3 \\ -2 \end{pmatrix} = 1 + 6\lambda + 6 - 6\lambda = 7$ so the line lies on the plane.

12 $39.5°$

13 $61.9°$

14 $110°$

15 $k = -7$

16 $\sqrt{54}$

17 $\mathbf{r} = 6\mathbf{i} - 21\mathbf{j} + 11\mathbf{k} + \lambda(\mathbf{i} - 6\mathbf{j} + 3\mathbf{k}) + \mu(3\mathbf{i} + 7\mathbf{j} + 4\mathbf{k})$

18 $\mathbf{r} \cdot \begin{pmatrix} 18 \\ -5 \\ 6 \end{pmatrix} = 59$

19 $\mathbf{r} \cdot \begin{pmatrix} -1 \\ -2 \\ 1 \end{pmatrix} = -8$

20 $2x + 6y - 13z = -37$

21 11.7 square units

Exercise 5.4A

1 6

2 3

3 $2\sqrt{2}$

4 $\dfrac{\sqrt{114}}{3}$

5 a $\sqrt{66}$
 b $3\sqrt{10}$
 c $\dfrac{4}{3}\sqrt{30}$
 d $\sqrt{14}$

6 a $\dfrac{7}{6}$ **b** $\dfrac{11}{14}\sqrt{14}$
 c $\dfrac{13}{15}\sqrt{3}$ **d** 0.4
 e $\dfrac{5}{11}\sqrt{11}$

7 2

8 Distance $= \left| \dfrac{n_1\alpha + n_2\beta + n_3\gamma + d}{\sqrt{n_1^2 + n_2^2 + n_3^2}} \right|$

$= \left| \dfrac{n_1 \times 0 + n_2 \times 0 + n_3 \times 0 - p}{\sqrt{n_1^2 + n_2^2 + n_3^2}} \right|$

$= \left| -\dfrac{p}{|\mathbf{n}|} \right|$

$= \dfrac{p}{|\mathbf{n}|}$ as required

Exercise 5.4B

1 $\dfrac{\sqrt{110}}{5}$

2 $\dfrac{\sqrt{30}}{6}$

3 a $\dfrac{\sqrt{3290}}{10}$
 b $x = \dfrac{2}{3}\sqrt{42}$

4 $\dfrac{8}{7}\sqrt{14}$

5 L_2 comes closer to the origin (by 2.85 units) than L_1.

6 B comes closer to the line (by 0.151) than line A.

7 $a = \pm 2$

8 a i $\left| \dfrac{-20\sqrt{51}}{51} \right|$ **ii** $\left| \dfrac{-20\sqrt{51}}{51} \right|$

 b Since A and B are the same distance from Π and are the same side of Π the line l must be parallel to the plane therefore it doesn't intersect it.

9 a i 1.42 **ii** 0.291 **iii** 3.13

 b Each pair of planes is parallel, hence they do not meet and we can find the perpendicular distance between them since it is constant.

10 a $\begin{pmatrix} 2 \\ -3 \\ 1 \end{pmatrix} + \lambda \begin{pmatrix} 3 \\ -5 \\ 2 \end{pmatrix} = \begin{pmatrix} 9 \\ -10 \\ -3 \end{pmatrix} + \mu \begin{pmatrix} 1 \\ 3 \\ 6 \end{pmatrix}$

 i: $2 + 3\lambda = 9 + \mu \Rightarrow \mu = 3\lambda - 7$
 j: $-3 - 5\lambda = -10 + 3\mu \Rightarrow -3 - 5\lambda = -10 + 3(3\lambda - 7)$
  $\Rightarrow \lambda = 2, \mu = -1$
 k: $1 + 2\lambda = 5$
  $-3 + 6\mu = -9$ so not consistent, therefore they do not intersect

 b They are not parallel and do not intersect so must be skew

 c $\begin{pmatrix} 3 \\ -5 \\ 2 \end{pmatrix} \cdot \begin{pmatrix} 18 \\ 8 \\ -7 \end{pmatrix} = 54 - 40 - 14 = 0$ so perpendicular to l_1

 $\begin{pmatrix} 1 \\ 3 \\ 6 \end{pmatrix} \cdot \begin{pmatrix} 18 \\ 8 \\ -7 \end{pmatrix} = 18 + 24 - 42 = 0$ so perpendicular to l_2

 d Π_1 is $\mathbf{r} \cdot \begin{pmatrix} 18 \\ 8 \\ -7 \end{pmatrix} = 5$

 Π_2 is $\mathbf{r} \cdot \begin{pmatrix} 18 \\ 8 \\ -7 \end{pmatrix} = 103$

e $\dfrac{98}{\sqrt{437}}$

f The lines lie on each of the two parallel planes so the shortest distance between them is the shortest distance between the planes which is 4.7

11 a $\begin{pmatrix} 3 \\ 4 \\ 20 \end{pmatrix} \cdot \begin{pmatrix} 0 \\ -5 \\ 1 \end{pmatrix} = 0 - 20 + 20 = 0$ so $\begin{pmatrix} 3 \\ 4 \\ 20 \end{pmatrix}$ and $\begin{pmatrix} 0 \\ -5 \\ 1 \end{pmatrix}$ are perpendicular

$\begin{pmatrix} 4 \\ 2 \\ -1 \end{pmatrix} \cdot \begin{pmatrix} 3 \\ 4 \\ 20 \end{pmatrix} = 12 + 8 - 20 = 0$ so $\begin{pmatrix} 3 \\ 4 \\ 20 \end{pmatrix}$ and $\begin{pmatrix} 0 \\ -5 \\ 1 \end{pmatrix}$ are perpendicular.

The normal to Π_2 is also perpendicular to two non–parallel vectors contained on Π_1, therefore the planes must be parallel. We know they aren't the same plane as $(5, 3, -2)$ does not lie on Π_2, since

$\begin{pmatrix} 5 \\ 3 \\ -2 \end{pmatrix} \cdot \begin{pmatrix} 3 \\ 4 \\ 20 \end{pmatrix} = 15 + 12 - 40 = -13$ (not 12).

b $\dfrac{25\sqrt{17}}{85}$

12 a $\mathbf{r} = \begin{pmatrix} -4 \\ 7 \\ -5 \end{pmatrix} + \lambda \begin{pmatrix} -4 \\ 7 \\ -6 \end{pmatrix} + \mu \begin{pmatrix} 2 \\ 5 \\ -1 \end{pmatrix}$

b 15.2

13 a $\dfrac{\sqrt{2}}{2}$

b $\begin{pmatrix} 2 \\ 1 \\ 0 \end{pmatrix} + t \begin{pmatrix} 1 \\ 1 \\ 0 \end{pmatrix} = \begin{pmatrix} 0 \\ 1 \\ -1 \end{pmatrix} + s \begin{pmatrix} 1 \\ 0 \\ 1 \end{pmatrix}$

$\mathbf{i} : 2 + t = s$

$\mathbf{j} : 1 + t = 1 \Rightarrow t = 0$

$\Rightarrow s = 2$

$\mathbf{k} : 0 + 0 = 0$

but $-1 + 2 = 1$

Therefore they never meet.

c $\dfrac{1}{4}$ square units

Review exercise 5

1 a $\dfrac{x-1}{2} = \dfrac{y+1}{-3} = \dfrac{z-4}{1}$

b $\mathbf{r} = \mathbf{i} - \mathbf{j} + 4\mathbf{k} + s(2\mathbf{i} - 3\mathbf{j} + \mathbf{k})$

2 a $\dfrac{x-3}{5} = \dfrac{y}{-4} = \dfrac{z-5}{-2}$

b $\mathbf{r} = 3\mathbf{i} + 5\mathbf{k} + s(5\mathbf{i} - 4\mathbf{j} - 2\mathbf{k})$

3 a Skew

b Parallel

c Intersect at $(17, 9, -2)$

d Intersect at $(10, -7, 7)$

4 a -13

b $60.6°$

5 $109.1°$

6 $\begin{pmatrix} -2 \\ 8 \\ -4 \end{pmatrix} \cdot \begin{pmatrix} -4 \\ 2 \\ 6 \end{pmatrix} = 8 + 16 - 24$

$= 0$ therefore they are perpendicular

7 $\mathbf{r} = \begin{pmatrix} 2 \\ 0 \\ -3 \end{pmatrix} + s \begin{pmatrix} -2 \\ 0 \\ 3 \end{pmatrix} + t \begin{pmatrix} 1 \\ -4 \\ -1 \end{pmatrix}$

8 a $\mathbf{r} \cdot \begin{pmatrix} 7 \\ -5 \\ 1 \end{pmatrix} = -15$

b $7x - 5y + z + 15 = 0$

9 a $32.3°$

b $\left(-\dfrac{1}{3}, -2, \dfrac{2}{3} \right)$

c $\dfrac{\sqrt{41}}{3}$

10 a $(-1, -1, -6)$

b $168°$

11 $\dfrac{11}{\sqrt{238}}$

12 $153°$

13 $\sqrt{19}$

14 $\sqrt{3}$

15 2.59

16 0.463

1 a $\mathbf{r} = 2\mathbf{i} + 4\mathbf{j} - \mathbf{k} + \lambda(\mathbf{i} + \mathbf{j} - 2\mathbf{k})$

b $-\mathbf{i} + \mathbf{j} + 5\mathbf{k} = 2\mathbf{i} + 4\mathbf{j} - \mathbf{k} + \lambda(\mathbf{i} + \mathbf{j} - 2\mathbf{k})$

Coefficients of $\mathbf{i}$: $-1 = 2 + \lambda \Rightarrow \lambda = -3$

Check coefficients of $\mathbf{j}$: $4 - 3 = 1$

Check coefficients of $\mathbf{k}$: $-1 - 3(-2) = 5$

So $(-1, 1, 5)$ lies on the line.

2 a $\mathbf{r} = \begin{pmatrix} 3 \\ 0 \\ 1 \end{pmatrix} + \lambda \begin{pmatrix} 1 \\ -1 \\ 1 \end{pmatrix}$

b $3 + \lambda = 8 \Rightarrow \lambda = 5$

Check: $0 - 5 = -5$ as required

$1 + 5 = 6$ not 7 so $(8, -5, 7)$ does not lie on the line.

3 $\mathbf{r} = \begin{pmatrix} 2 \\ 1 \\ 0 \end{pmatrix} + \lambda \begin{pmatrix} 8 \\ -5 \\ 3 \end{pmatrix}$

4 a -3

b $87.3°$

5 $\dfrac{1}{6}\sqrt{3}$

6 $(-4\mathbf{i} + \mathbf{k}) \cdot (2\mathbf{i} - \mathbf{j} + 8\mathbf{k}) = -8 + 0 + 8 = 0$

Therefore they are perpendicular.

7 a Intersect at $(1, -2, 6)$

b Parallel.

c Skew.

8 a $\begin{pmatrix} 3 \\ 1 \\ -2 \end{pmatrix} \cdot \begin{pmatrix} 0 \\ 4 \\ -2 \end{pmatrix} = 0 + 4 + 4$

$= 8$ so lies on this plane

b $2(3) + 3(1) - (-2) = 11$

so lies on this plane

9 a $r = \begin{pmatrix} 1 \\ -5 \\ 2 \end{pmatrix} + s\begin{pmatrix} 0 \\ -3 \\ -1 \end{pmatrix} + t\begin{pmatrix} -2 \\ 4 \\ 0 \end{pmatrix}$

b $\begin{pmatrix} 2 \\ 1 \\ -3 \end{pmatrix} \cdot \begin{pmatrix} 0 \\ -3 \\ -1 \end{pmatrix} = 0 - 3 + 3 = 0$

So perpendicular to $\begin{pmatrix} 0 \\ -3 \\ -1 \end{pmatrix}$

$\begin{pmatrix} 2 \\ 1 \\ -3 \end{pmatrix} \cdot \begin{pmatrix} -2 \\ 4 \\ 0 \end{pmatrix} = -4 + 4 + 0 = 0$

So perpendicular to $\begin{pmatrix} -2 \\ 4 \\ 0 \end{pmatrix}$

Hence perpendicular to plane

c i $r \cdot \begin{pmatrix} 2 \\ 1 \\ -3 \end{pmatrix} = -9$

ii $2x + y - 3z = -9$

10 a e.g $r = \begin{pmatrix} 7 \\ 12 \\ -14 \end{pmatrix} + s\begin{pmatrix} 1 \\ 4 \\ -2 \end{pmatrix} + t\begin{pmatrix} 2 \\ 1 \\ -3 \end{pmatrix}$

b $\begin{pmatrix} -1 \\ 0 \\ 4 \end{pmatrix} = \begin{pmatrix} 7 \\ 12 \\ -14 \end{pmatrix} + s\begin{pmatrix} 1 \\ 4 \\ -2 \end{pmatrix} + t\begin{pmatrix} 2 \\ 1 \\ -3 \end{pmatrix}$

$7 + s + 2t = -1 \Rightarrow s + 2t = -8$

$12 + 4s + t = 0 \Rightarrow 4s + t = -12$

$\Rightarrow s = -\dfrac{16}{7}$

$\Rightarrow t = -\dfrac{20}{7}$

Check for **k** component:

$-14 - \dfrac{16}{7}(-2) - \dfrac{20}{7}(-3) = -\dfrac{6}{7}$ not 4

So does not lie on the plane

11 a $\dfrac{x-2}{-1} = \dfrac{y-0}{4} = \dfrac{z+3}{3}$

b They are not parallel as the direction vectors are not equal.

So it is not the same line.

12 a $a = 2, b = 7, c = -3$

b $r = \begin{pmatrix} 2 \\ 7 \\ -3 \end{pmatrix} + t\begin{pmatrix} 2 \\ -3 \\ 1 \end{pmatrix}$

13 B and C

14 $\dfrac{1}{3}\sqrt{137}$

15 $\dfrac{\sqrt{42}}{14}$

16 $63.1°$

17 a $67.4°$

b $(5, -1, -2)$

18 $\dfrac{7}{\sqrt{14}}$

19 a $48.2°$

b $\left(\dfrac{-1}{3}, \dfrac{19}{3}, \dfrac{-4}{3}\right)$

20 $r = \begin{pmatrix} 5 \\ -2 \\ 7 \end{pmatrix} + \lambda\begin{pmatrix} 3 \\ -1 \\ 6 \end{pmatrix} + \mu\begin{pmatrix} -1 \\ -6 \\ 8 \end{pmatrix}$

21 3.32 square units

22 Closer to second line

23 $\dfrac{4}{9}\sqrt{26}$

Index

2D (two dimensions) 114, 118
3D *see* three dimensions

A
addition
    matrices 74
    vectors 14
algebra 35–56
    proof by induction 35, 48–51
    roots of polynomials 36–41
    series 35, 42–50
    summing powers 42–7
    uses 35
angles
    intersecting lines/planes 123, 130–2
    radians 19, 21, 58, 64
    of rotation 85, 87
    scalar products 121–5
areas under curves 58–9, 64
Argand diagrams 14–18, 19, 21
arguments (arg z) 19–27
associative law 79, 103

B
bisectors 23
brackets
    complex numbers 6, 10
    matrices 79
    proof by induction 51
    roots of polynomials 41

C
calculators
    complex numbers 5, 17
    determinants 94
    equation solver 16
    matrices 76, 78, 94, 97
    modulus-argument 20, 21
    polynomial equations 9
    scalar products 122
Cartesian equations
    distances 135
    modulus-argument 20, 25
    vector equations 114–15, 126–7, 130, 132
circles 21–2
coefficients 9, 11, 12, 91–2
column of matrix 78, 81
column vectors 118, 119
combined transformations 86
common factors 49
commutative law 79, 124
completing the square 9
complex conjugate pairs 9
complex conjugate (z^*) 5, 8–12, 14
complex numbers 3–34
    $a + bi$ 4, 6, 12, 17
    Argand diagrams 14–18, 19, 21
    arithmetic 4–7
    loci 19, 21–6
    modulus-argument format 19–27
    notation 4, 26
    polynomial equations 8–13
    properties 4–7

results needing proofs 24
    uses 3
    written form 11
consistency 100, 102–3, 118–19
constants 42–3, 74
cosine (cos)
    angles of rotation 85, 87
    $\cos (A - B)$ 24
    intersecting lines/planes 130–2
    scalar products 121–5
cube numbers 45–6
cubic equations 9, 11, 37–8, 40–1

D
data analysis 80
denominators 5
determinants 94–7, 101–3
differences method 43–4
differentiation 136–7
direction vectors 117–19, 123, 131–2
discriminants 8
distances 134–8
    point to line 134, 136–7
    point to plane 135
distributive law 79
dot product *see* scalar product

E
element of matrix 74
enlargements 85, 95
equating coefficients 9, 11, 12, 91–2
equations
    cubic 9, 11, 37–8, 40–1
    inconsistent 100
    linear 94–105
    parametric 114–15
    polynomial 8–13, 36–41
    quadratic 7, 8, 10–11, 36–7
    quartic 9, 11–12, 16, 38–9
    vectors 114–20, 126–33
    see also Cartesian equations; simultaneous equations
equivalent vectors 117
expanding brackets
    complex numbers 6, 10
    matrices 79
    proof by induction 51
    roots of polynomials 41

F
factorisation 16, 79
factors 8, 10–12, 49
factor theorem 10–11

G
gradients 16
graphs 14–18, 60–2, 66

H
half-lines 22

I
identity matrices 84, 96
imaginary numbers 4–7, 8, 14, 24
inconsistency 100, 102–3, 118–19

index laws 49
induction *see* proof by induction
inequalities 26
inspection method 16
integers
    positive 48, 80
    proof by induction 50–1
    summation 42, 45–6
integration 57–72
    uses 57
    volumes of revolution 58–67
intersecting lines
    distances 134
    equations of lines 118–19
    equations of planes 130–1
    scalar products 123
    see also points of intersection
intersecting planes 130–2
invariant points/lines 91–2
inverse matrices 96–7, 100–2

K
kites 16

L
laws of indices 49
like terms 49
linear equations 94–105
linear independence 130
linear transformations 83, 91, 119
    see also transformations
lines
    determining if point lies on line 115–16
    distance to point 134, 136–7
    equations representing same line 117
    invariant 91–2
    parallel 118–19, 134–5, 137
    skew 118–19
    vector equations 114–20
    $y = mx + c$ 92
    $y = x/-x$ 83, 86, 91
    see also intersecting lines
loci 19, 21–6
long division 9, 11, 12, 16

M
magnitude of vector 122, 123, 137
matrices 73–112
    arithmetic 74–82
    conformable for multiplication 75
    linear equations 94–105
    properties 74–82
    transformations 83–93, 119
    uses 73, 74
matrix of minors 97
minors 95, 97
mirror planes 86–7
modulus-argument format 19–27
multiplication
    complex numbers 4–5
    matrices 74–6, 78–81, 83, 86, 103
    pre-/post-multiplication 83, 85, 88, 91, 100–3, 119

N

natural numbers 48
negative determinants 95
negative numbers 4
normal to plane 127, 131–2

O

order of matrix 74, 75, 78, 86

P

parallel lines 118–19, 134–5, 137
parallel vectors 117, 118, 126, 130
parametric equations 114–15
perpendicular bisectors 23
perpendicular distance 134–5, 137
perpendicular lines 16, 23
perpendicular vectors 121, 124, 127, 134
pi (Π) notation 127
planes
    Cartesian equations 126–7, 130, 132
    distance from point 135
    equations representing same plane 130
    intersections with lines/planes 127–8,
        130–2
    linear equations 100–2
    normals to 127, 131–2
    reflection 86–7
    scalar products 126–7, 130
    vector equations 126–33
points
    determining if point lies on line 115–16
    distance to line 134, 136–7
    distance to plane 135
    equations representing the same line 117
    invariant 91–2
    sets of 26
points of intersection
    distances 134
    equations of lines 118–19
    equations of planes 127–8
    volumes of revolution 60–1, 66
    see also intersecting lines
polynomial equations 8–13, 36–41
position vectors
    Argand diagrams 14
    distances 134–5
    matrices 85
    vector equations 114–15, 126–7, 130
positive integers 48, 80
post-multiplication 100, 103
powers 42–7
pre-multiplication 83, 85, 88, 91, 100–3, 119
principal arguments (arg z) 19
prisms 100, 102
products of matrices 75–6
proof by induction 35, 48–51
    expressions divisible by integers 50–1
    key steps 48
    matrices 79–80
    sums of series 48–50

Q

quadratic equations 7, 8, 10–11, 36–7
quadratic formula 8–9
quartic equations 9, 11–12, 16, 38–9

R

radians 19, 21, 58, 64
rationalising the denominator 5
real numbers 4–7, 8, 14, 24
reflections 83, 86–7, 91
regions 25–6, 61, 64, 66
revolution, volumes of 58–67
roots 8–11, 16, 36–41
rotation 58–67, 84–8
row of matrix 78, 81

S

scalar products 121–5
    angle between intersecting lines
        123
    distances 134
    equations of planes 126–7, 130
    properties 124
scalar quantities 114–19, 121–5, 127–8
scale factors 85, 95
self-inverse matrices 97
series 35, 42–50
    proof by induction 48–50
    summing powers 42–7
sets of points 26
sheaf 101
sigma notation 42
simultaneous equations
    complex numbers 6, 7
    linear equations 100, 102–3
    matrices 78, 91–2, 100, 102–3
    transformations 91–2
    vector equations 118–19
sine (sin)
    angles of rotation 85, 87
    intersecting lines/planes 130–2
    sin (A − B) 24
singular matrices 94, 96
skew lines 118–19
solids 58–67
sphere, volume 62
square matrices 76
square numbers 43, 45–6
square roots 4
stretches 84–5, 95
subscript matrix notation 78, 79
subtracting matrices 74–5
summation 42–7
    powers 42–7
    proof by induction 48–50
    required formulas 44
    sigma notation 42
sum of n terms 42, 48–9
superscript matrix notation 76

T

tabular problems 80–1
tangent (tan) ratio 25
three dimensions (3D)
    intersecting lines 118
    parallel lines 118
    skew lines 118
    transformations 86–7
    vectors 114, 118, 122, 124
    volumes of revolution 58–67
transformations 83–93
    complex numbers 16, 17
    enlargements 85, 95
    matrices 83–93, 119
    reflections 83, 86–7, 91
    rotations 84–8
    stretches 84–5, 95
    vector equations 119
transforming equations 40–1
transposed matrices 97
triangular prisms 100, 102
two dimensions (2D) 114, 118

U

unit vectors 114, 121

V

vector equations
    lines 114–20, 127–8
    planes 126–33
vectors 113–43
    addition 14
    distances 134–8
    matrices 80–1, 83–6, 88, 91–2
    scalar products 121–5
    transformations 83–6, 88, 91–2
    uses 113
    vector equations 114–20, 126–33
volumes of revolution 58–67
    around x-axis 58–63
    around y-axis 64–7
    formulae 58, 60–2, 64, 66
    integration 58–67
    steps to finding 60, 66

X

x-axis 58–63, 83, 87

Y

y-axis 64–7, 83, 87

Z

z* see complex conjugate
z-axis 86–7
zero matrices 75